METABOLOME ANALYSIS

METABOLOME ANALYSIS

An Introduction

SILAS G. VILLAS-BÔAS
AgResearch Limited
Grasslands Research Centre
New Zealand

UTE ROESSNER
Australian Centre for Plant Functional Genomics
School of Botany, University of Melbourne, Australia

MICHAEL A. E. HANSEN
JORN SMEDSGAARD
JENS NIELSEN
Center for Microbial Biotechnology, BioCentrum-DTU
Technical University of Denmark

BICENTENNIAL 1807 WILEY 2007 BICENTENNIAL

WILEY-INTERSCIENCE

A John Wiley & Sons, Inc., Publication

Published by John Wiley & Sons, Inc., Hoboken, New Jersey
Published simultaneously in Canada

Library of Congress Cataloging-in-Publication Data:

Metabolome analysis : an introduction / Silas G. Villas-Bôas ... [et al.].
p. ; cm.
Includes bibliographical references.
ISBN-13: 978-0-471-74344-6
1. Metabolites. 2. Genomics. I. Villas-Bôas, Silas G. (Silas Granato)
[DNLM: 1. Metabolism. 2. Cell Physiology. 3. Genomics–methods.
4. Systems Biology–methods. QU 120 M587973 2007]
QP171.M48 2007
572.8'6–dc22 2006022114

Printed in the United States of America

10 9 8 7 6 5 4 3 2 1

To
our colleagues,
families and
friends

CONTENTS

PREFACE

It has been less than a decade the word "metabolome" was first used referring to all low molecular mass compounds synthesized and modified by a living cell or organism. As a consequence, metabolomics emerged as a new field in the biological science, achieving tremendous development and popularity in the last couple of years. Many would say that metabolomics is a new word for an old science, because it revives the classical biochemical concepts and studies what became "unfashionable" during the genomics era, and it makes extensive use of analytical techniques idealized much earlier than the massive genome sequencing programmes. But, the applicability of metabolomics combined with genomic information or other system wide approaches make this field unique in modern science, both because of its multidisciplinary requirement, where biologist, chemists, engineers, physicists, mathematicians, and statisticians have to join forces to solve common problems; or by its ambition in connecting the different levels of biological information at the molecular level.

As a postgenomics tool, metabolomics is a young field in science but in an exponential growth phase. There is already a peer reviewed journal in its second year of publication, totally dedicated to publish works in the metabolomics field (*Metabolomics, Springer*), an international Metabolomics Society that was formed in 2004 (www.metabolomicssociety.org), and six annual international conferences focused entirely on metabolomics developments and studies (the International Conference on Plant Metabolomics and the Scientific Meeting of the Metabolomics Society).

Despite of all the advances in the metabolomics area, there has been a lack of a concise and basic literature focused on metabolome analysis, particularly an introductory text that can be used as a general guide for a novice interested to start exploring this new field or as a textbook for graduate and undergraduate students attending specialized courses. We, professionals with different scientific backgrounds, therefore joint efforts to write this textbook, aiming to guide the reader to the main steps involved in metabolite analysis, and covering different biological materials (e.g., from plant and animal tissues to microbial and cell cultures, body fluids, and extracellular media), as well as presenting and discussing the principles of the most used methodologies for sample preparation, separation techniques, and detection methods.

The reader will find the book divided into two parts: Part I presents and discusses the concepts and methodology behind metabolite analysis. We first introduced the metabolomics field and its new terminologies (Chapter 1), followed by a general introduction to the diverse biochemical world of small molecules, where the basic concepts of cell metabolism are presented and the differences between primary and secondary metabolites as well as the dynamics of biochemical reactions and metabolite turnover are discussed (Chapter 2). Then, progressively, the reader is taken through the several steps of metabolome analysis, starting with reviewing the diversity of techniques used for sampling and sample preparation (Chapter 3), followed by a global overview of modern analytical methods used in the separation, detection, and identification of metabolites (Chapter 4) and ending with Chapter 5 that is fully dedicated to the most challenging aspect of metabolomics—the data analysis.

Part II of the book is aimed to illustrate the applicability of metabolomics and to discuss specific particularities and requirements of metabolomics in certain groups of organisms. Thereby, we review successful cases of metabolome analysis, illustrating yeast metabolomics (Chapter 6); reviewing specialized sampling devices for microbial metabolomics (Chapter 7); discussing the plant systems and reviewing the major achievements in plant metabolomics (Chapter 8); illustrating the applicability of metabolomics in the classification of filamentous fungi (Chapter 9); and finishing the book with a complete review of metabolomics applied to human and other mammals (Chapter 10).

Our goal as authors was to write a concise and practical focused book as an introduction to metabolome analysis. A book focused on an integrated analytical approach combining the whole analytical chain from sampling over extraction and separation to state-of-the art mass spectrometry and data processing. Although we included a few review chapters in the second part of the book, it is important to emphasize that this book was not intended to be a review book but a textbook that introduces the principles rather than the latest results. The readers will find in the next pages bits of biochemistry, bits of molecular biology, bits of analytical chemistry, bits of mathematics and statistics, and even bits of chemical engineering. That was the challenges that we faced when decided to write this book: to organize the work-flow in metabolome analysis covering all different biological systems and all interdisciplinary aspect. We believe in metabolomics as a field *per se* rather than an additional tool in science. We borrow tools from different sciences to build this new field: METABOLOMICS. Now we invite you to try it.

LIST OF CONTRIBUTORS

Dr. David Wishart, Deptments of Biological Sciences & Computings Sciences, 2-21 Athabasca Hall, University of Alberta, Edmonton, AB Canada, T6G 2E8

Dr. Jens Nielsen, Center for Microbial Biotechnology, Building 223, BioCentrum-DTU, Technical University of Denmark, DK-2800, Kongens Lyngby, Denmark

Dr. Jørn Smedsgaard, Center for Microbial Biotechnology, Building 221, BioCentrum-DTU, Technical University of Denmark, DK-2800, Kongens Lyngby, Denmark

Dr. Michael Adsetts Edberg Hansen, Center for Microbial Biotechnology, Building 223, BioCentrum-DTU, Technical University of Denmark, DK-2800, Kongens Lyngby, Denmark

Dr. Silas Granato Villas-Bôas, AgResearch Limited, Grasslands Research Centre, Tennent Drive, Private Bag 11008, Palmerston North, New Zealand

Dr. Ute Roessner, Australian Centre for Plant Functional Genomics, School of Botany, the University of Melbourne, 3010 Victoria, Australia

PART I

CONCEPTS AND METHODOLOGY

1

METABOLOMICS IN FUNCTIONAL GENOMICS AND SYSTEMS BIOLOGY

By Jens Nielsen

This chapter gives a brief introduction to the field of metabolomics and puts this in perspective of the current development in molecular biology, where genomics have resulted in a move from a reductionistic analysis of biological systems (or even sub-systems) to a systems (or global) view on the function of biological systems. Thus, the chapter serves as an introduction to the textbook.

1.1 FROM GENOMIC SEQUENCING TO FUNCTIONAL GENOMICS

In 1992 the first nucleotide sequence of a complete chromosome was obtained, namely the DNA sequence of chromosome III of the yeast *Saccharomyces cerevisiae*, and around the same time efforts to sequence the human genome were initiated. In 1995 the first complete genome was sequenced, namely that of the pathogenic bacterium *Haemophilus influenzae*, and in 1996 the complete genomic sequence of the yeast *S. cerevisiae* was released. Since then there has followed genomic sequences of many different organisms (Figure 1.1), and currently the number of sequences entered into GenBank is doubled every 10 months. Genomic sequences provide the blueprint for cellular function, and the complete set of genes within a genome basically defines a functional space for the organism. However, in order to further define this functional space it is necessary (1) to know the function of all the proteins and (2) to know the relationship between which genes are expressed (or which proteins are present) at different environmental conditions. Since the first complete genome was released,

Metabolome Analysis: An Introduction, by Silas G. Villas-Bôas, Ute Roessner, Michael A. E. Hansen, Jorn Smedsgaard and Jens Nielsen

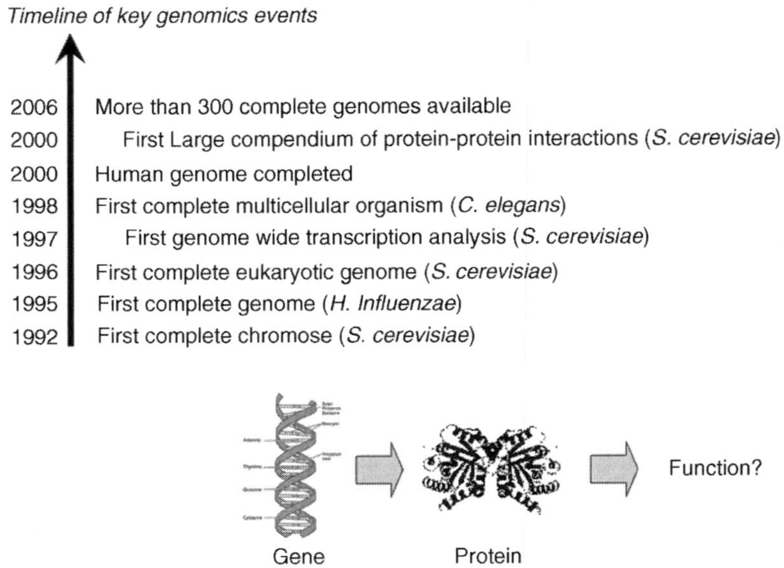

Figure 1.1 A timeline of key developments in the genomics and postgenomics era. The availability of complete genomic sequence raises the question of the function of the individual genes as illustrated in the figure.

the costs of sequencing has steadily decreased and new technologies offer the possibility to dramatically decrease the costs further, opening up for complete sequencing as a tool in diagnostics. With this development, focus has shifted from genomic sequencing toward understanding the function of the individual genes (Figure 1), referred to as functional genomics. The availability of complete genomic sequences and requirement for identification of function for a large number of genes basically resulted in a paradigm shift in biology, as traditionally function was known (or studied) and research was focused on identification of the gene(s).

Bioinformatics has played a central role in functional genomics, but still experimental techniques are essential, and following the availability of complete genomic sequences a number of high-throughput experimental techniques have been developed that enables analysis of a large number of components within a living cell. These include DNA arrays for analysis of all (or a very high fraction) mRNAs, 2D-gel electrophoresis and advanced mass spectrometry for analysis of a large number of proteins, and yeast-two hybrid and other technologies for mapping of protein–protein interactions. These techniques are often referred to as omics techniques (derived from genomics), and terms such as transcriptomics, proteomics, and interactomics are used to describe these different analytical approaches. Even though all high-throughput techniques enable analysis of a large number of components (or interactions), it is, however, currently only transcriptomics that enables measurement of all the relevant components (in this case the mRNAs). Metabolomics is one of the more recently introduced "omics" technologies and as the word indicates it focus on analysis of all the metabolites within the cell under study. Similar to the use of

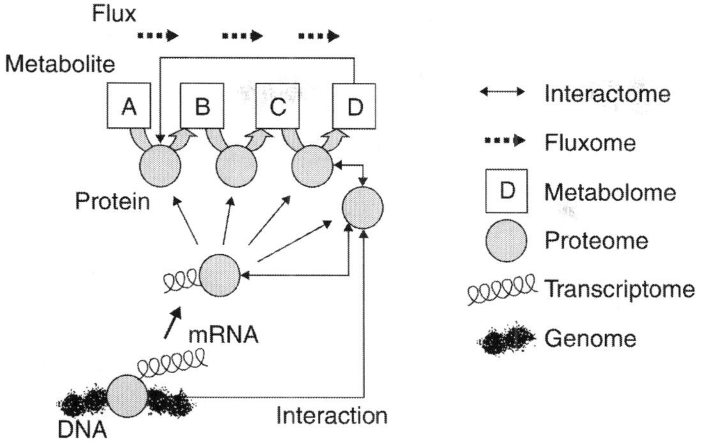

Figure 1.2 An overview of some key "omes" within a cell. The overview captures the central dogma of biology where genes are transcribed into mRNA, which is further translated into proteins. Proteins serve many different functions within the cell, but some acts as enzymes that catalyze the interconversion of metabolites. The interconversion rates of metabolites are given as a set of fluxes through the different biochemical pathways operating in the cell. The different components of the cell may interact with each other resulting in the appearance of complex control loops imposed on many key functions in the cell.

"omics" the term "ome" is often used to describe all the components in a given group of compounds (or interactions). Figure 1.2 gives an overview of the different "omes" in the context of cellular function; and Table 1 gives our definition of some of the most frequently analyzed "omes."

TABLE 1.1 Definitions of Frequently Analyzed "Omes".

Genome	The complete nucleotide sequence in the genetic material of a living cell and further the complete list of all open reading frames (ORFs) that encode proteins.
Transcriptome	The complete set of all mRNA present in the cell.
Proteome	The complete set of all proteins present in the cell. The pool includes different forms of the same protein, e.g. a protein can be present in different states (phosphorylated/non-phosphorylated), and the proteome may therefore include many more components than the transcriptome and the number of ORFs.
Metabolome	The complete set of all metabolites formed by the cell in association with its metabolism.
Fluxome	The complete set of all fluxes through the different biochemical reactions that are involved in the interconversion of metabolites.
Interactome	The complete set of interactions between different components within the cell. These interactions include protein-protein interactions, protein-DNA interactions, protein-metabolite interactions as well as other possible interactions.

1.2 SYSTEMS BIOLOGY AND METABOLIC MODELS

A fundamental problem in interpreting results from the analysis of the different "omes" is that the individual components in all the "omes" are complex functions of a large number of different cellular components (see Figure 1.2). This has called for integrated analysis, where several "omes" are measured in parallel, and mathematical models are used for the analysis of the data. This approach is referred to as systems biology, and in recent years there has been a major shift toward integrated analysis, and in particular building detailed mathematical models describing different parts that forms the basis for the complete biological system that makes up a living cell.

As an illustration of the interaction of the different components in a living cell, the transcription of a given gene is a function of the level of transcription factors, and also the activities of upstream kinases and receptors. Similarly, the level of any given protein is determined, not just by the level of its corresponding mRNA, but also by the activity of the translational apparatus, protein kinases, phosphatases, and proteases. Whereas the levels of metabolites are determined directly by the activities of many different enzymes (parts of the proteome), the individual components of the metabolome are generally far more complex functions of other components in the cell than is the case for mRNAs or proteins. Thus, the level of any metabolite in the cell is determined by the activity of all the enzymes that are involved in the synthesis and conversion of that metabolite. Detailed metabolic models (see Table 1.2 and text below) have shown that less than 30% of the metabolites are involved in only two reactions, whereas about 12% of the metabolites participate in more than 10 reactions and about 4% of the metabolites even participate in more than 20 reactions. Furthermore, most reactions in a living cell involve more than a single substrate and a single product (more than 67% in the yeast *S. cerevisiae*) and this ensures a high degree of connectivity in the metabolic network (see Figure 1.3). Thus, the metabolic network operating in a living cell is a complex myriad of reactions that are tightly connected. Due to this coupling of many different reactions within the metabolic network, even small perturbations in the proteome (e.g., an alteration in the level of a few enzymes) may result in a significant change in the levels

TABLE 1.2 Some Data from a Few Detailed Metabolic Models (From Borodina and Nielsen, 2005).

Organism	Reactions	Metabolites	Metabolic ORFs	Total ORFs
H. pylori	444	340	268	1638
H. influenzae	477	343	362	1880
E. coli	720	436	695	4485
S. coelicolor	700	501	769	8042
S. cerevisiae	1175	584	708	5773
M. musculus	1220	872	—	—

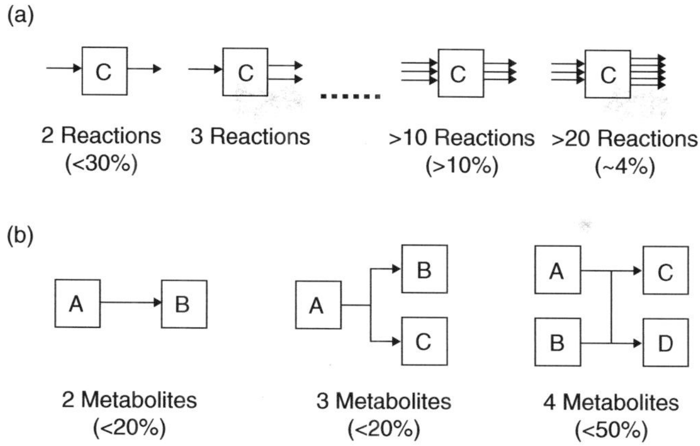

Figure 1.3 Illustration of the tight coupling of the different reactions in the metabolic network operating in a living cell. (a) Distribution of the number of reactions spanning the different metabolites. (b) Distribution of the number of metabolites being involved in the different reactions in the metabolic network.

of many metabolites. The biological reason for this may well be that this ensures a stable operation of the metabolic network with respect to the occurrence of mutations, i.e., upon a decrease in the activity of a particular enzyme, the response may be an increase in the level of the substrates of that enzyme, thereby ensuring that the change in the flux may only be slightly altered. Thus, evolution may have favored the establishment of metabolic networks that are tightly coupled and hence are robust to different kinds of perturbations.

As mentioned above the objective of systems biology is to represent cellular function through mathematical models, and many different types of mathematical models have been developed for the description of a wide range of cellular processes. Due to the conserved nature of the central metabolism in different biological systems, the function of metabolism has been extensively studied, and also the genes encoding enzymes involved in the central metabolism are very well annotated for most organisms. This has formed the basis for reconstruction of complete metabolic networks of several different organisms (see Table 1.2). This reconstruction process relies on genomic information and biochemical information of the studied organism (Palsson, 2006). These reconstructed metabolic networks serve as scaffolds for metabolic models that can be used to predict cellular function and study the role of individual reactions, and also for analysis of "omics" data (Borodina and Nielsen, 2005; Palsson, 2006). In the context of metabolomics these models are particularly useful as they provide links between the different metabolites in the metabolic network. They can also be used to calculate the fluxes through different parts of the metabolism, and through combination with metabolome analysis; it is hereby possible to correlate metabolite levels and fluxes, which enables identification of key control points in the metabolism.

1.3 METABOLOMICS

Being the intermediates of biochemical reactions, metabolites play a very important role in connecting the many different pathways that operate within a living cell. As mentioned above the level of metabolites represents integrative information of the cellular function, and, hence, defines the phenotype of a cell or tissue in response to genetic or environmental changes. Analysis of cellular function at the molecular level requires recruitment of several different analytical techniques. Whereas comprehensive methods for analysis at the transcriptional level (transcriptome) and at the translational level (proteome) are currently in a rapid state of development, and high-throughput analytical methods are already in use, methods for analysis of the metabolomics approaches are, however, so far less common, and currently there is no single method that enables analysis of the metabolome. Although metabolite profiling has long been applied for medical and diagnostic purposes as well as for phenotypic characterization, it is not until recently that increasing efforts have been undertaken to develop methods to screen of a high number of intracellular metabolites in the context of functional genomics (Fiehn, 2001).

Metabolome analysis covers the identification and quantification of all intracellular and extracellular metabolites with molecular mass lower than 1000 Da[1], using different analytical techniques. In common with the transcriptome and the proteome, the metabolome is context-dependent, and the levels of each metabolite depend on the physiological, developmental, and pathological state of a cell, tissue, or organism. However, an important difference is that, unlike mRNA and proteins, it is difficult or impossible to establish a direct link between genes and metabolites. The convoluted nature of cell metabolism, where the same metabolite can participate in many different pathways, complicates the interpretation of metabolite data.

The genome, transcriptome, and proteome elucidations are based on target chemical analyses of biopolymers composed of four different nucleotides (genome and transcriptome) or 22 amino acids (proteome). Those compounds are highly similar chemically, and facilitate high-throughput analytical approaches. Within the metabolome, there is, however, a large variance in chemical structures and properties. Thus, the metabolome consists of extremely diverse chemical compounds from ionic inorganic species to hydrophilic carbohydrates, volatile alcohols and ketones, amino and nonamino organic acids, hydrophobic lipids, and complex natural products. That complexity makes it virtually impossible to simultaneously determine the complete metabolome (Chapter 2). To further add to the complexity of metabolome analysis is the very rapid turnover of metabolites, i.e., many metabolites are present in low concentrations and there are very high fluxes through the metabolite pools. It

[1]This cut-off molecular weight is obviously not very strict as many secondary metabolites have molecular weights above 1000 Da, and still they are considered to be metabolites. However, it is necessary to have some kind of discrimination between metabolites and macromolecules that are the major constituents of the cell, i.e., proteins, DNA, RNA, lipids, etc.

is therefore important to quench the metabolism rapidly and this calls for dedicated methods for quenching and extraction of metabolites from living cells. Therefore, the metabolomics encompass sample preparation (Chapter 3), sample analysis (Chapter 4), and date analysis (Chapter 5). Basically each metabolome study requires an evaluation of the sample preparation and the extraction procedure and how they couple to a combination of different analytical techniques in order to achieve as much information as possible, and we will illustrate this in a number of examples at the end of the textbook (Chapters 6–9).

As there are no single analytical method for analysis of the metabolome, different terms are often used in the field of metabolomics (see Table 1.3). There is a general consensus that the term metabolome describes the total sum of metabolites a given biological system can either use or form by its metabolism. The metabolome is often divided into the exometabolome and the endometabolome, where the former represents metabolites outside the cell and the latter represents intracellular metabolites. Whereas this distinction between exo- and endometabolome is quite useful for microbial systems where it is easy to separate the cells from the extracellular medium, it is less useful for multicellular systems where it may be difficult to isolate the cells from complete tissue. However, still it is conceptually important to differentiate between these two as the exometabolome often plays a very different

TABLE 1.3 Some Definitions Used in Metabolome Analysis (Adapted from Nielsen and Oliver, 2005).

Metabolome	The complete set of all metabolites used by or formed by the cell in association with its metabolism. The metabolome comprises both the endometabolome (the complete set of intracellular metabolites) and the exometabolome (the set of metabolites excreted into the growth medium or extracellular fluid).
Metabolomics	Approaches to analyze the metabolome or a fraction of the metabolome. Metabolomics involves sampling, sample preparation, chemical analysis, and data analysis.
Metabolic fingerprinting	Spectra from NMR or MS analysis that provides a fingerprint of metabolites produced by a cell. The fingerprint typically does not provide information about specific metabolites.
Metabolic footprinting	Analysis of the exometabolome. This may be either through analysis of specific metabolites or through spectra that do not provide information about specific metabolites (in analogy with metabolite fingerprinting).
Metabolite profiling	Analysis of a group of specific metabolites, e.g. a class of metabolites such as carbohydrates or amino acids. The analysis does not need to be quantitative, but often it is at least semiquantitative.
Metabolite target analysis	Quantitative analysis of metabolites participating in a specific part of the metabolism.

physiological role than the endometabolome. Two terms that are often used to describe analysis of a part of the metabolome are metabolite profiling and metabolic fingerprinting. These two terms are often used as synonyms with no clear distinction, but here we will use the definitions given in Table 1.3, which is adapted from Fiehn (2001) (see also Nielsen and Oliver, 2005). According to these definitions, metabolite profiling is the analysis of a given set of metabolites, e.g., a set of amino and organic acids, whereas metabolic fingerprinting is an unspecific analysis of a sample, e.g., a range of mass peaks obtained by mass spectrometry. The former provides direct physiological information, and the data can be integrated into metabolic models, whereas the latter provides a fingerprint that only can be used for grouping of different samples, e.g., using cluster analysis. As one may use nonspecific analysis of both the exo- and the endometabolome, the term metabolic footprinting has been introduced to describe analysis of the exometabolome in microbial cultures (Allen et al., 2003). The term footprinting indicates that the microbial cells leaves a footprint in the extracellular medium when they take up nutrients and secrete metabolites in connection with their growth process. Even though metabolic fingerprinting (or footprinting) does not provide information about the levels of specific metabolites, these analysis techniques may still be used for classification of mutants (or growth conditions) and permit the assignment of functions to orphan genes through the concept of guilt-by-association. It is, however, difficult to integrate this kind of data with other types of data, e.g., transcriptome data, and even though the concept of guilt-by-association is useful for classification of and hence can be used in functional genomics, it is less useful in systems biology where quantitative data are required. There are basically two solutions to this fundamental problem: (1) one may identify the peaks (or metabolites) that are playing a key role in distinguishing the different mutants (e.g., by using MS–MS) or (2) one may restrict the analysis to a group of metabolites which can be measured quantitatively (e.g., by CE–MS, LC–MS, or GC–MS), i.e., using metabolite profiling. Whereas the first solution provides some insight into the qualitative response of metabolism to the genetic change, it is associated with the risk of not identifying the quantitative effects of a given mutation. The other solution may produce a quantitative phenotype for a given mutation, but miss metabolites that are the key to the analysis. Some new developments in CE–MS (Soga et al., 2003) and GC–MS (Roessner et al., 2000; Weckwerth et al., 2004; Villas-Boas et al., 2005) do, however, enable true quantitative analysis of a relatively large number of metabolites.

Mass spectrometry (MS) and nuclear magnetic resonance (NMR) are the most frequently employed methods of detection in the analysis of the metabolome (Chapter 4). NMR, in particular, is very useful for structure characterization of unknown compounds and has been applied for the analysis of metabolites in biological fluids and cells extracts. However, in certain circumstances, the ^{1}H NMR spectrum is insufficient on its own to provide information that will fully characterize a metabolite, but it may still provide a valuable metabolic fingerprint. This is obvious the case where analytes contain functional groups that are deficient in protons or where the protons can readily chemically exchange with the solvent, the signals thus being broadened beyond detection. Alternatively, other nuclei can also be used, such as ^{13}C NMR. However, ^{13}C NMR spectroscopy presents relatively low sensitivity, i.e., in the range of μmol to mmol. In addition, ^{13}C NMR analysis may take several hours for a single sample, as a consequence of its low sensitivity, and the equipment costs are much higher compared to MS-based techniques.

The most important advantages of MS is its high sensitivity, and high-throughput in combination with the possibility to confirm the identity of the components present in the complex biological samples as well as the detection and, in most of the cases, the identification of unknown and unexpected compounds. Furthermore, the combination of separation techniques (e.g., chromatography) with MS tremendously expands the capability of the chemical analysis of highly complex biological samples. The basic information of mass spectra is characterized by its simplicity. The spectrum displays masses of the ionized molecule and its fragments, and those masses are simply the sums of the masses of the component atoms. In some cases, a mass spectrum contains a wealth of specific analytical and structural information, much more information than the expert in the field can currently utilize; unfortunately that abundance of information can discourage the novice who turns to compendia of mass spectrometric information for help. Nevertheless, it is comparatively simple to handle the mass spectra and there are several available software applications that make the interpretation of mass spectrometric data relatively easy.

1.4 FUTURE PERSPECTIVES

From the recent past it became obvious that metabolomics is a scientific field which develops with an enormous speed which makes it already difficult to follow the increasing numbers of scientific publications presenting the development of novel instrumentation, methodologies, or exciting applications in biology. With this development metabolomics has attracted increasingly interests, not only by biologists but also by the public and politicians as its value has been convincing from many successful applications. In near future, many institutions and laboratories worldwide will have established the physical and intellectual capacities to apply metabolomics in their research programs. Metabolomics will become more and more advanced, which will concurrently lead to certain confidence in the way it is applied and in the validity of the data obtained. In plant research, potential applications for metabolomics are enormous as described in Chapter 8, and for this reason the Plant Metabolomics Society has been founded some years ago (www.plantmetabolomics.nl) and four international conferences so far were held by the society, which has given the opportunity to share exciting new developments in the field. This society has been followed by the recently founded Metabolomics Society (www.metabolomicssociety.org).

As discussed above, the strength of metabolome analysis is that metabolite levels present a high degree of integrative information. This is, however, also a drawback as it is inherently difficult to interpret the results. In those cases where the levels of many different metabolites have been measured, it is often difficult to bring the data into a physiological context that matches our current understanding of metabolism (measurement of many metabolites is, however, valuable for discovery of new pathways). Some studies have succeeded in mapping measurements of several metabolites onto metabolic charts and have hereby demonstrated how metabolite profiling can be combined with transcriptome analysis for mapping responses when the cells are exposed to different environmental conditions (Hirai et al., 2004; Villas-Bôas et al., 2005). However, as mentioned above, metabolism is far more connected than is shown by maps downloaded from KEGG (www.genome.jp/kegg) or other databases. Therefore, if a large number of metabolites are measured, it is necessary to adopt a more structured approach to data analysis. This is provided through the integration of experimental data with mathematical models, and as metabolism has been particularly well described for many microorganisms (Kell, 2004), it makes sense to start such model-driven data analyses using such single-celled systems. Recently, it has been demonstrated how a detailed metabolic model for *E. coli* could form the basis for integrating transcriptome data with computational data (Covert et al., 2004). Furthermore, by converting a genome-scale metabolic model to a metabolic graph, it has shown possible to use genome-scale metabolic models for identification of parts of the metabolic network that are transcriptionally coregulated (Patil and Nielsen, 2005), and this concept can easily be extended to the integration of transcriptome, proteome, and metabolome data.

As has been shown in a number of cases and will be shown in this textbook, metabolome analysis has proven successful for phenotypic mapping of cells, and thereby for the clustering of different mutants. However, as pointed out recently by Nielsen and Oliver (2005), it is a requirement for a wider use of metabolome analysis, and particularly for integration of these data with mathematical models as mentioned above, that there is a shift toward *truly quantitative analysis of specific metabolites obtained under well-defined conditions.* By "true quantitative analysis" they mean not only measurement of relative levels, but also measurement of actual concentrations of the different metabolites. This calls for

- Definition of appropriate data standards
- Development of standard analytical methods
- Development of appropriate libraries of mass spectra of GC–MS and LC–MS for standard analytical methods.

Definition of data standards is important for enabling comparison of data from different experiments, and from transcriptome analysis the true value of accumulating large data-sets has been demonstrated in several cases. Thus, in analogy with the MIAME standards for transcription analysis, it is interesting to define data standards for metabolome analysis, and there are already movements in this direction (Jenkins et al., 2004), and obviously the above-mentioned Metabolomics Society will play an important role in defining standards and building libraries. This is not an easy task because, for example, many different synonyms are used for one and the same metabolite and many different methodologies are used to analyze metabolites. Therefore, ways for the standardization of metabolomics experiments have to be defined and accepted by the community, and anthologies have to be determined and used commonly. The driving force behind these initiatives is the desire of each metabolomics user to increase the number of identified metabolites and hereby increase the amount of information extractable from measurements. In addition, a functional database for public metabolomics data will attract computer scientists and bioinformaticians to develop novel methods for analysis of these huge data-sets leading, for example, to the development of new and useful software packages for data visualization, mining, and information extraction. This again will be of great help and use for the biologists. In recent years, there have been some reports on standard analytical methods that enable quantitative analysis of a large number of metabolites and there is a trend toward defining mass spectral libraries for these methods (Villas-Bôas et al., 2005; Halket et al., 2005; Schauer et al., 2005), which will clearly support further advancement of the research field.

In conclusion, it is an extremely exciting time for metabolomics as a new, rapidly growing scientific field. Most interestingly in near future will be the development of a common language among biologists, biochemists, geneticists, molecular biologists, analytical chemists, bioinformaticians, and computer scientists for best and most satisfactory outcomes of any metabolomics approach. We hope that our textbook will assist in this development and spur further developments in metabolomics.

REFERENCES

Allen J, Davej HM, Broadhurst D, Heald JK, Rowland JJ, Oliver SG, Kell DB. 2003. High-throughput classification of yeast mutants for functional genomics using metabolic footprinting. Nature Biotechnol 21:692–696.

Borodina I, Nielsen J. 2005. From genomes to *in silico* cells via metabolic networks. Curr Opin Biotechnol 16:1–6.

Covert MW, Knight EM, Reed JL, Herrgard MJ, Palsson BØ. 2004. Integrating high-throughput and computational data elucidates bacterial networks. Nature 429:92–96.

Fiehn O. 2001. Combining genomics, metabolome analysis and biochemical modelling to understand metabolic networks. Comp Funct Genomics 2:155–168.

Halket JM, Waterman D, Przyborowska AM, Patel RKP, Fraser PD, Bramley PM. 2005. Chemical derivatization and mass spectral libraries in metabolic profiling by GC/MS and LC/MS/MS. J Exper Bot 56:219–243.

Hirai MY, Yano M, Goodenowe DB, Kanaya S, Kimura T, Awazuhara M, Arita M, Fujiwara T, Saito K. 2004. Integration of transcriptomics and metabolomics for understanding of global responses to nutritional stresses in *Arabidopsis thaliana.* Proc Nat Aca Sci USA101:10205–10210.

Jenkins H, Hardy N, Beckmann M, Draper J, Smith AR, Taylor J, Fiehn O, Goodacre R, Bino RJ, Hall R, Kopka J, Lane GA, Lange BM, Liu JR, Mendes P, Nikolau BJ, Oliver SG, Paton NW, Rhee S, Roessner-Tunali U, Saito K, Smedsgaard J, Sumner LW, Wang T, Walsh S, Wurtele ES, Kell DB. 2004. A proposed framework for the description of plant metabolomics experiments and their results. Nat Biotechnol 22:1601–1606.

Kell DB. 2004. Metabolomics and systems biology: Making sense of the soup. Curr Opin Microbiol 7:296–307.

Nielsen J, Oliver S. 2005. The next wave in metabolome analysis. Trends Biotechnol 23:544–546.

Palsson BO. 2006. *Systems Biology*, Cambridge University Press, New York, NY, USA.

Patil K, Nielsen J. 2004. Uncovering transcriptional regulation of metabolism using metabolic network topology. Proc Natl Acad Sci USA 102:2685–2689.

Roessner U, Wagner C, Kopka J, Trethewey RN, Willmitzer L. 2000. Simultaneous analysis of metabolites in potato tuber by gas chromatography-mass spectrometry. Plant J 23:131–142.

Schauer N, Steinhauser D, Strelkov S, Schomburg D, Allison G, Moritz T, Lundgren K, Roessner-Tunali U, Forbes MG, Willmitzer L, Fernie AR, Kopka J. 2005. GC-MS libraries for the rapid identification of metabolites in complex biological samples. FEBS Lett 579:1332–1337.

Soga T, Ohashi Y, Ueno Y, Naraoka H, Tomita M, Nishioka T. 2003. Quantitative metabolome analysis using capillary electrophoresis mass spectrometry. J Proteome Res 2:488–494.

Villas-Boas SG, Moxley JF, Åkesson M, Stephanopoulos G, Nielsen J. 2005. High-throughput metabolic state analysis: The missing link in integrated functional genomics. Biochem J 388:669–677.

Weckwerth W, Loureiro ME, Wenzel K, Fiehn O. 2004. Differential metabolic networks unravel the effects of silent plant phenotypes. Proc Natl Acad Sci USA 101:7809–7814.

2

THE CHEMICAL CHALLENGE OF THE METABOLOME

By Ute Roessner

This chapter focuses on the description of the chemistry behind metabolism and why metabolites from the analytical point of view can be treated as chemicals in a constantly dynamical environment. A metabolite is synthesized to fulfill a finite biological function. Metabolites undergo chemical reactions carried out by enzymes, which change the chemical properties of the metabolites. These chemical reactions in a series are called pathway and the sum of all pathways is called metabolism. Metabolites are determined by specific characteristics, which are described in detail. When all metabolite-connecting reactions are transformed into a linear matrix, a metabolic network can be reconstructed, which is in fact a subnetwork within all interactions of various types of cellular molecules, such as proteins, RNA, and DNA. The analyses of the structure and architecture of these cellular networks have not only increased our understanding of life's complexity but also pointed the importance of determining the identity and function of each component in a cell.

2.1 METABOLITES AND METABOLISM

All living cells derive energy and building blocks required for growth and maintenance from the conversion of small chemical compounds to another set of chemical compounds with lower free energy content. This conversion or transformation of chemicals involves a large number of chemical reactions with many chemical intermediates, the completeness of these reactions is called metabolism, and the chemicals involved in metabolism are called metabolites. The word metabolism

Metabolome Analysis: An Introduction, by Silas G. Villas-Bôas, Ute Roessner, Michael A. E. Hansen, Jorn Smedsgaard and Jens Nielsen

comes from the Greek *metabolē* and means change or transformation. The complexity of life processes requires that the number of metabolites that participate in the metabolism is quite large, but still there is a high degree of organization of the different interconversion processes. Thus, in any living cell, the carbon and energy source for the cell is first converted to a set of so-called precursor metabolites, and these precursor metabolites are subsequently converted to metabolites that serve as building blocks for biomass synthesis and other metabolites that are secreted by the cells. The properties of metabolites and their functionality as they interact within their natural environment determine the chemistry of life. Metabolites are the products of enzyme-catalyzed reactions that occur naturally within living cells. A molecule has to meet certain properties and characteristics before it is called a metabolite. First of all, a metabolite is synthesized by the cell for the purpose of performing a useful, if not indispensable, function in the maintenance and survival of the cells by, for example, contributing to the infrastructure or energy requirement of the cell. If it does not directly perform a biological function, it will, after a structural modification, serve as a precursor for further conversion into a biologically active compound. Another important feature of a metabolite is that it is recognized and acted upon by enzymes, which will change its properties by means of a chemical reaction.

The many different reactions within a living cell are normally organized into a series of reactions that serve a coordinated function within the cell. Such series of reactions are called pathways, and pathways may have a varying number of metabolites as intermediates. In some pathways, metabolites retain many of the properties of their parent metabolite, which are at the start of the pathway, until its carbon structure forms larger constructions or reduces to smaller structures. Examples of this are the conversion of free amino acids into proteins; the conversion of glucose moieties into high molecular weight carbohydrate structures such as starch; and the conversion of free fatty acids into complex lipids. Smaller metabolites are produced if the parent compound undergoes systematic degradation, for example, during oxidation reactions, which may eventually result in the formation of water and/or carbon dioxide. In this process the cell is, however, capturing much of the free energy in the parent metabolite and in other metabolites as will be described later.

A major characteristic of metabolites is that they have a finite half-life, which means they are constantly taken up, produced, degraded, or excreted by the cell. Last but not the least, many metabolites can serve as regulators of carbon flow in competing and interacting pathways to control their own and other metabolites' pace of conversion.

These features of metabolites have to be borne in mind when their comprehensive determination, identification, and quantification are aimed by a metabolomics approach. The fast turnover and modification of metabolites require specific and especially quick extraction methodologies, and the enormous chemical diversity requires a range of different separation and detection techniques. Chapter 3 will give detailed descriptions of applicable and feasible approaches to extract, and Chapter 4 gives an outline of the currently applied analytical technologies to measure compounds from different biological sources.

As described above, metabolites are molecules, which are constantly transformed and changed in chemical reactions within a living cell. A series of these reactions are called pathways, and the sum of all pathways is called metabolism. In general, a few important points can be summarized to describe the concept of metabolism:

(i) All chemical reactions of life are organized and linked into a network of metabolic pathways.
(ii) Metabolism is maintained and regulated to ensure constant supply of resources for the living cell and hence for survival of the cell and is highly dependent on the environment.
(iii) The free energy of cells is stored in chemical substances, which are metabolites themselves, whereas other metabolites are bound in structural components of the cell.
(iv) Metabolic reactions are influenced by metabolites by a number of specific control mechanisms.
(v) Metabolism can be segregated into central (or primary) metabolism and secondary metabolism. The central metabolism is primarily related to energy and production of core structures in the cell, e.g., proteins and structural components and mostly influenced by the nutritional environment. The central metabolism share many similarities across species, and most metabolites of the central metabolism are widespread in nature. The secondary metabolism relates to production of far more specialized metabolites, some that are unique to a single species and require many genes to be produced. These metabolites are often of unknown function but may act as, for example, signal compound, for defense and other purposes that improve function or survival in a multicellular environment (organism).
(vi) Metabolism can be divided into anabolic and catabolic metabolic reactions. Anabolism means the synthesis of complex molecules from simple compounds to store energy whereas the degradation of complex molecules for energy release is called catabolism. In general, anabolic reactions require energy whereas catabolic reactions release energy. Metabolic energy capture occurs largely through the synthesis of ATP, NADH, or NADPH, molecules that are designed to provide energy for biological work, which is one of the most important metabolites itself.

Chemical reactions are carried out to transform and change the chemical nature of metabolites. Often these reactions only proceed because of the presence of specific catalysts, which are called enzymes and are highly specialized protein structures. A catalyst increases the rate or velocity of a chemical reaction without being changed itself in the overall process. They change the rates of reactions, but do not affect the equilibrium of a reaction. These enzymes work simply by lowering the energy barrier of a reaction and by doing so, the catalyst increases the fraction of molecules that have enough energy to attain the transition state, thus making the reaction go faster in both directions. Details of different working principles of enzymes and their mode of action is described by most biochemistry textbooks (see, e.g., Stryer, 1995 and Voet and Voet, 2004).

2.2 THE STRUCTURAL DIVERSITY OF METABOLITES

Metabolome analysis presents one of the most exciting and also challenging investigations compared with the other cell product analyses, the "omes" such as the genome and transcriptome. This is because of the fact that each metabolite is characterized by its individual chemical structure determining the physical and chemical properties of the compound. Therefore, each metabolite is unique and their features are specific, and metabolites from the same pathway can present very different chemistry. The properties and chemistry of metabolites and their occurrence in the metabolism are determined by two major properties: the chemical and physical properties and the dynamics by which a metabolite is converted, both strongly dependent on the environment at any one time. And indeed, this great diversity in chemical and physical properties of metabolic compounds requires an assortment of procedures allowing the accurate and comprehensive measurement of metabolites within a metabolomics approach. An example of different metabolites and their chemical structures is represented in Figure 2.1.

2.2.1 The Chemical and Physical Properties

Text box 2.1 illustrates a few of the features determining the chemical properties of a metabolite. Altogether, there are a range of objectives resulting in the enormous variety of chemical and also physical properties, which determine the behavior of each metabolite and concurrently its ability to be analyzed.

(i) Molecular weight—The weight of a molecule is calculated by the sum of the weights of all atoms making the molecule. It is therefore a specific value for each molecule. Exceptions for molecules are made by the same number of certain atoms resulting in the same sum (e.g., isomers). Metabolites are, by definition, small molecular weight compounds (in comparison with polymers such as proteins or starch) and their weight ranges from as low as 18 g/mol (H_2O) to more than 1000 g/mol for lipid structures.

(ii) Molecular size—The molecular size of a molecule is represented by its special volume and tridimensional structure. These depend on the molecular structure and how many other molecules like water are attracted to have noncovalent binding on the surface of the molecule. Thereby, the efficient volume of the molecule is increased. The unit in which molecular size is calculated is Å.

(iii) Polarity—The polarity of a molecule is a physical property of a compound, which in the context of metabolomics, is related to the ability to form polar interactions (noncovalent bonds in particular hydrogen bonds) with water molecules and

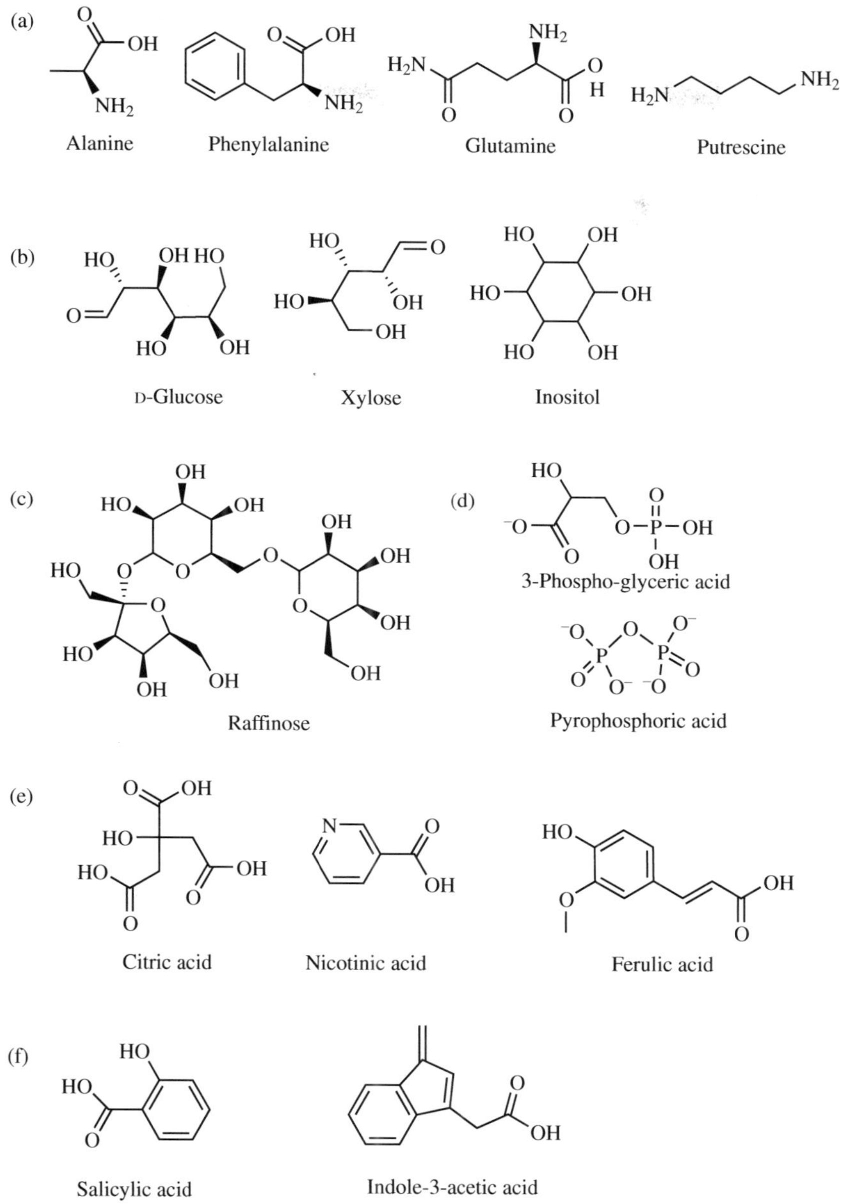

Figure 2.1 A selection of metabolites from different chemical classes. (A) amino acids and amines, (B) monosaccharides, (C) trisaccharide, (D) important very small phosphorylated compounds, (E) primary and secondary organic acids, (F) phytohormones, (G) fatty acids, (H) lipid, (I) sterol, (J) acyclic diterpene, (K) vitamins.

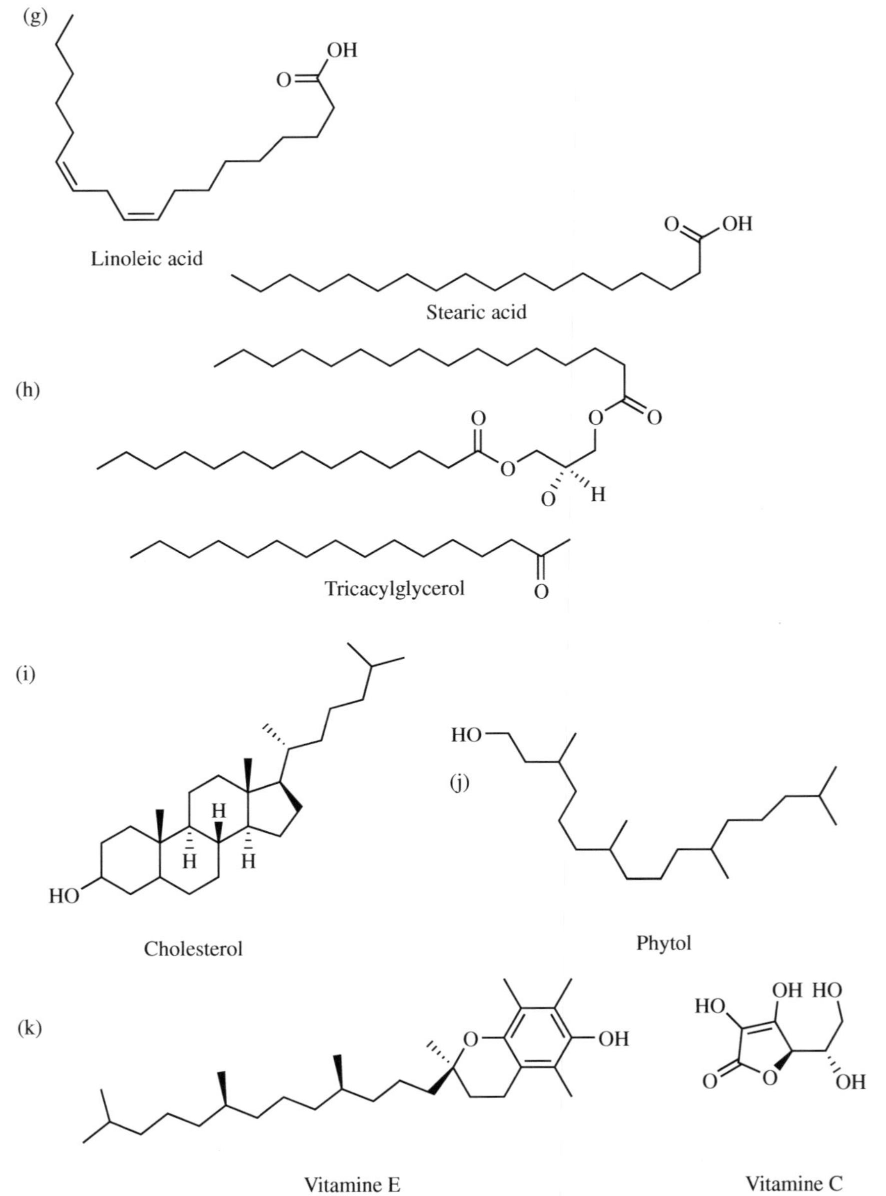

Figure 2.1 (*Continued*)

other polar compounds. This again relates to other physical properties such as melting and boiling points, solubility, and intermolecular interactions between different molecules. In most cases, there is a close correlation between the polarity of a molecule and the number and types of polar or nonpolar covalent bonds, which are present in the molecule. In general, with some exceptions, the greater the electronegativity

◊ Text box 2.1 Chemical diversity of metabolites.

This text box represents selected example demonstrating different characteristics resulting in a huge chemical diversity of metabolites.

(1) *Molecular size—molecular weight*

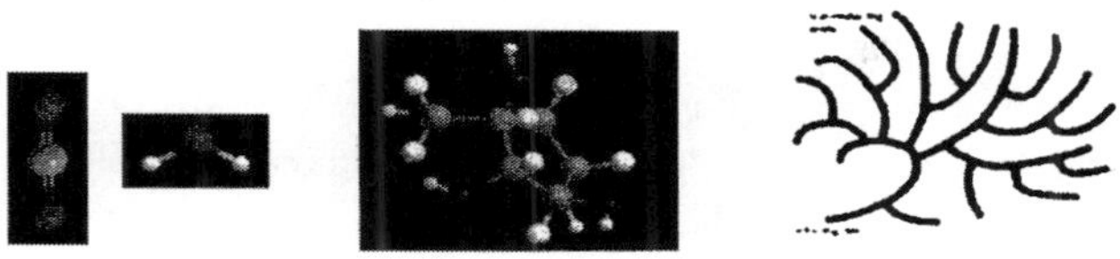

Formular	CO_2	+	H_2O	⟶	glucose	⟶	glycogen
Molecular weight	44		18		180		$n \times 180$

(2) *Polarity*

Highly apolar ⟵				⟶ Highly polar
Lipids	Carotenoids	Phenolics	Amino acid	Sugars
Fatty acids	Chlorophylls	Alcohols	Organic acids	Nucleotides
Waxes	Steroids		Organic amins	Phosphates
Terpenes	Flavenoids		Alkaloids	Metals
			Nucleosides	Salts

(3) *Isomers*

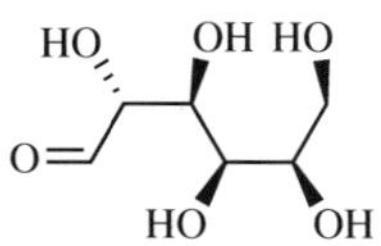

D-Glucose

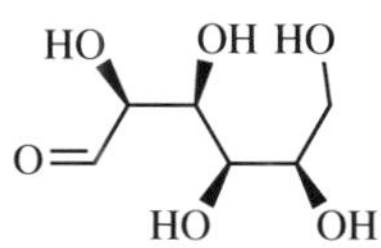

D-Mannose

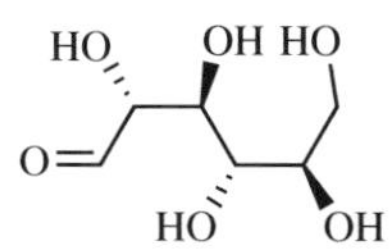

D-Galactose

(4) *Examples for additional modifications*

(A) *Hydroxylation*; (B) *Phosphorylation*; (C) *Reduction*; (D) *Amidation*; (E) *Acetylation*

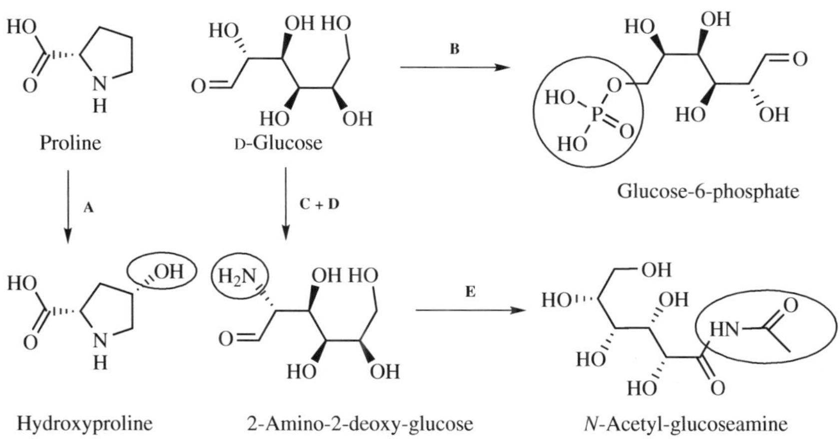

differences between atoms in a bond, the more polar is the bond. For example, the presence of an oxygen atom makes the compound more polar than a nitrogen atom, because oxygen is more electronegative than nitrogen. The catch is that these effects can be pH dependent so that amines can be very polar (ionic) at low pH and apolar at higher pH. Similarly, for organic acids, they can be very polar at higher pH (ionic) and lesser polar at low pH. However, in both cases, the compounds are somewhat polar because of their ability to form hydrogen bond with water, and oxygen with two lone-pairs can form better hydrogen bond network than nitrogen with only one lone-pair. Depending on the functional groups positioned at the molecule and the pH of its environment, a ranking in polarity is possible, the most polar being on the left:

Acid > Amide > Alcohol > Ketone ~ Aldehyde > Amine > Ester > Ether > Alkane

In addition, the polarity determines the forces of interaction between the molecules in the liquid state. Polar molecules are attracted by the opposite charge effect (the positive end of one molecule is attracted to the negative end of another molecule). Molecules have different degrees of polarity as determined by the functional group present. The general principle is as follows: The greater the forces of attraction, the higher the boiling point or the greater the polarity, the higher the boiling point.

(iv) Volatility—The volatility of a compound depends on its boiling or melting point, meaning the temperature at which it changes from solid or liquid to gaseous state. As described above, there is a strong correlation between the polarity and boiling point of a compound and therefore between the polarity and volatility of the molecule as well: Greater polarity means less volatility.

(v) Solubility—The solubility of a solute is the maximum quantity of solute that can dissolve in a certain quantity of solvent or solution at a specified temperature. This feature is mostly related to polarity, pK_a, temperature, solvent, and size. There are a few major factors, which have to be considered as they affect the solubility and also the time until a solute is dissolved. First, the nature of the solute and the solvent is the main factor determining the solubility. For a solvent to dissolve in a solute, the particles of the solvent must be able to separate the particles of the solute and occupy the intervening spaces. Polar solvent molecules can effectively separate the molecules of other polar substances. This happens when the positive end of a solvent molecule approaches the negative end of a solute molecule. For example, ammonia, water, and other polar substances do not dissolve in solvents whose molecules are nonpolar. However, nonpolar substance such as fat will dissolve in nonpolar solvents. On the contrary, polar solvents can generally dissolve solutes that are ionic. The negative ion of the substance being dissolved is attracted to the positive end of a neighboring solvent molecule. The positive ion of the solute is attracted to the negative end of the solvent molecule. Secondly, the size of the solute particles affects the solubility and rate of solution. When a solute dissolves, the action takes place only at the surface of each particle. When the total surface area of the solute particles is increased, the solute dissolves more rapidly. Breaking a solute into smaller pieces increases its surface area and hence its rate of solution; therefore, breaking apart a cell into very small parts will increase the solubility of many metabolic compounds. Thirdly, an increase in the temperature of the solution increases the solubility of a solid solute. On the contrary, for all gases, solubility decreases as the temperature of the solution rises. Fourthly, changes in the pressure have a strong effect on the solubility of gaseous solutes: An increase in the pressure increases the solubility and a decrease in the pressure decreases the solubility. In addition, stirring of the solvent containing the liquid or solid solutes brings fresh portions of the solvent in contact with the solute, thereby increasing the rate of solution, and of course, when there is little solute already in solution, dissolving takes place relatively more rapidly. As the solution approaches the point where no solute can be dissolved, dissolving takes place more slowly until it reaches saturation.

(vi) pK_a is an important parameter to describe many metabolites. The pK_a describes at what pH an equal number of the acidic or alkaline functional group will be protonated and at what pH they will not. Hence above or below the pK_a value, the metabolites may be ionized or neutral.

(vii) Stability—The stability of a chemical is defined by its resistance to chemical reactions, changes, or degradation due to internal or external reactions. There are two factors affecting stability: the thermodynamics and the kinetics. A substance that is thermodynamically unstable (or energetically unstable) has a more negative Gibbs free energy (ΔG). A substance or mixture that would be mostly converted into something else at equilibrium is said to be thermodynamically unstable. On the contrary, the substance or mixture is said to be kinetically unstable when it reacts extremely fast. The time a substance takes for a reaction to occur is a measure of its kinetic stability. The *slower* the reaction, the *greater* the kinetic stability. This is especially important with respect to metabolite analysis. Many metabolic compounds are extremely unstable, particularly when removed from their cellular environment. Therefore, the right conditions for increasing the thermodynamic and kinetic stability have to be chosen in the extraction process. There are different types of unstability to consider. The highest impact with respect to metabolite analysis may well be thermo-unstability. Many metabolic compounds degrade when exposed to higher temperatures, which may be already room temperature. Another factor influencing stability is photodegradation caused by too much light. Lastly, some compounds are sensitive to oxidative or reductive conditions. Therefore, the right conditions for the extraction of cellular compounds and sample preparation for metabolite analysis using any analytical method have to be carefully chosen. More detail on appropriate extraction methods and sample handling of unstable compounds are discussed in Chapter 3.

2.2.2 Metabolite Abundance

There are several factors that affect the concentration levels (abundance) of each metabolite in a cell at any one time. The most important factors influencing the cellular concentration and excretion of metabolites are the environment (medium), the uptake, turn-over rate, the number of pathways in which the metabolites take part, whether it is an intermediate or end product, cell status, and so forth. Even though a cell can perform millions of metabolic reactions, they all are not running simultaneously at any given moment. Also, some metabolites play roles in many different pathways where some may have a very low or even zero fluxes whereas other metabolites are very active, channeling a lot of metabolites through them. Finally, some metabolites are intermediates and are never released from the enzyme complex where they are used. Clearly, the level of the fluxes will strongly affect the actual amount of metabolite present in the cell at a given time. Thus, some metabolites will be highly abundant and others will be present in only trace amounts. In many cells, glucose, for example, is present in millimolar concentrations whereas certain signaling molecules may be present only with a few molecules per cell. This has an important impact on the analytical method an investigator needs to apply for coping with this huge dynamic range in which metabolite levels exist in biological systems.

2.2.3 Primary and Secondary Metabolism

The compounds in a living organism are divided into primary and secondary metabolites. Primary metabolites are generally distributed within all living organisms and are intimately connected with essential life processes and include ubiquitous compounds, such as sugars, amino acids, or organic acids. These are produced by and involved in primary metabolic processes, such as glycolysis, respiration, or photosynthesis. In addition, the universal building blocks and energy sources like proteins, nucleic acids, or polysaccharides belong to primary metabolism although they differ in structural detail from one organism to another. In contrast, secondary metabolites have only restricted distributions and are often a specific characteristic of individual organisms and species. In general, it can be noted that primary metabolites participate in nutrition and essential metabolic processes inside each cell. On the contrary, secondary metabolites do not appear to participate directly in growth and development and therefore are nonessential to life although they are important to the organism which produces them to influence ecological interactions between the organisms and their environment. Primary and secondary metabolisms are intimately related with secondary metabolites depending on precursors and energy generated through primary metabolism. Secondary metabolites are produced by pathways derived from primary metabolic routes and characterized by an enormous chemical diversity. It is interesting to note that despite this diversity, secondary metabolites are synthesized essentially from only a small number of key primary metabolites, which is the basis of a general classification of secondary metabolites into three major groups. Terpenoids are derived from the five-carbon precursor isopentenyl diphosphate (IPP), alkaloids are synthesized principally from amino acids, and phenolic compounds are originated from either the shikimic acid pathway or the malonate/acetate pathway.

As the set of secondary metabolites in each organism is specific and also a comprehensive analysis of these compounds within a metabolomics context is organism-specific, a more detailed description of secondary metabolites is given in the case studies (Chapters 8–10).

2.3 THE NUMBER OF METABOLITES IN A BIOLOGICAL SYSTEM

There have been many attempts to estimate the number of metabolites in a biological system. The size of the metabolome varies greatly, depending on the organism studied. The completion of whole genome sequences of many different species has enabled estimation of the number of metabolites, but owing to the lack of complete gene annotations in sequenced genomes, not all possible metabolic reactions can be predicted. For example, the well studied model of eukaryotic organism *Saccharomyces cerevisiae* contains more than 6000 genes of which only approximately 70% have been studied so far, and hence there are almost 2000 genes whose function is unknown. Therefore, the number of metabolites estimated is uncertain and only represents a rough estimate. In general, it has been stated that the number of possible metabolites in a cell is lower than the number of all genes and proteins in a cell. There are several reasons for this assumption. First, there is no one-to-one relationship between a gene and a chemical reaction in the same way as there is no direct linkage among genes, transcripts, and proteins. Secondly, quite a few metabolites participate in several pathways, and thus act on different enzymes that again are coded by different gene. Thirdly, some more complex metabolites, in particular the secondary metabolites, require many genes for their productions, often carried out by large enzyme complexes. An example is found within the polyketides, which are synthesized from long chains of acetyl moieties assembled, folded and modified in large enzyme complexes. These enzymes are oligomeric complexes, which contain more than one protein chain coded by different genes. Complexes are formed by noncovalent bonds or static or transient association of several different protein molecules. In most cases, these protein complexes are responsible only for very specific reactions and therefore may involve only two metabolic molecules, the substrate and the product, but on the contrary, it has to be noted that a number of enzymes can catalyze more than one chemical reaction resulting in the transformation of different metabolic structures whereas the type of reactions tend to be very similar. For example, some nonspecific glycosyltransferases are able to transfer the glucose moiety, in most cases, of UDP-glucose into different acceptors, always resulting in a glycosylated structure as their product, and fourthly, many key metabolites are involved in a large number of metabolic reactions which involve many different enzymes and therefore genes.

In reality, it is extremely difficult to determine the number of metabolites and also other cell products, such as transcripts and proteins, at a given time in a given cell because of the lack of analytical techniques to measure all cellular components in a comprehensive manner. In many bacteria and also some eukaryotes such as baker's yeast, detailed wide genome analyses have made great progress to get more information about the real complexity of these comparatively simpler cellular organisms. For example, in the well studied bacterium *E. coli*, there are about 4400 genes and it is estimated that only about 442 metabolic compounds are produced (Edwards and Palsson, 2000, PNAS 97, 5528-5533) whereas for the eukaryote *S. cerevisiae*, which contains about 6200 genes, it has been estimated that it contains slightly more than 700 metabolites (Forster et al., 2003, Genome Res. 13, 244-253). Most metabolites in these two relatively simpler organisms are related to the central metabolism responsible for energy turn-over, cell life cycle, and reproduction. None of these organisms produce more complex metabolites and relatively fewer, if any, produce secondary metabolites. In both cases these numbers represent all metabolic components ever capable of being made within the life cycle of these microorganisms. In higher organisms, the situation becomes much more complex. Additional dimensions, such as tissue specificity or organ structures, make correct estimations extremely difficult. For example, it has been estimated that the whole plant kingdom might be capable of producing between 200,000 and 400,000 primary and secondary metabolites and a similar number within the fungal kingdom. However, a single specie may use and produce many of the well-known metabolites from the central metabolism but may not produce all possible secondary metabolites. However, only about 5000 might be actually present in the well-studied plant model *Arabidopsis thaliana* at a given time point.

Finally, it is important to remember that the pool of metabolites in any organism also reflects the surrounding; thus, all metabolites that are taken up by the cell or organism will be a part of the metabolome even if they are not used in any way, and metabolites originating from cellular degradation also add to the complexity of the metabolome. As described in Section 2.2, given the large number of structural differences between metabolic compounds together with the enormous qualitative variety of the metabolomes, it is difficult to analyze all metabolites by one method.

2.4 CONTROLLING RATES AND LEVELS

Thousands of metabolic reactions can occur even in the simplest living cell. Each reaction needs a specific enzyme, which catalyses this reaction. However, it has to be noted that not all possible reactions that can occur within a living cell will typically operate at the same time. In reality, only a small fraction of the reactions operate at one given point of time, and it is essential for efficient functioning of living cells that the enzymatic activity and therefore the rate of interconverting the different metabolites is highly coordinated and regulated. There are different levels of regulating metabolic events. The three major levels are as follows:

(i) control of enzyme level
(ii) control of enzyme activity
(iii) control of uptake and transport

The concentrations of different enzymes vary widely in cellular extracts. Enzyme levels are controlled partly by regulating the enzyme's rate of synthesis, but the rate of enzyme degradation can also be a factor in controlling enzyme levels. Enzyme synthesis involves transcription of the gene that encodes the enzyme and further translation of the mRNA. There can be control at several different points in protein synthesis, and this may involve induction or repression by the presence or absence of certain metabolites. The control of protein synthesis is complex and involves many different biological processes, but we will not discuss this further here as our focus is at the level of metabolism and, hence, control of enzyme activity. The regulation of the enzyme activity is archived by a reversible interaction of the enzyme with ligands and by covalent modification of the enzyme itself. Low molecular weight ligands, which are metabolites themselves, can interact with enzymes and exert positive and negative controls. Indeed, pathway intermediates can influence the rate or their own conversion as well as the conversion of other metabolites in a pathway of which they are a member. In the following sections we will discuss different mechanisms involved in regulation of enzyme activity.

2.4.1 Control by Substrate Level

The concentration of a reactant in a given enzymatic reaction can regulate the catalytic activity of the enzyme performing the transformation. This type of control of enzyme activity is called cooperativity. Often the first step of a pathway is controlled by these stimuli and is in principle simple: The more the substrate available, the higher is the rate of conversion and hence, feeding into that particular pathway resulting in an increased amount of product being formed.

2.4.2 Feedback and Feedforward Control

Feedback control mechanisms usually involve inhibition of specific enzymes, and often a metabolite formed in a pathway inhibits the action of an earlier step in the pathway. In most cases, the level of the end product of a particular pathway inhibits the starting reaction, the first step at which the pathway begins. By this regulation mechanism, entire pathways may be down regulated when the end product is present in sufficient amounts. The inhibition of the enzyme activity can be reversible or irreversible. Another mechanism of regulation, but in this case in a positive manner, is feedforward, which occurs when a molecule in a reaction series activates the activity of an enzyme that is involved in a reaction downstream in the pathway.

2.4.3 Control by "Pathway Independent" Regulatory Molecules

Many biological processes require catalytic functions beyond those provided by the protein making up the enzyme, i.e., the enzyme requires the help of other small organic molecules or ions to carry out the reaction. Molecules which can bind to enzymes and regulate their activation level are called coenzymes. It has to be noted that some of these are metabolites itself, which have to be synthesized specifically for this purpose in independent pathways. A coenzyme may either be attached by covalent bonds to a particular enzyme or exist freely in solution, but in either case it participates intimately in the chemical reactions catalyzed by the enzyme. Often a coenzyme is structurally altered in the course of reaction, but it is always regenerated to its original form in a subsequent reaction catalyzed by other enzyme systems. The most abundant and known coenzymes are used for energy transfer and in redox (electron transfer processes) reactions, e.g., adenosine triphosphate (ATP), nicotinamide adenine dinucleotide (NAD), and nicotinamide adenine dinucleotide phosphate (NADP), whereas others are crucial in catabolism of metabolites and key structures including DNA, e.g., coenzyme A (CoA) (structure see Figure 2.2), riboflavin mononucleotide (FMN) and flavin adenine dinucleotide (FAD), biotin, pyridoxal phosphate, thiamine pyrophosphate, or tetrahydrofolic acid (THFA).

ATP is a coenzyme of vast importance in the transfer of chemical energy derived from biochemical oxidations and its importance will be discussed in more detail in Section 2.7. NAD^+ and its phosphorylated form $NADP^+$ are derived from adenine, ribose, and nicotinic acid or niacin (a vitamin of the B complex) and are important intermediates in biochemical oxidations and reductions within the cell. Both NAD^+ and $NADP^+$ can be reduced by accepting a hydride ion (H^-, a proton with two electrons) from an appropriate donor; the resulting NADH and NADPH can then be oxidized back to their original states by transferring their hydride ions to various acceptors. In this fashion, electron pairs (and protons) are shuttled around in the cell from high-energy donors to low-energy acceptors. CoA is another coenzyme that has been shown to participate in a variety of biochemical reactions, all involving acyl groups such as the acetyl unit; it is, for instance, associated with the pivotal first step of the tricarboxylic acid cycle, in which an acetyl unit (the breakdown product of carbohydrates) is introduced into the cycle to be converted eventually into carbon dioxide, water, and chemical energy. CoA is derived from adenine, ribose, and pantothenic acid (a vitamin of the B complex). Other functions of acetyl-CoA are acting as a donor of acetate for the synthesis of fatty acids, ketone bodies, or cholesterol. Here a classical regeneration occurs; i.e., following the transfer of the acetyl group onto its acceptor, CoA is released. The regeneration is carried out by the pyruvate dehydrogenase complex, which catalyzes the oxidative decarboxylation of pyruvate to form acetate which is further attached to the CoA to form acetyl-CoA. The process is simplified in Figure 2.3.

Another class of regulators for enzymatic reactions are inorganic substances or metal ions, which are called cofactors. Many enzymes require the presence of these cofactors to catalyze their reactions; in other cases, the presence of the cofactor may increase the rate of the catalysis of the reaction. Some examples of common cofactors are presented in Table 2.1.

2.4.4 Allosteric Control

Many enzymes exist in active and inactive conformation. These enzymes are invariably multisubunit proteins, with specific allosteric sites for binding an activation molecule. The binding of the activator will transform the inactive enzyme into its active conformation and *vice versa*. There are two forms of allosteric regulation: first, if the substrate of the reaction itself is the activator (homoallostery) and second,

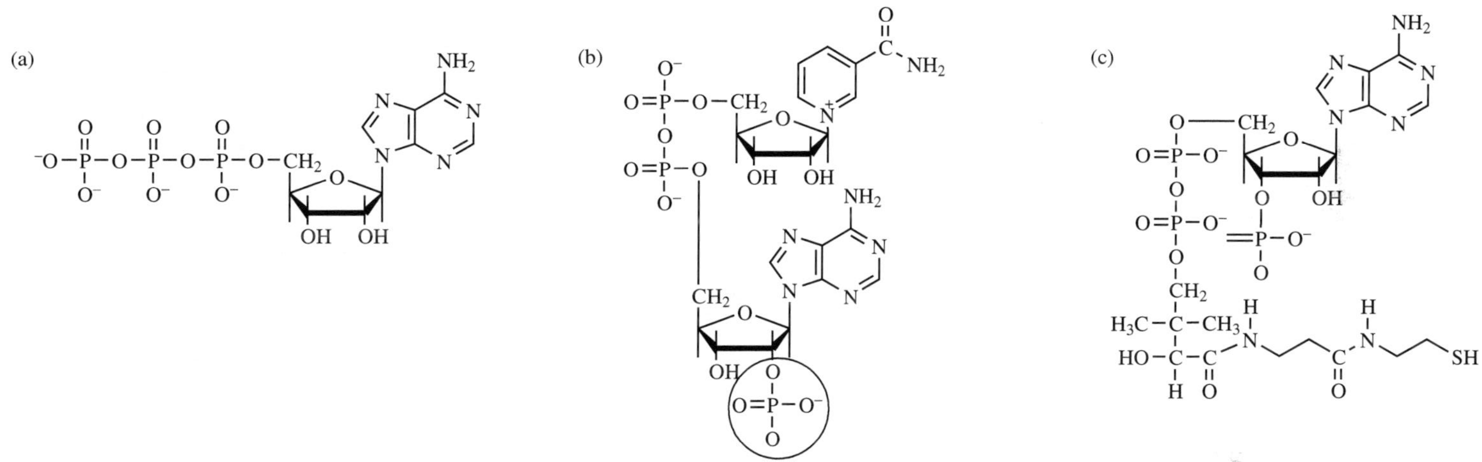

Figure 2.2 Molecular structure of (a) ATP, (b) NAD(P)$^+$, and (c) CoA.

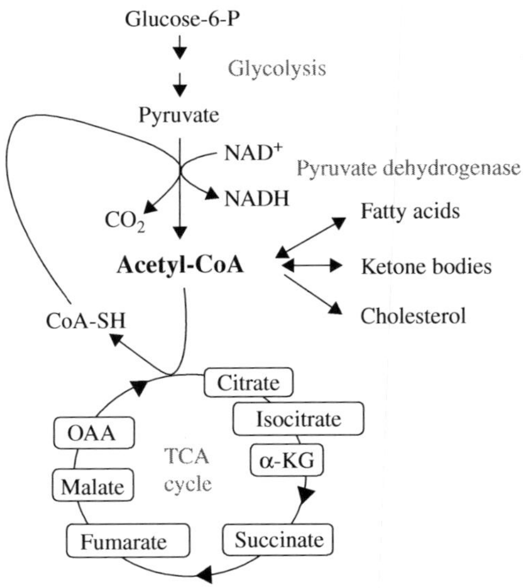

Figure 2.3 The role of acetyl-CoA as a primary acetyl-group donor and its production and generation.

if another molecule, the effector, which is not being transformed in this particular pathway, is bound to the enzyme (heteroallostery).

2.4.5 Control by Compartmentalization

A major way in which cells control the flow of metabolites in relation to the bioenergetic status of a cell is by separating metabolic reactions into different compartments, which not only allows a spatial but also temporal regulation of enzyme activities, and hereby the rate metabolites undergo various metabolic reactions. One of the most well known and simplest examples is the process of starch biosynthesis in heterotrophic plant tissues (Figure 2.4). Sucrose as the energy source, which is produced in the photosynthetic green "source" tissues, is delivered via the apoplastic

TABLE 2.1 Common Cofactors with Examples of Enzymes and Proteins that Require Them for Their Functionality.

Cofactor	Enzyme
Fe^{3+} or Fe^{2+}	Ferredoxin
Zn^{2+}	Alcohol dehydrogenase
Cu^{2+} or Cu^{+}	Cytochrome oxidase
K^{+} and Mg^{2+}	Pyruvate phosphokinase

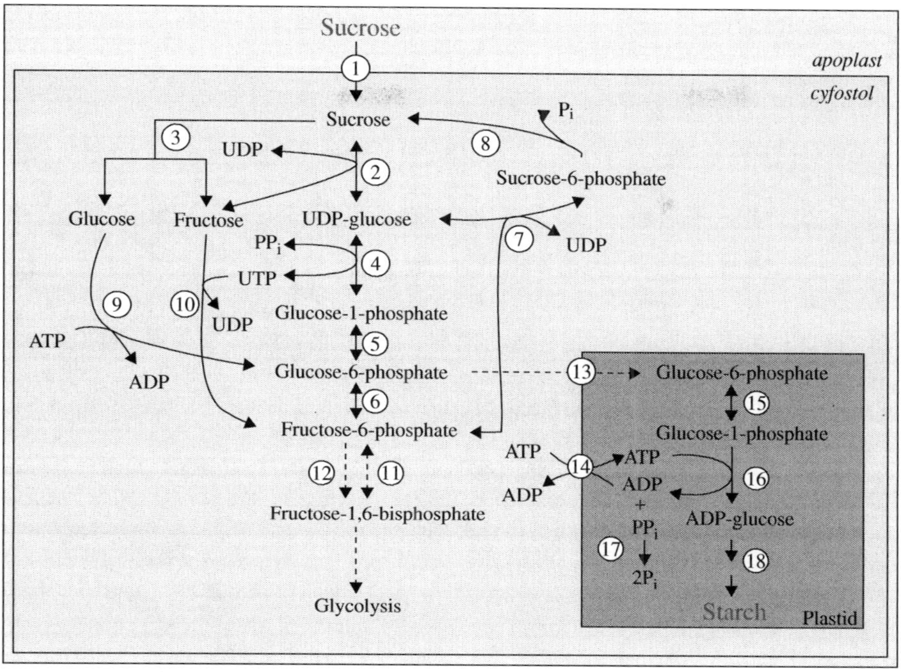

Figure 2.4 Compartmentalization of the sucrose to starch metabolism in heterotrophic plant cells. The numbers denote the following enzymes: (1) sucrose transporter; (2) sucrose synthase; (3) alkaline invertase; (4) UDPglucose pyrophosphorylase; (5) cytosolic phosphoglucomutase; (6) phosphoglucose isomerase; (7) sucrose phosphate synthase; (8) sucrose phosphate phosphatase; (9) hexokinase; (10) fructokinase; (11) pyrophosphate:fructose-6-phosphate phosphotransferase; (12) phosphofructokinase; (13) plastidial glucose-6-phosphate transporter; (14) plastidial ATP/ADP translocator; (15) plastidial phosphoglucomutase; (16) ADPglucose pyrophosphorylase; (17) pyrophosphatase; (18) starch synthetic enzymes.

stream and taken up by the heterotrophic "sink" cells (e.g., roots or tubers). It is degraded to glucose-6-phosphate, which either enters the glycolytic pathway or is transported by a plastidial glucose-6-phosphate transporter into the amyloplast, a nonphotosynthetic form of plastids. Glucose-6-phosphate serves there as the precursor for starch synthesis by an initial transformation into ADP–glucose.

2.4.6 The Dynamics of Metabolism—the Mass Flow

As described above, metabolites are under constant transformation; thus, once formed they may be used immediately. The levels of many metabolites change in half a minute or second, or even faster, in any case far faster than the turn-over for nucleic acids or proteins. Therefore, not only the concentration of metabolites provides information on the status of the cell but also the flow through the many different pathways provides important information on the cellular state. It is important to distinguish between reactions and the fluxes through reactions. As an example, a reaction can be described as a one-to-one relationship and can be described by defined values:

1 molecule glucose + 1 molecule ATP → 1 molecule glucose-6-P + 1 molecule ADP

The fluxes through pathways are, however, the rates of the reaction at which the amount of material (atoms) is going through in a given time. Therefore, flux values represent the amount of substrate that is being converted to a product in a unit time.

Several different approaches have been developed to quantify metabolic fluxes through the different pathways operating within living cells. This includes the measurement of the consumption rate of a substrate or the accumulation rate of a product. This, however, does not provide information on how the fluxes distribute within the many different pathways inside the cell. Information on this can be obtained by the use of labeled metabolites, i.e., metabolites containing enrichment in certain isotopes like ^{13}C. In these experiments, a specifically stable or radioactive isotope-labeled substrate is provided to the biological system (*in vivo* to whole cells or organisms or *in vitro* to, e.g., tissue slices). Over a certain time frame, the label is then distributed all over the network until, finally, the enrichment of label in intracellular metabolite structures is measured either by determination of radioactivity or by the stable isotopic pattern using NMR or mass spectrometry. When the distribution of label is quantified per time unit, the actual fluxes can be calculated. It is very important to distinguish between steady-state and kinetic labeling. In steady-state labeling experiments, it is assumed that the equilibrium of labeled and unlabeled molecules of a certain metabolite is reached. In kinetic labeling, a steady-state is not reached, but the kinetics of the changes in labeling enrichment of different metabolite pools is determined.

Metabolic flux analysis (MFA) is a global approach to quantify metabolic fluxes through the entire biochemical reaction network of a cell or organism. This results in a flux map that shows the distribution of fluxes over the complete network (or at least a reasonable representation of this). In this method, intracellular fluxes are calculated from a few measured fluxes, e.g., fluxes in and out of the cell, by using a mathematical model for the metabolic network. A key assumption in these calculations is a steady-state level in all intracellular metabolites, but owing to the low half-times, this is generally a reasonable assumption. This approach is quite valuable as it is not (yet) possible to determine fluxes through all metabolic pathways comprehensively by other methods, mainly owing to major limitations in the ability to determine all metabolic compounds and their isotope enrichment simultaneously.

The major application of metabolic flux analysis is in the field of metabolic engineering which aims at the overproduction of high-value metabolites (e.g., essential amino acids in feeding crops, ethanol in yeast) preventing side effects in the overproducing organism.

For further reading see Christensen et al. (2002); Schwender et al. (2004); Fernie et al. (2005).

2.4.7 Control by Hormones

A higher level of regulation of reactions and transport processes can be achieved by the action of specific signaling substances, e.g., hormones. Hormones are metabolites synthesized in one type of cells and then transported to another type of cells, where they trigger a specific effect. They are therefore considered as metabolites having an important biological function to transfer information from one cell to another. Classes of compounds involved in this type of regulations are hormones, growth factors, neurotransmitters, and pheromones. The examples are steroid hormones, such as testosterone and estradiol, well known as sex hormones, which are bound to a hormone receptor that will undergo a conformational change either initiating a complex signaling cascade or directly interacting with DNA to control the transcription of selected genes. The cascade initiated by binding of the extracellular substance (the first messenger, the hormone) is based on the action of second messengers. In addition to their function in relaying information from the first messenger to the control point (e.g., DNA transcription), they importantly serve as an amplifier of the strength of the signal. The binding of a first messenger to a single receptor at the cell surface may result in massive changes in the biochemical activities within the cell. There are three major types of second messengers: (i) cyclic nucleotides (e.g., cAMP, cGMP); (ii) inositol triphosphates (IP_3); and (iii) calcium ions, where the first two classes are by definition metabolites themselves. The analysis of hormones from a metabolomics point of view is challenging because their concentrations in biological tissues are very low. Special enrichment and purification procedures have to be applied allowing the detection and also quantification of these messenger molecules. Potential methods aiming at enrichment of low-abundant metabolites are described in Chapter 3.

2.5 METABOLIC CHANNELING OR METABOLONS

The interior of a cell is very crowded and owing to dense packing of its molecular contents, the mobility of solutes is limited. To overcome the hindered diffusion of molecules, the cell needs to compartmentalize metabolic pathways. As described in Section 2.1.4, one way is to accomplish different pathways in different cell compartments, such as the mitochondrion or the Golgi apparatus. Another possibility is to facilitate the direct transfer of metabolic intermediates to the active site of the subsequent transforming enzyme in the pathway without release of the metabolite to a free aqueous phase. This can be accomplished by building aggregates of the relevant enzymes involved in a given pathway. The association of the various cooperating enzymes belonging to a pathway in large complexes is called metabolons. The enzyme clusters fall into two different classes: (i) the static association, where the set of enzymes belonging to the metabolon exists in the absence of the starting substrate and/or any intermediate and (ii) the dynamic association, which only assembles when a certain metabolic component is bound to one of the enzymes in the pathway. In most cases, this initiator of the assembling is the metabolite that is involved in the metabolon.

The enzyme complexes allow the direct transfer of the series of biosynthetic intermediates between catalytic sites of enzymes belonging to the pathway without releasing them into the bulk solvent of the cell. An intermediate, which is formed by one catalytic site of one enzyme, can then be directly transferred to the catalytic site of the following enzyme. There are a number of advantages of metabolic channeling, for example, the intermediates are (i) not diluted, (ii) contaminated by other molecules, (iii) the transition time between catalytic sites is dramatically reduced, and most importantly, (iv) competing site reactions are excluded. In addition, regulatory aspects of metabolism are enhanced by, for example, remaining an optimal local substrate concentration for maximal enzyme activity and regulating the competition of other pathways for common metabolites. Another important feature of metabolic channeling is that highly reactive or toxic intermediates are separated from other components of the cell or directly sequestered for excretion. In many cases, metabolons are associated with structural elements in the cells such as membranes, which may facilitate the transport of the final product through the membrane.

The state of the association of a metabolon often provides a rapid and powerful mechanism for regulating metabolic activity. Although all components of the metabolon may be present, but as long they are not associated, the channeling process and therefore metabolic action is not possible. Specific mechanisms sensing the metabolic status or energy demands of the cell lead to activating the association process of the metabolon enzymes by, for example, phosphorylation of one or more of the proteins involved in the metabolon.

The *in-vitro* and *in-vivo* investigation of multienzyme formations is very difficult. Therefore, only a few numbers of metabolons are studied in detail so far. A well known, detailed and characterized example is the Calvin cycle in green tissues. During this cycle, which consists of a serious of various reactions, CO_2 is incorporated into a five-carbon sugar named ribulose-1,5-bisphosphate by an enzyme called ribulose-1,5-bisphosphate carboxylase/oxygenase (called Rubisco). The product of the reaction is a six-carbon intermediate which immediately splits into half to form two molecules of 3-phosphoglycerate. In further reactions, ATP and $NADPH_2$, delivered from the photosynthetic light reactions, are used to convert 3-phosphoglycerate to glyceraldehyde 3-phosphate, the three-carbon carbohydrate precursor to glucose and other sugars which are then transported through the cell for other biosynthetic reactions or storage. In the third phase, more ATP is used to convert some of the pool of glyceraldehyde 3-phosphate back to RuBP, the acceptor for CO_2, thereby regenerating and completing the cycle. This complex is loosely associated with the tylakoid membranes in the chloroplasts of the green tissues, such as leaves, near the sites of ATP and NADPH production within photosynthesis. The assembly of the complex mainly enhances the step of carbon fixation by Rubisco, and also the activity of other enzymes involved in the cycle is dependent on their complex formation. It could be demonstrated that by enzyme association, a mechanism for enhanced intermediate channeling and the flux through the cycle is controlled by modifications of individual enzymes for additional regulation of activity.

For scientists who aim to identify and quantify all small molecules in biological system, i.e., the metabolome, it is important to have in mind that all cells contain many different organelles, microcompartments, and possible metabolons. Therefore, the analysis of metabolite concentrations in tissue parts or single cells results only in average cellular concentrations but does not provide the actual concentration of a substrate around the active site of its transforming enzyme. There is a lot of developmental potential for highly sensitive, extremely spatial resolved metabolite detection assays, which also enable accurate quantification at any place in the cell. A prerequisite for these methodologies is that the cell or parts of the cell have to be fixed to stop any further flux of metabolites and arrest all enzymatic activities. Then the compounds have to be visualized and quantified, for example, by using colorimetric assays or some sort of imaging technique. This technique has been successfully applied to determine the distribution of ATP in legume embryos during development (Borisjuk et al., 2003, Plant J., 36, pp. 318–329). For further readings see Winkel (2004); Jørgensen et al. (2005).

2.6 METABOLITES ARE ARRANGED IN NETWORKS THAT ARE PART OF A CELLULAR INTERACTOME

With increasing knowledge about metabolites and their transformation, it is now possible to analyze the structure and the behavior of the networks on the basis of the connection between two metabolites by the chemical reaction forming one from the other. On the basis of the knowledge about a (nearly) full set of transforming chemical reactions and associated transport processes, which become available for more and more organisms, the reconstruction of the underlying metabolic networks *in silico* is possible. In this, the biochemistry of the reaction networks is directly translated into the realm of linear algebra in the form of a stoichiometric matrix. As metabolites are connected by reactions and therefore enzymes, however the questions raised, which metabolites play key roles within the network structure or if there are particular well-suited metabolites keeping the network in its structure. In the past, increasing information from genome sequencing, advanced protein and metabolite analyses, gave the opportunity to map a picture of the complex relationships between all components of the network. The simplest measurement of network complexity is to measure the node degree; this determines how many neighbors each node has. This determination of the neighborhood of each network components is also described as the connectivity of the components (Dandekar and Schmidt, 2004). Pathway-genome wide databases have been developed and can be used to reconstruct organism-specific connectivity maps of metabolites and their connecting reactions. The degree of connectivity of a metabolic network can be characterized by the network diameter, defined by the shortest biochemical pathway averaged over all pairs of metabolites. The diameters of a range of metabolic networks from different organisms are very similar, irrespective of the number of metabolites found in the given species. The reason for this might be that with increasing complexity of the organism, individual metabolites are increasingly connected. It has been found that the average number of possible reactions, in which a metabolite participates, increases with the number of metabolites in the system. Very important to note is that only a few well connected metabolites ("hubs") dominate the overall connectivity of the network. Once one of these "hub" metabolites is removed from the network, the network diameter increases dramatically, demonstrating the importance of these metabolites (Jeong et al, 2000). As the large-scale architecture of the network is determined by these well-connected compounds, it is interesting to investigate if in all organisms the same "hub" metabolites are functional or whether there are organism-specific differences in the identity of the highly connected nodes. A general feature of many complex networks is their "small world" character, meaning that any two nodes in the system are connected by relatively shorter paths along existing nodes, which enables messages to reach every node in the network in a very rapid way and therefore optimizes the reaction efficiency of metabolism to any kind of perturbations (Wagner and Fell, 2001).

It could be demonstrated that the ranking of most connected metabolites is similar for 43 analyzed organisms, meaning that the network structure is highly conserved within species. The species-specific differences were only for very lowly connected metabolites. The majority of metabolites are rarely used whereas only a few are used very frequently. Interestingly, these highly connected metabolites belong to energy-capturing metabolites or to cofactors; however in general, it was determined that many small hydrophilic compounds are selected (see Figure 2.5). The most used molecule in nearly all networks is water, which is not surprising as it is needed and released by a huge number of enzymatic reactions. The most frequently used metabolites are ATP and ADP, the reduction equivalents NAD+ and NADP+, and their reduced form NADH and NADPH. The "small world" behavior of the

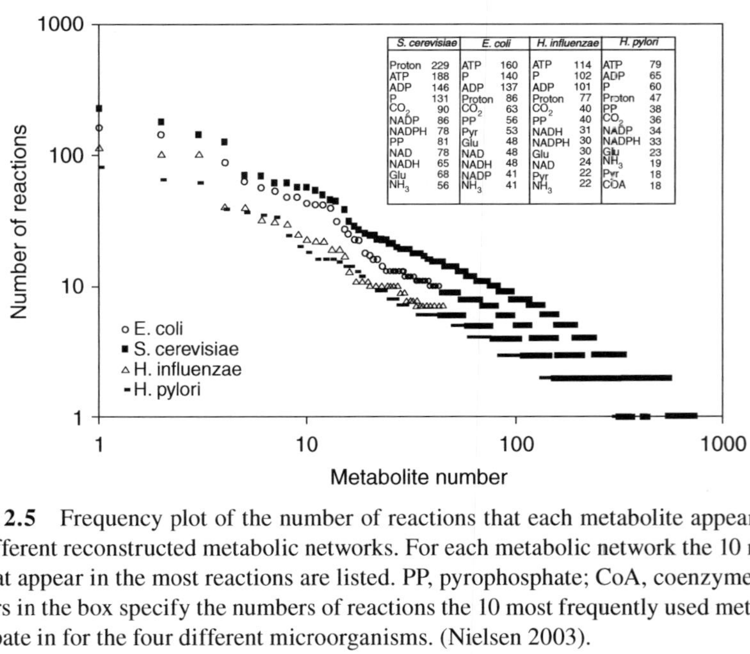

Figure 2.5 Frequency plot of the number of reactions that each metabolite appears in for four different reconstructed metabolic networks. For each metabolic network the 10 metabolites that appear in the most reactions are listed. PP, pyrophosphate; CoA, coenzyme A. The numbers in the box specify the numbers of reactions the 10 most frequently used metabolites participate in for the four different microorganisms. (Nielsen 2003).

network and the reason why ATP is the major "hub" metabolite is extremely obvious. When ATP levels are high, there is less need for energy generation, e.g., by carbon oxidized in the citric acid cycle. At such times, the cell can store carbon as fats and carbohydrates; so fatty acid synthesis, gluconeogenesis, and related pathways come into play. When ATP levels are low, the cell must mobilize carbon storages to generate substrates for energy metabolism, and carbohydrates and fat are therefore broken down. The information of the actual ATP levels therefore has to be distributed fast through the network to regulate and activate the right pathways.

Other well linked metabolites such as pyruvate, phosphoenolpyruvate, glutamate, α-ketoglutarate, AMP, acetyl CoA, and glutamine all belong to very central metabolic pathways, namely glycolysis, TCA cycle, or transamination reactions. This is again not surprising as these are of central importance for the cell survival by belonging to the energy metabolism or representing so-called precursor metabolites for synthesis of all carbon structures synthesized within a cell. In general, key metabolites are always those compounds that link two or even more different pathways. Interestingly, by detailed characterization of the metabolic network structures, it is now possible to not only identify the key metabolites generated by catabolism to be used in anabolism *per se*, but also define the center of metabolism dividing the degrading and the synthesizing metabolism.

Metabolic networks are only one way to model and describe a living cell. In fact, most biological characteristics are based on complex interactions of the numerous constituents of the cell, such as metabolites, proteins, mRNA transcripts, and also the genome. Therefore, it becomes extremely important to increase our understanding as to how this enormous complex machinery works and is regulated not only within a single isolated cell but also as an integrated system surrounded by other cells. Till date, the development of advanced analytical technology to determine cell products simultaneously and the application of powerful computing techniques have enabled scientists to construct and compare cellular networks. Various types of networks could be identified including metabolic, protein–protein-interaction, signaling, and transcription regulation networks, but none of these networks function on their own; they rather form a "network of the networks," also called the interactome. Detailed comparisons of the different networks in between and within the interactome could reveal that there is a high degree of common features in the architectural organization and structure of the networks. These include the small-world behavior mentioned above, conserved connectivity degree of nodes, the presence of well connected "hubs," preferential attachment of nodes (nodes prefer to attach to nodes that have already many links), the robustness of the network structure against perturbations, and the rapidity and efficiency to react to changes in external conditions. Interestingly, the activity of metabolic reactions or molecular interaction differs; some are highly active throughout the life cycle of the cell whereas others switch on only at certain environmental conditions. This goes in agreement with the known fact that some reactions have small or even zero flux coexisting with other reactions exhibiting very high fluxes. To increase the ability to analyze and understand network structure and topology completely, data collection skills have to be enhanced. This will require the optimization and development of highly sensitive methodologies for detection, identification, and quantification of the various types of molecules in a cell at extremely high resolution in both space and time. Finally, it becomes especially challenging to integrate the different types of networks and to look how the interactome contributes to the performance of the cell and finally understand the biological system as a whole. For further reading see Jeong et al. (2000); Wagner and Fell (2001); and Nielsen (2003); Barabasi and Oltvai (2004).

REFERENCES

Barabasi AL, Oltvai ZN. 2004. Network biology: Understanding the cell's functional organization. Nat Rev Gen 5:101–113.

Borisjuk L, Rolletschek H, Walenta S, Panitz R, Wobus U, Weber H. 2003. Energy status and its control on embryogenisis of legumes: ATP distribution within Vicia faba embryos is developmentally regulated and correlated with photosynthetic capacity. Plant J 36:318–329.

Christensen B, Gombert AK, Nielsen J. 2002. Analysis of flux estimates based on (13)C-labelling experiments. Eur J Biochem 269:2795–2800.

Dandekar T, Schmidt S. 2005. Metabolites and pathway flexibility. In Silico Biol 5:103–110.

Edwards JS, Palsson BØ. 2000. The Escherichia coli MG1655 in silico metabolic genotype: its definition, characteristics, and capabilities. PNAS 97:5528–5533.

Fernie AR, Geigenberger P, Stitt M. 2005. Flux an important, but neglected, component of functional genomics. Curr Opin Plant Biol 8:174–182.

Forster J, Famili I, Fu P, Palsson BO, Nielse J. 2003. Genome-scale reconstruction of the saccharomyces cerevisiae metabolic network. Genome Res 13:244–253.

Jeong H, Tombor B, Albert R, Oltvai ZN, Barabasi AL. 2000. The large-scale organization of metabolic networks. Nature 407:651–654.

Jorgensen K, Rasmussen AV, Morant M, Nielsen AH, Bjarnholt N, Zagrobelny M, Bak S, Moller BL. 2005. Metabolon formation and metabolic channeling in the biosynthesis of plant natural products. Curr Opin Plant Biol 8:280–291.

Nielsen J. 2003. It is all about metabolic fluxes. J Bacteriol 185:7031–7035.

Schwender J, Ohlrogge J, Shachar-Hill Y. 2004. Understanding flux in plant metabolic networks. Curr Opin Plant Biol 7:309–317.

Stryer L. 1995. *Biochemistry* (5th edition), W.H. Freeman, New York, USA.

Voet D, Voet J.G. 2004. *Biochemistry* (3rd edition), John Wiley & Sons, New York, USA.

Wagner A, Fell DA. 2001. The small world inside large metabolic networks. Proc R Soc Lond B 268:1803–1810.

Winkel BS: 2004. Metabolic channelling in plants. Ann Rev Plant Biol 55:85–107.

3

SAMPLING AND SAMPLE PREPARATION

By Silas G. Villas-Bôas

As a result of the complexity of the metabolome in both the diversity of chemistry and its wide dynamic range, adequate methods for sampling and sample preparation are of outmost importance in analysis of metabolites. Therefore, this chapter guides the reader through the main steps involved in harvesting and preparing the samples for metabolite analysis, covering the most important techniques to stop the cellular metabolism and to extract metabolites from different biological matrices.

3.1 INTRODUCTION

The metabolome is complex both in terms of chemical diversity and in terms of a wide dynamic range, and adequate methods for sampling and sample preparation are therefore of outmost importance in analysis of metabolites. Sample preparation is generally considered the limiting step in metabolome analysis because it is an important source of variability in the analysis. Because of the differences in cell structures, sample preparation from eukaryotes and prokaryotes is quite different, and even within the eukaryotic kingdom it is not possible to establish a general method for sample preparation in metabolome analysis. Sample preparation protocols in metabolomics are organism-dependent or, more precisely, cell-structure-dependent.

Figure 3.1 summarizes the general steps involved in sample preparation for analysis of metabolites. Since metabolome studies aim to relate metabolite levels with the response of biological systems to a genetic or environmental changes, the first step in

Metabolome Analysis: An Introduction, by Silas G. Villas-Bôas, Ute Roessner, Michael A. E. Hansen, Jorn Smedsgaard and Jens Nielsen

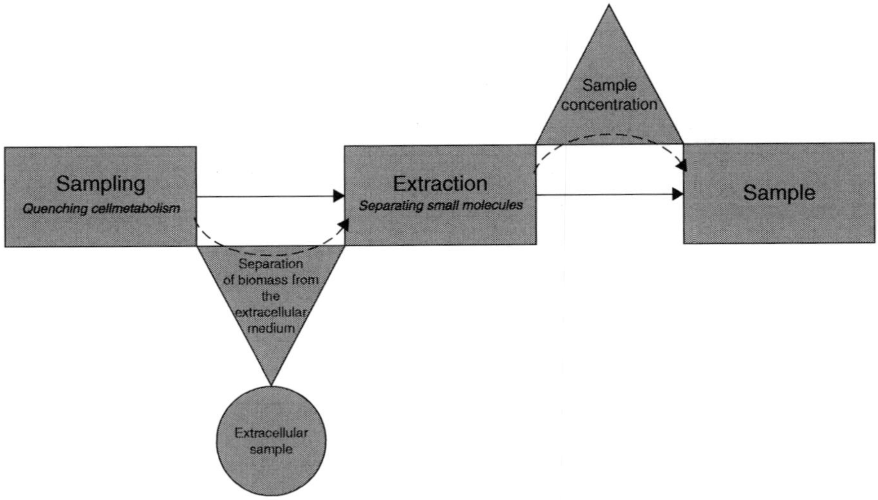

Figure 3.1 General steps involved in sample preparation. Full arrows indicate the sequence of the main events in sample preparation, and dashed arrows point alternative steps to improve analysis.

sample preparation is a rapid quenching of all biochemical processes concomitantly or immediately after sample harvesting. We have already discussed in Chapter 2 that metabolite concentrations change very rapidly induced by any (unnoticed) variation in the environment of the cells or organism. The metabolite turnover will depend mainly on the metabolite species (e.g., if primary or secondary metabolites), and its localization (e.g., intra- or extracellular). However, most primary metabolites have an intracellular half-life in the order of seconds or less, i.e., cytosolic glucose is converted to glucose-6-phosphate at an approximate rate of 1 mM/s and ATP is used in many different reactions at a rate of about 1.5 mM/s (Table 3.1). Quenching of metabolism is, therefore, an

TABLE 3.1 The Intracellular Turnover Value for Some Metabolites.

Metabolite	Turnover rate mM/s	Determined on	Reference
Glucose	1.0	*Saccharomyces cerevisiae*, aerobic cultivation on glucose	De Koning and van Dam, 1992
Glucose	0.3	Isolated adipocytes previous treated with insulin	Marshall et al., 2004
ATP	1.5	*Saccharomyces cerevisiae*, aerobic continuous cultivation on glucose ($D = 0.1$/h)	Rizzi et al., 1997
ADP	2.0	*Saccharomyces cerevisiae*, aerobic continuous cultivation on glucose ($D = 0.1$/h)	Rizzi et al., 1997

extremely important step for metabolome analysis, and it should be seriously considered during establishment/development of the sample preparation method.

Following the quenching step, it is necessary to make the metabolites accessible to the analytical method that will be used achieving minimal losses because of chemical degradation or further biochemical conversions. This second step usually involves the extraction of metabolites from the intracellular media by disrupting the cell envelop and subsequent separating the low molecular mass compounds from the biological matrix. In addition, several biological samples (i.e., microbial and cell cultures, blood, and others) will require separation of cells from the extracellular medium and distinct analysis of intra- and extracellular metabolites is often desirable. This step is the most time consuming, and it is virtually impossible to avoid losses mainly because of the high chemical diversity and the wide dynamic range of metabolite concentrations. Choices have to be made concerning which metabolites should be measured, and often analysis of some classes of compounds has to be sacrificed in favor of a good reproducibility of other metabolites. Alternatively, multiple extraction procedures should be applied to enable analysis of as many metabolites as possible, but still keeping the variability sufficiently low to allow reliable comparisons between samples and batches of samples.

Furthermore, many metabolites are present at fairy low levels in the samples and additional sample dilution is often observed during sample preparation procedures, which impose a requirement for sample concentration prior to the analysis in order to improve detection. However, losses by degradation and metabolite-class discrimination are also observed at this stage and again choices will need to be made guided by the objectives of the study that is being carried out.

3.2 QUENCHING—THE FIRST STEP

3.2.1 Overview on Metabolite Turnover

The turnover of metabolites and dynamics of cellular metabolism are discussed in details in Chapter 2. Here we will briefly review this important issue to permit the reader to understand, independently of Chapter 2, the necessity of quenching the cellular metabolism prior to any other procedure during sample preparation.

In analogy with taking a photography, which captures a static image from a dynamic environment, metabolome analysis represents snapshots of the *in vivo* metabolic state of a cell or organism in a specific developmental stage and environmental condition. The cellular metabolism is dynamic and the level of the measurable metabolites is the result of the ratio between the specific formation rates of each metabolite and their specific conversion rates to other metabolic products, as specified in Equation (3.1):

$$\mathrm{Met}_{\mathrm{level}} = \mathrm{Met}_{\mathrm{formed}} - \mathrm{Met}_{\mathrm{consumed}} \tag{3.1}$$

The rates of metabolic reactions depend mainly on the enzyme concentrations and the substrate availability (including availability of cofactors) and frequently also on different effectors, e.g., activators and inhibitors. Therefore, the rate of metabolic reactions not only determines the turnover of metabolites but also depends on the levels of the metabolites, and hence on the development stage of the cells or organism, and the environmental conditions. In the following we will look specifically at intracellular and extracellular turnover of metabolites.

3.2.1.1 Intracellular Turnover. For cellular cultures grown in suspension, the turnover of metabolites intracellularly is much faster than the turnover in the extracellular medium, simply because the cells generally account only for a relatively small fraction of the volume in the system. However, the intracellular metabolite concentration is usually much higher than the extracellular concentration. Table 3.1 lists a few metabolites and their intracellular turnover rates.

The primary metabolites, which are metabolites related to biochemical reactions involved in cellular synthesis and hence play a key role in cellular function (e.g., fuelling reactions), are intermediates of several different reactions, and they, therefore, usually have very rapid intracellular turnover (Box 3.1). On the contrary, metabolites formed via secondary metabolism usually accumulate in the cells or are secreted to the extracellular medium and, therefore, have a much slower turnover (Box 3.1). Thus, the primary metabolic reactions are the most critical part of the metabolic network in terms of rapid quenching. Furthermore, most primary metabolites participate in a large number of reactions and this means that most environmental or genetic alteration results in alterations in the levels of these metabolites. Primary metabolites are, therefore, often the main focus of metabolome studies, and measuring the intracellular levels of these compounds requires a rapid sampling with simultaneous inactivation of metabolic enzymes in a time window of seconds.

3.2.1.2 Extracellular Turnover. Extracellular metabolites are usually metabolites that have been secreted by the cells or resulted from degradation of polymers, but they may also appear due to cell lyses. The extracellular medium is more diluted than the intracellular and, therefore, the turnover of extracellular metabolites is slow if not absent. The main source of variability in the extracellular metabolite levels are the presence of living cells in the medium, which are responsible for metabolite uptake and secretion, cell lyses, and secretion of extracellular enzymes. However, turnover of extracellular metabolites is typically relatively slow due to relatively high concentrations compared with the uptake/secretion rates. For some cases, e.g., when microbial cells are grown at low limited substrate concentrations, e.g., at conditions with low glucose concentrations, but still with a high rate of substrate uptake the turnover can be in the order of seconds. In these cases, it is also important to rapid quench the cellular activity, but otherwise it is sufficient to simply separate the cells from the extracellular media to ensure a low variability on measurement of extracellular metabolite levels. However, there are still three other main potential sources of variability in the extracellular samples: (i) extracellular enzyme activities, (ii) chemical degradation, and (iii) chemical interactions.

Extracellular enzymes are a particular important source of variability in samples containing complex substrates or biopolymers that can be further degraded,

◊ Text box 3.1 Turnover of secondary metabolites.

The secondary metabolites are mainly produced at the stationary growth-phase when the biomass has reached its maximum. These compounds are usually the end product of a metabolic pathway and tend to be accumulated inside the cells or be secreted to the extracellular medium because they have a very slow turnover. Usually, they are stable chemically and can resist to heating and hard sample workup. However several secondary metabolites also exhibit photo- and thermo-lability, which can lead to great variability on the profile of these metabolites. Therefore, special care should also be taken to avoid chemical degradation and chemical interactions, when handling samples containing secondary metabolites that will be used within a metabolome context. Low temperatures and protection against light must be the guidelines during processing these samples.

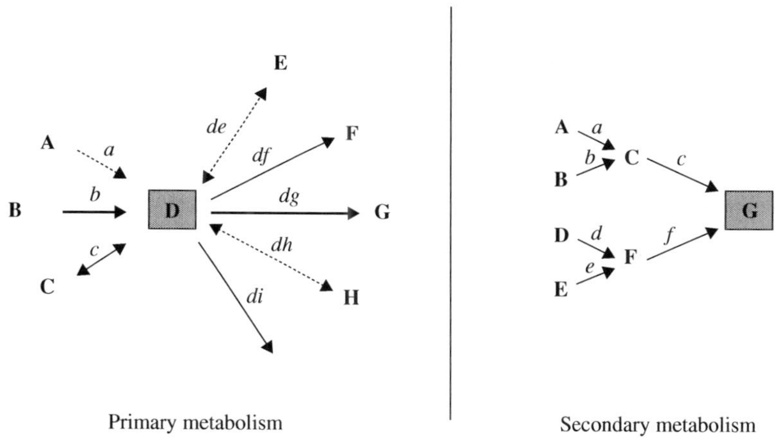

A sketch illustrating the main differences between a primary and secondary metabolism: on primary metabolism the primary metabolite "D" can be formed from the precursors A, B, or C, with B being its main source. However, metabolite "D" can also be reversely converted to C and is a precursor to several other metabolites (E, F, G, H, and I). H and E can also be converted back to "D." On secondary metabolism, the metabolites A and B are converted to C, and D and E is converted to F. The secondary metabolite "G" can be formed from the precursors C or F, but it is not an intermediate to any other reaction, therefore it accumulates inside the cell or it is secreted.

i.e., starch, glycogen, peptone, yeast extract, xylan, cellulose, pectin, and others. For such cases, the extracellular enzymes must be inactivated right after sampling and biomass separation. Losses by chemical degradation are another important source of variability in analysis of extracellular metabolite levels. Particularly, thermo- and photo-labile metabolites can be degraded quickly if kept for long time at room temperature or exposed to light. For instance, phosphorylated compounds, some sulphur derivatives, and some reduced metabolites can be degraded or oxidized rapidly at room temperature. Similarly, photo-degradation is a process that may result in high variability in the level of certain metabolites sensitive to light. For example, *S*-adenosyl-L-methionine, which is a methyl donor metabolite; a cofactor for enzyme-catalyzed methylations, including catechol *O*-methyltransferase (COMT) and DNA methyltransferases (DNMT), is a very unstable compound that can degrade very rapidly at temperatures above 0°C when exposed to light. Therefore, a quick storage of extracellular samples at low temperature (< −20°C) and preferably in the dark is highly recommended. The same procedure will also avoid further chemical interactions between active metabolites in the extracellular sample. Phosphorylated compounds are likely to exchange phosphate groups and oxido-reductive reactions are typically chemical interactions occurring in a mixture of different metabolite species. Box 3.2 provides some guidelines for handling samples of extracellular metabolites.

3.2.2 Different Methods for Quenching

A rapid inactivation of metabolism is usually achieved through rapid changes in temperature or pH. There are two general strategies depending on the objective. (i) Quenching and extraction of intracellular metabolites are combined, typically when the quenching procedure results in partial extraction of the intracellular metabolites because of disruption of the cellular envelope. In this case, intracellular and extracellular metabolites will be analyzed together. (ii) Quenching followed by separation of the biomass from the extracellular medium. This second option is particularly interesting for sampling microbial or cell cultures because it eliminates the interference of extracellular compounds, but it requires a reliable quenching method that avoids leakage of intracellular metabolites.

The quenching process itself consists of sampling the biological material (e.g., microbial and cell cultures, plant and animal tissues, body fluids) with simultaneous inactivation of the cellular metabolism and enzymatic activities. This is usually done by placing the biological sample in contact with a cold (< −40°C) or hot (>80°C) solution or with an acidic (pH < 2.0) or alkaline (pH > 10) solution. This process must be sufficiently fast to avoid changes in metabolite levels caused by alteration in the environment of the cells, ideally in a time window of a second.

Different biological samples require different techniques to achieve a proper quenching. We are, therefore, going to discuss the quenching techniques applied to specific class of samples in the following sections.

3.2.3 Quenching Microbial and Cell Cultures

Microbial or cell cultures are generally characterized by a high dilution ratio between biomass and extracellular medium, and this affects the quenching process. The most common quenching methods for this kind of samples are based on aqueous solutions containing an organic solvent, usually methanol or ethanol, buffered or nonbuffered, set to an extreme temperature (very cold or very hot), or acidic

◊ Text box 3.2 Handling samples of extracellular metabolites.

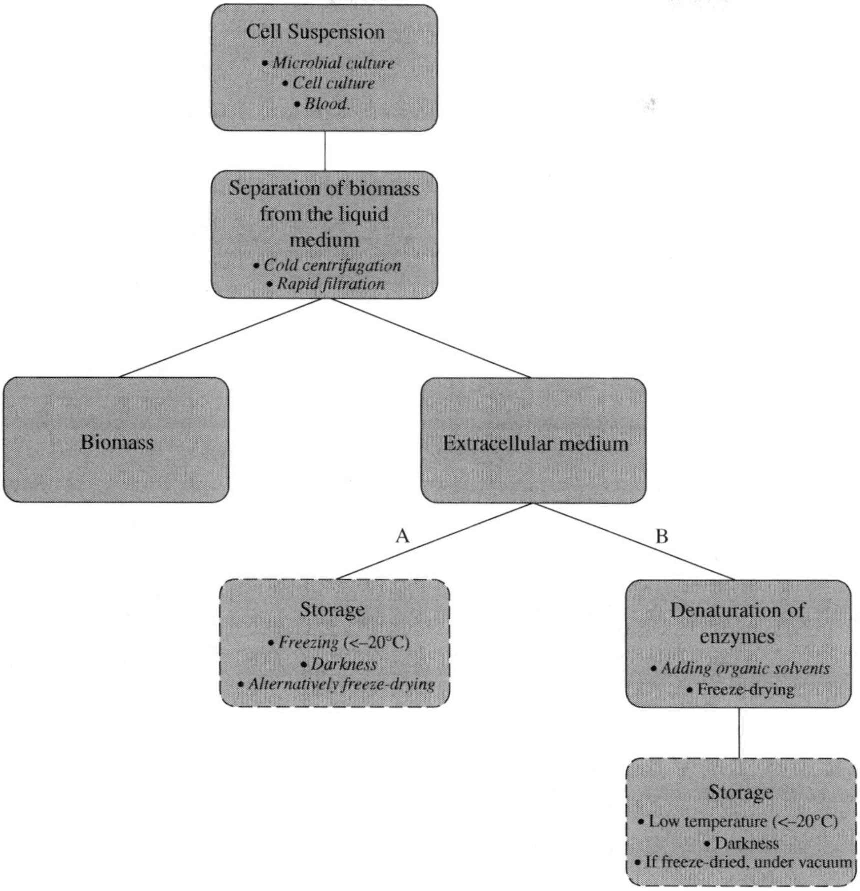

Samples containing extracellular metabolites must be rapidly separated from the cells, which are usually achieved by centrifugation at low temperature (1–4°C) or fast filtration under vacuum. The low temperature during centrifugation is necessary to slow down the secretion of metabolites and uptake of medium components and even decrease extracellular enzymatic activity, without disrupting the cell envelops (avoiding freezing). (A) Once separated the extracellular medium from the biomass it can be divided in small portions and frozen. The samples must be stored at low temperature (< -20°C) and in the dark to avoid any chemical alteration of the metabolites. Alternatively, the samples can be freeze-dried and stored at low temperature (< -20°C), under vacuum and in the dark. (B) However, if the extracellular medium free of cells still contains high enzymatic activity, mainly related to substrate breakdown such as hydrolases and oxidases,

◊ Text box 3.2 (*Continued*)

it will be extremely necessary to quench the enzyme activities, which can be done by adding organic solvents (e.g., chloroform, ethyl acetate, acetonitrile, and others) into the samples and rapid mixing to denaturate the enzymes. Alternatively, the samples can be frozen and freeze-dried. They must be stored similarly to samples obtained in "A."

solutions, typically perchloric acid. Sometimes, liquid nitrogen is also used as a quenching agent.

There are several techniques for a fast transferring of cultivation samples from the cultivation flasks or reactor to the quenching solution and the different techniques vary with respect to speed and practicability. Again, choices have to be made to achieve good reproducibility between sample replicates, keeping in mind that the quenching efficiency is maximized by a high sample-quenching solution surface area, e.g., by spraying the sample into the quenching solution.

Batch cultivations using shake flasks or similar vessels are typically sampled manually using automatic pipettes or syringes. A fixed volume of culture is quickly harvested and sprayed into sample flasks containing the quenching solution. The analyst must be trained to be quicker enough to quench all samples in a short time window, which usually takes 3–6 s per sample. One faster alternative is to fill a syringe with quenching solution before harvesting the cultivation sample. The time window obtained via manual sampling is acceptable for a wide range of purposes, and the amount of sample harvested is usually controlled by weighting the quenching flask before and after quenching, because a quick sampling process usually results in considerable variability in the sample volume taken. However, pipettes are not suitable to harvest samples from bioreactors and syringes generally results in too slow sampling. For this reason, several specialized techniques and devices have been developed to harvest and quench cultivation media from bioreactors and they are discussed in details in Chapter 7.

Most quenching agents or solutions (e.g., perchloric acid, trichloroacetic acid, boiling ethanol, boiling water, liquid nitrogen) disrupt the cell envelopes and, therefore, impede a reliable separation between intra- and extracellular metabolites. Only the cold methanol solution seems to be less aggressive for certain cells, but it does not completely prevent intracellular metabolite leakage. The effect of different quenching procedures on the different types of microbial cells will be discussed in the following sections.

3.2.3.1 Bacterial Cells. Recent research on method development for quenching microbial cultures containing bacterial cells is scarce. What is known today is that bacterial cells are sensitive to any quenching techniques developed until present date, including cold methanol, and, therefore, cell separation from the quenching solutions should not be done and analysis of intracellular and extracellular metabolites must be combined. Usually, the extracellular metabolites are determined separately in the samples of spent culture media and their levels are subtracted from the pool (intra + extra) in order to get an estimation of the intracellular levels, but this approach may give rise to large standard deviations for intracellular metabolites that typically make up a small fraction of the total metabolite pool.

According to Britten and McClure (1962), the levels of intra- and extracellular metabolites in *Escherichia coli* are in an osmotic equilibrium. Addition of distilled water completely removes the free amino acids from the cells, and a relative mild osmotic shock, such as a 30% reduction in the osmotic strength, removes 40% of the amino acids. However, solutions with the same osmotic strength of the culture medium or hyperosmolarity have little effect on the amino acid pool. Other classes of metabolites are also subjected to similar osmotic equilibrium and, therefore, leak from the intracellular medium during quenching or cell wash but at highly varying rates. Similar behavior has also been observed in Gram positive bacteria such as *Bacillus subtilis* (Smeaton and Elliott, 1967).

Aqueous solutions containing organic solvents, such as methanol, ethanol, butanol, acetone, and others, remove most of intracellular metabolites from bacterial cells (Britten and McClure, 1962; Jensen et al., 1999; Letisse and Lindly, 2000; Wittmann et al., 2004), and cold methanol solution has even been suggested as an efficient extraction agent for intracellular metabolites of bacterial cells (Maharjan and Ferenci, 2003).

E. coli cells quenched/washed with cold iso- or hyposmotic solution tend to present a greater leakage of intracellular metabolites than if quenched/washed with the same solution at room temperature (Leder, 1972). However, the leakage can be prevented or minimized if the cells are subjected at the moment of cold shock to a simultaneous hyperosmotic transition. It is suggested that iso-osmotic cold shock causes crystallization of the liquid-like lipids within the membrane. The hydrophilic channels created in this process would facilitate the rapid efflux of metabolites. The imposition of a simultaneous hyperosmotic transition by dehydrating the cell periphery would cause increased lipid interaction, thus, preserving the integrity of the cell membrane.

Wittmann et al. (2004) proposed a protocol for fast separation of bacterial cells from extracellular media using fast filtration under vacuum and washing the biomass with four volumes of cold saline solution (0.9%) at −0.5°C (the whole filtration step including the washing can be finished in less than 45 s). This method seems to permit authentic quantification of intracellular amino acid pools. However, this procedure does not seem to be suitable for precise analysis of metabolites with a faster turnover, e.g., phosphorylated intermediates.

Key references describing protocols for quenching bacterial cell cultures are listed in Table 3.2.

3.2.3.2 Yeast Cells. The most widely spread method for quenching yeast cell cultures makes use of cold methanol solution as the quenching agent and was originally proposed by de Koning and van Dam (1992). This method was developed for the determination of changes of glycolytic metabolites in yeast at the subsecond time scale. In their original application of the method, samples of incubated yeast

TABLE 3.2 Literature Sources for the Main Protocols for Quenching Bacterial Cell Cultures.

Quenching agent	Main conditions	Organism quenched	Reference
Perchloric acid	0.85 M in water 1:2 sample: $HClO_4$ sol. room temperature	*Alcaligenes eutrophus*	Cook et al., 1976
Hot sodium hydroxide	0.25 M in water 4:1 sample: NaOH sol. 85°C	*Alcaligenes eutrophus*	Cook et al., 1976
Cold perchloric acid	35% (w/w) in water ~1:1 sample: $HClO_4$ sol −40°C	*Zymomonas mobilis*	Weuster-Botz, 1997
Cold methanol	60% (v/v) in water 1:3 sample: methanol −50°C	*Escherichia coli*	Schaefer et al., 1999
Cold methanol	60% (v/v) in buffer 1:3 sample: methanol −35°C	*Lactococcus lactis*	Jensen et al., 1999
Cold ethanol	75% (v/v) in buffer 1:5 sample: ethanol sol. −75°C	*Xanthomonas campestris*	Letisse and Lindley, 2000
Liquid nitrogen	~1:3 sample: liquid N_2 −196°C	*Escherichia coli*	Buziol et al., 2002
Cold NaCl sol.	0.9% (w/w) in water 1:40 sample: saline −0.5°C	*Corynebacterium glutamicum*	Wittmann et al., 2004

suspension are rapidly transferred (sprayed) into a 60% (v/v) cold methanol solution kept at −40°C in a proportion of one part of sample for four parts of cold methanol solution. After quenching, the cells are separated by centrifugation at −20°C and the drained pellet is resuspended in 2.5 mL of 100% cold methanol (−40°C). For complete denaturation of proteins, 1 mL of precooled chloroform is added to the samples and additional 20 μL of 200 mM EDTA (pH 7.0) is added to inhibit Mg^{+2}-dependent partly chloroform-resistant enzyme activities. The sample tubes are stored at −80°C for further metabolite extraction.

This method gained great popularity due to its ability in separate cells from extracellular metabolites without apparent damage of the yeast cell envelope. However, it was demonstrated recently that yeast cells, similarly to bacterial cells, are also sensitive to cold methanol solution either buffered or nonbuffered and leakage of some intracellular metabolites has been observed after quenching *S. cerevisiae* cultures with cold methanol solution following the original protocol proposed (Villas-Bôas et al., 2005a). Several organic and amino acids are practically washed out of the yeast cells after being in contact with the cold methanol solution. However, no evidence for leakage of phosphorylated sugars and nucleotides (NADP and NAD) has been found (Villas-Bôas et al., 2005a). By decreasing the time the yeast cells stay in contact with the methanol solution (e.g., applying quicker centrifugation), the leakage of intracellular metabolites can be minimized significantly. However, a few metabolites may present a higher leakage under faster centrifugation, e.g., lactate, citramalate, myristate (Villas-Bôas et al., 2005a).

Nonetheless, the cold methanol method for quenching yeast cells still represents the only alternative where the biomass can be separated from the extracellular medium with good efficiency, but precautions must be taken to achieve minimal losses of intracellular metabolites. Since the longer the cells are in contact with the quenching solution the higher the leakage, the common practice of washing the cell pellet with cold methanol solution to eliminate interference of extracellular metabolites should be reconsidered and probably avoided. Alternatively, the method proposed by Wittmann et al. (2004) for fast separation of bacterial cells from extracellular media by fast filtration under vacuum and washing the biomass with cold saline solution ($0.9\%_{w/w}$, $-0.5°C$) can probably be adapted to yeast cells, but as mentioned before, it is not possible to achieve a very fast quenching using this procedure.

Yeast cells can also be quenched with perchloric acid, boiling ethanol and liquid nitrogen, but all these alternatives will release the intracellular metabolites to the quenching suspension during the quenching process. Table 3.3 lists the literature source of main protocols used for quenching yeast cell cultures.

3.2.3.3 Filamentous Fungi. The physiology and the morphology of filamentous fungi are quite different from those of yeast, and, therefore, different quenching methods must be considered. The cultures of filamentous fungi are usually highly viscous and heterogeneous, and it is, therefore, difficult to obtain a representative sample from a fermentation process. The easiest methods for quenching this kind of samples are using either liquid nitrogen or cold methanol solution (Hajjaj et al., 1998).

TABLE 3.3 Literature Sources for the Main Protocols for Quenching Yeast Cell Cultures.

Quenching agent	Main conditions	Organism quenched	Reference
Perchloric acid	0.66 M in water 1:1 sample: $HClO_4$ sol. room temperature	*Saccharomyces cerevisiae*	Larsson and Törnkvist, 1996
Cold methanol	60% (v/v) in water 1:4 sample: methanol sol. −40°C	*Saccharomyces cerevisiae*	De Koning and van Dam, 1992
Cold methanol	75% (v/v) in water/buffer 1:2 sample: methanol sol. −40°C	*Saccharomyces cerevisiae*	Villas-Bôas et al., 2005a,b
Boiling ethanol	75% (v/v) in buffer 1:4 sample: ethanol sol. 80°C	*Saccharomyces cerevisiae*	Gonzales et al., 1997
Liquid nitrogen	−196°C	*Saccharomyces cerevisiae*	Mashego et al., 2003

Quenching by liquid nitrogen allows rapid and repeated sampling under short periods of time, but it does not allow separation between intra- and extracellular metabolites. On the contrary, quenching in cold methanol allows separation of intra- and extracellular metabolites, but no study has been reported investigating whether or not leakage of intracellular metabolites takes place by quenching filamentous fungi with cold methanol. In addition, technical adaptations of the protocol developed for quenching yeast cells are needed to perform the sampling on short timescales and to separate the biomass from the extracellular medium at low temperatures.

3.2.4 Quenching Plant and Animal Tissues

When determining the metabolite levels from plant or animal tissues, the analyst must be aware that the obtained metabolite profile are originated from a heterogenic mixture of differentiated cells, which are at different stages of their development. Another important issue to be considered is the size of the sample that should be compatible with the quenching technique used. Cell tissues are, most of the times, distributed in several layers, where the peripheral cells tend to be quenched before the central ones, increasing the sample variability. Therefore, the tissue thickness as well as a reproducible sample size should be seriously considered when planning the experiments.

The process for quenching plant or animal tissues can be divided into four basic steps as illustrated in Figure 3.2. The first and most critical step is removing the

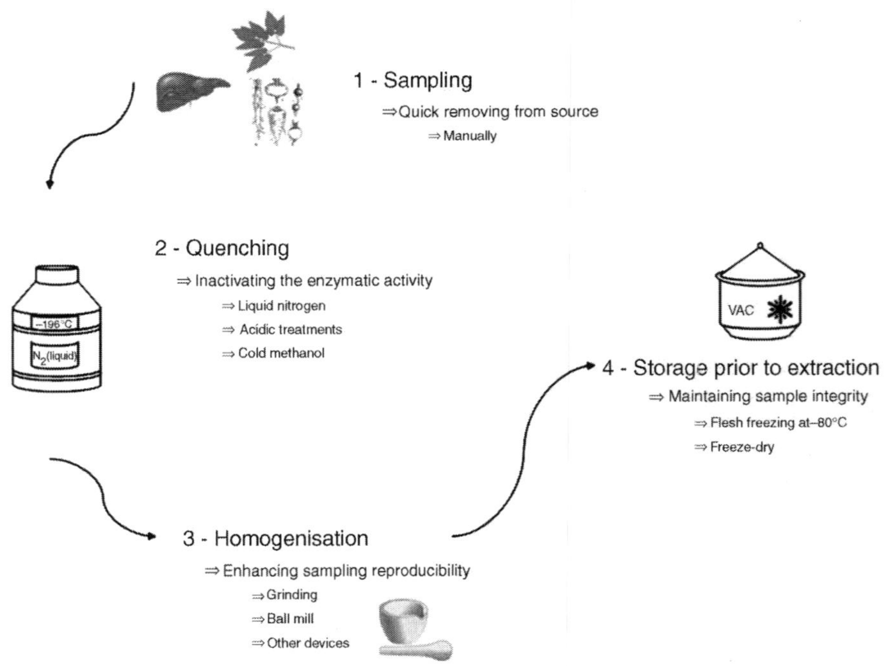

Figure 3.2 Main steps during sampling animal and plant tissues for metabolome analysis.

target tissues from the whole organism. This step should be very quick but it has to be done manually. This is a critical step because during cutting plant tissues or sacrificing a living animal, an immediate alteration of cellular metabolism is induced modifying the original *in vivo* levels of the metabolites.

Once the targeted tissues are removed from the original organism, the cellular metabolism must be quenched immediately. The most reasonable way to achieve an efficient quenching of plant or animal tissue is by rapid freezing in liquid nitrogen. As liquid nitrogen is an inert substance (boiling point at −196°C) it can be rapidly eliminated from the sample by evaporation. Liquid CO_2 has been considered as an alternative for liquid nitrogen but it should be avoided because CO_2 can oxidize a series of metabolites. Alternatively, cold methanol solution or acidic treatments using perchloric or nitric acid can be used as quenching agents; however, their efficiency is controversial and no validation of these methods to quench plant or animal tissues has been reported so far.

In order to enhance the sample reproducibility and extraction efficiency, the quenched tissue samples must be homogenized and the sample surface must be increased. Different types of homogenization can be used, which vary according to the type of tissue, but all process must be done at low temperature to avoid metabolite degradation or further metabolic conversions. Usually, the samples are grounded under liquid nitrogen using a mortar and pestle as illustrated in Figure 3.2. Alternatively, the frozen tissues can be grounded using a ball mill with prechilled holders (Fiehn, 2002), but harder tissues such as plant roots will require specialized devices such as ultraturax (Orth et al., 1999).

The last step in sampling/quenching plant or animal tissues is their storage prior to the metabolite extraction. There are two alternatives for storage of quenched plant/animal tissue samples: (i) shock freezing at −80°C or (ii) freeze-dry and storage under vacuum at low temperature. Shock freezing at −80°C is advantageous for metabolome analysis because it improves the sample integrity, but depending on the number of the samples being handled this method could limit the physical space for sample storage, and great care must be taken to avoid partially thawing samples before extracting metabolites. On the contrary, freeze-dried samples ensure the inactivation of cellular metabolism because enzymes and transporters are unable to work in complete absence of water. However, freeze-dried samples must be stored in dry environment such as evacuated desiccators and at low temperatures to avoid absorption of water and degradation of metabolites. But, according to Fiehn (2002), freeze-drying may potentially lead to irreversible adsorption of metabolites on cell walls and membranes, decreasing the extraction efficiency.

Extracellular metabolites present in biofluids from animal tissues, such as milk, urine, and plasma, are an important source of metabolic information and can be handled easily than the samples from solid tissues. For instance, the metabolites present in the blood provide metabolic information on all tissues that deliver metabolites to the blood and obtain metabolites from it. When extracellular metabolites are concerned, the basic guidelines for quenching samples containing these compounds are applied as shown in Box 3.2.

3.3 OBTAINING METABOLITES FROM BIOLOGICAL SAMPLES

The biological samples contain three general classes of metabolites: (1) water soluble metabolites or polar compounds, (2) water insoluble metabolites or nonpolar compounds, and (3) volatile metabolites. All these three classes of metabolites can be found both intra- and extracellularly. There is no single method able to extract and group all the three classes of metabolites simultaneously, and, thus, different techniques are usually applied to extract the different classes of compounds, and they will vary according to the nature of the biological sample (e.g., if cells or extracellular media).

3.3.1 Release of Intracellular Metabolites

A large part of the metabolome is located in the interior of cells in a highly diverse range of concentrations (i.e., from pmol to mmol). The extraction of these intracellular metabolites is inevitably a time-consuming step and the extraction solvent or conditions should be able to prevent any further physical and chemical alterations of the molecules as well as the whole entire extraction process should ensure minimal loss of the metabolites to be extracted. The extraction procedure aims to disrupt the cell structures liberating all or the maximum number of metabolites in their original state and in a quantitative manner to a defined extraction medium. The choice and development of efficient methods for extraction of intracellular metabolites requires an understanding of: (i) the cell wall structures, which are the first and main barrier to be broken; (ii) the chemical nature of the metabolites (i.e., physical and chemical form, solubility, stability); and (iii) the sources of losses (especially their impact on subsequent recovery of metabolites).

The alterations in metabolic composition and ration of metabolites after extraction of intracellular metabolites that are expected to be provoked by any extraction procedure are illustrated in Figure 3.3. It is impossible until present date to extract all intracellular metabolites keeping their original state and original intracellular ratio. First, all extraction procedures dilute the metabolite concentrations and change the original ratio of several compounds as a result of incomplete extraction of many metabolites in addition to chemical modifications or partial degradation of labile molecules. Furthermore, artifacts are usually introduced into the samples during extraction procedures such as chemical contaminants from solvents and vessels, polymer degradation, and many others. Therefore, we need to be able to extract meaningful information from metabolome data regardless the alterations introduced into the samples, and, hence, appropriate data analysis procedures always play an important role.

3.3.2 Structure of the Cell Envelopes—the Main Barrier to be Broken

The cell envelope basically consist of a *cytoplasmic membrane* and for many organisms also a rigid outer supporting structure the *cell wall*. The cytoplasmic membrane is primarily composed of lipids and proteins and its basic structural function is to

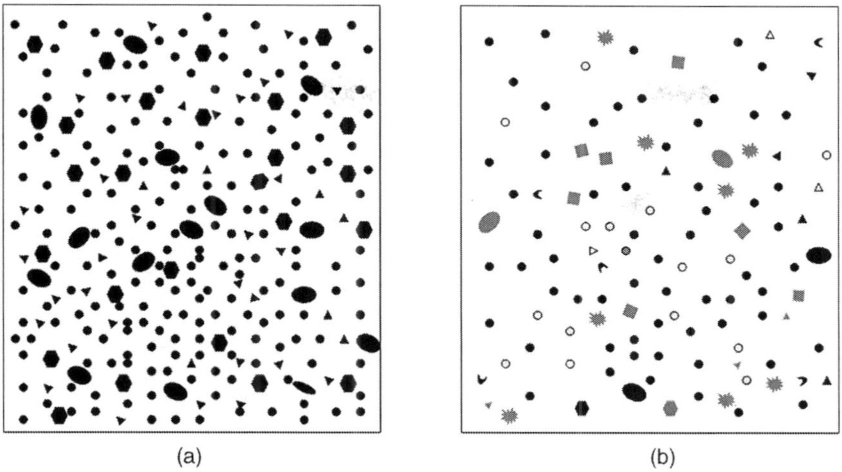

Figure 3.3 Schematic figure illustrating the alterations expected to be provoked by any extraction procedure in the metabolic composition and ratio of metabolites after intracellular metabolite extraction. It is impossible until present date to extract all intracellular metabolites keeping their original state and original intracellular ratio. (a) Illustrates symbolically the state of different metabolites inside the living cells (black symbols). (b) Illustrates symbolically the state of different metabolites and chemical compounds in an optimal extracted sample, showing: a clear dilution of the metabolite concentrations and change the original ratio of several compounds as a result of expected incomplete extraction of many metabolites; chemical modifications or partial degradation of labile molecules (changing in the color pattern of the symbols or lost of original shape); and introduction of artefacts into the samples expected to occur during extraction procedures such as chemical contaminants from solvents and vessels, polymer degradation, and many others (e.g., symbols not present in (a)).

maintain the osmotic balance within the cell. The interior of a cell contains very high protein and metabolite concentrations and in the absences of a cell wall it is very susceptible to osmotic shock. The cell wall, present in many organisms and cells, offers the primary resistance to disruption and its strength is related to many factors. A huge diversity of wall structures and compositions exist in nature, but, nevertheless, there are some gross similarities (e.g., Gram-positive and Gram-negative bacteria, yeasts and other fungi, plant cells).

3.3.2.1 Cell Wall Structures of Bacteria. The rigid wall matrix of nearly all bacteria is a continuous bag-like molecule completely encapsulating the cell, providing both shape and strength, and protects the cell from bursting due to the osmotic pressure that exists within the cell. Two different types of walls exist among the bacteria (Figure 3.4). Bacteria possessing a single, but thick, cell wall can be stained using the Gram stain procedure and are hence called Gram-positive bacteria, whereas, bacteria that contain two, but relatively thin, cell walls do not stain using the Gram stain procedure and are, therefore, called Gram-negative bacteria (for further details on Gram staining procedure and mechanism consult any basic microbiology book). The strength

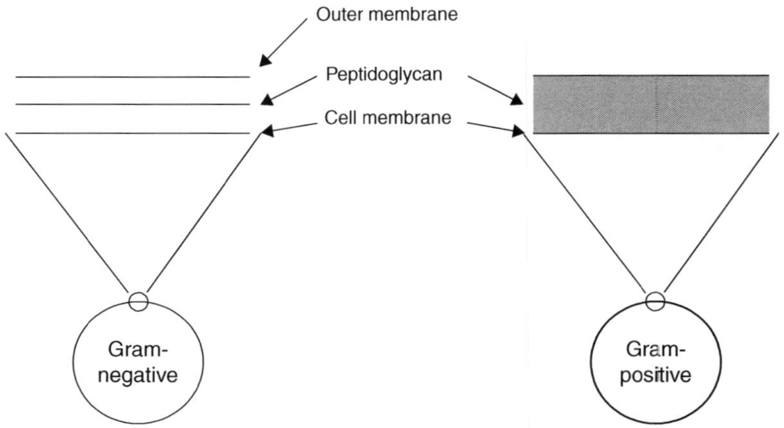

Figure 3.4 Schematic figure comparing the cell wall structure of Gram-negative and Gram-positive bacteria. Gram-positive bacteria present a thicker layer of peptidoglycan in their cell wall, conferring greater strength and resistance to mechanical disruption comparing to Gram-negative cells.

and rigidity of bacterial cell walls are due to a glycopeptide called peptidoglycan or murein, which consists of glycan chains cross-linked by peptides (Figure 3.5). The polysaccharide (glycan) chains consist of alternating *N*-acetylglucosamine (NAG) and *N*-acetylmuramic acid (NAM) units linked by β-1,4 glycosidic bonds. The peptides that cross-link the glycan chains to each other are basically two short peptide units: a tetrapeptide of variable composition (with rare D-amino acids) linked to NAM residues via the lactyl side chain and a bridging pentapeptide $(Gly)_5$. The degree of cross-linking varies considerably, e.g., ~50% in the Gram-negative bacterium *E. coli* and ~90% in the Gram-positive bacterium *Lactobacillus acidophilus.*

The major resistance to disruption of bacterial cell walls is offered by the peptidoglycan layer. The extent of cross-linking of peptidoglycan affects the wall strength and therefore the ease of disruption. There are some important differences between the peptidoglycan in Gram-positive and Gram-negative bacteria. The peptidoglycan of Gram-negative bacteria can be isolated as a sac of pure peptidoglycan that surrounds the cell membrane in the living cell. It is called the *murein sacculus*. The sacculus is elastic and believed to be under stress *in vivo* because of the expansion due to osmotic pressure against the cell membrane. In contrast, the peptidoglycan from Gram-positive bacteria is covalently bonded to various polysaccharides and teichoic acids and it cannot be isolated as a pure murein sacculus. The cross-linking in the peptidoglycan is usually direct in Gram-negative bacteria, whereas there is usually a peptide bridge in Gram-positive bacteria providing more strength and resistance to disruption.

3.3.2.2 Structure of Yeast Cell Envelopes. The basic structural components of the yeast cell envelopes are glucans, mannans, and proteins. The overall wall

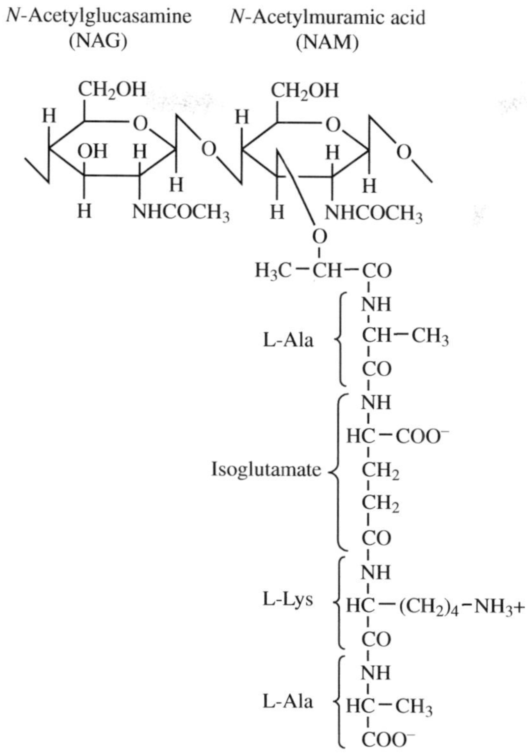

Figure 3.5 The repeating unit of peptidoglycan present in bacterial cell walls. The major resistance to mechanical disruption of bacterial cell walls is offered by the peptidoglycan layer.

structure is generally thicker than that in Gram-positive bacteria, and the thickness increases with age. The inner part of the cell wall is composed of glucan fibrils, which constitute a rigid matrix that assists in providing the cellular shape (Figure 3.6). Covering the fibrils is a layer of glycoprotein and beyond this is a mannan mesh cross-linked by 1,6-phosphodiester bonds. The majority of proteins in yeast cell walls are within the mannan mesh, existing as mannan–enzyme complexes, some of which are covalently attached to the mesh. The glucan structure is moderately branched, and glucose units are linked by β-1,3 and β-1,6 glycosidic bonds.

The mannan backbone consists of mannose units linked by α-1,2 and α-1,3 configurations. As with bacterial cells, resistance of yeast cell walls to disruption appears to be a function of how tightly cross-linked and how thick the structural portion is, but usually yeast cell wall is more resistant to disruption than bacterial cell walls.

3.3.2.3 Envelopes of Other Fungi. Generalizations about the cell envelopes of other fungi are not possible due to very diverse cell wall compositions. The structure of hyphal walls is the most widely studied. In most filamentous fungi the cell wall

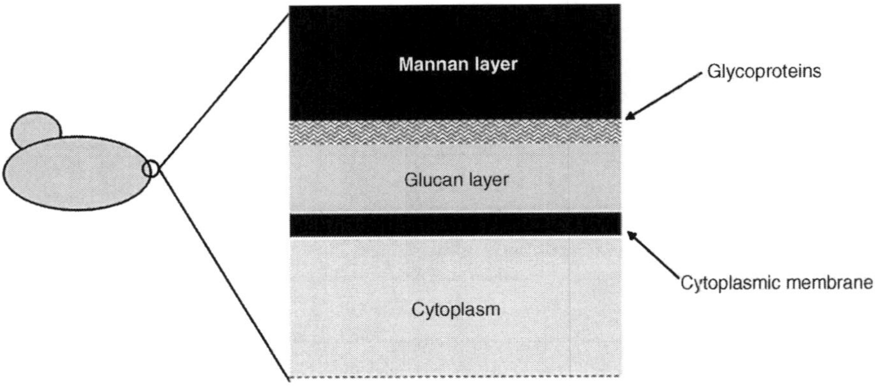

Figure 3.6 Schematic illustration of the yeast cell envelope. The overall yeast cell wall structure is generally thicker than in Gram-positive bacteria, and yeast cells are more resistant to mechanical disruption than bacterial cell walls.

is more resistant to disruption than in yeast cell walls and is primarily composed of polysaccharides with lesser amounts of proteins and lipids. As for bacteria and yeasts, shape and strength of the wall is provided by the amount of polysaccharides. Chitin (*N*-acetylglucosamine polymer linked by β-1,4 bonds) and β-glucan polymers are most common and are constructed in layers. Mature walls of *Neurospora crassa* consist of concentric layers arranged from the interior outwards as illustrated in Figure 3.7.

3.3.2.4 Structure of Plant Cell Envelopes. In plant cell envelopes, the cell wall is a rigid multilayered structure that lies outside the cytoplasmic membrane (Figure 3.8). The thickness as well as the composition and organization of plant cell

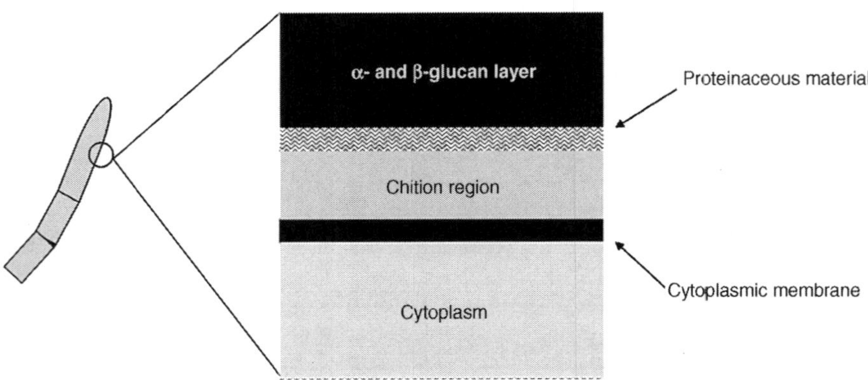

Figure 3.7 Schematic illustration of the envelope of *Neurospora crassa*. Generalizations about the cell envelopes of other filamentous fungi are not possible due to very diverse cell wall compositions.

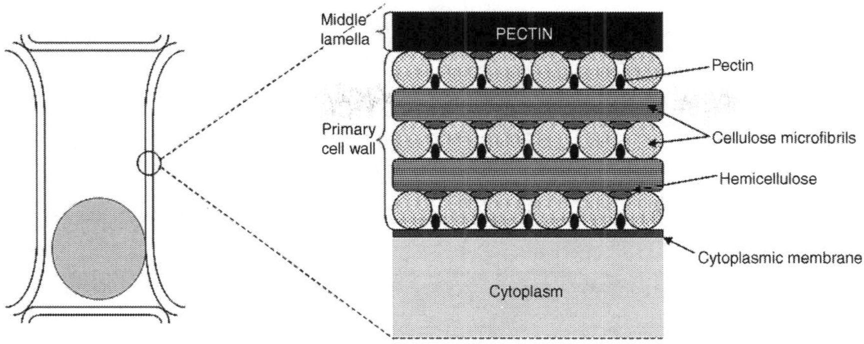

Figure 3.8 Schematic illustration of the multilayered primary cell wall structure of plant cell envelopes. The secondary plant cell wall, which is often deposited inside the primary cell wall as a cell matures, sometimes has a composition nearly identical to that of the earlier-developed wall. More commonly, however, additional substances, especially lignin, are found in the secondary wall.

walls can vary significantly. Many plant cells have both a primary cell wall, which accommodates the cell as it grows, and a secondary cell wall, which develops inside the primary cell wall after the cell has stopped growing. The primary cell wall is thinner and more pliant than the secondary cell wall, and it is sometimes retained in an unchanged or slightly modified state without the addition of the secondary wall even after the growth process has ended.

The main chemical components of the primary plant cell wall include cellulose (in the form of organized microfibrils; see schematic Figure 3.8), a complex carbohydrate made up of several thousands of glucose molecules linked end to end. In addition, the cell wall contains two groups of branched polysaccharides the pectins and cross-linking glycans or known as hemicellulose. Organized into a network with the cellulose microfibrils, the cross-linking glycans increase the tensile strength of the cellulose, whereas, the coextensive network of pectins provides the cell wall with the ability to resist compression. In addition to these networks, small amount of protein can be found in all plant primary cell walls. Some of this protein is thought to increase the mechanical strength and part of it consists of enzymes, which initiate reactions that form, remodel, or breakdown the structural networks of the wall.

The secondary plant cell wall, which is often deposited inside the primary cell wall as a cell matures, sometimes has a composition nearly identical to that of the earlier-developed wall. More commonly, however, additional substances, especially lignin, are found in the secondary wall. Lignin is the general name for a group of polymers of aromatic alcohols that have a very hard structure and provide considerable strength to the structure of the secondary wall. Lignin makes plant cell walls less vulnerable to attacks by fungi or bacteria as do cutin, suberin, and other waxy materials that are sometimes found in plant cell walls.

A specialized region associated with the cell walls of plants, and sometimes considered an additional component of them, is the middle lamella (see Figure 3.8).

Rich in pectins, the middle lamella is shared by neighboring cells and cements them firmly together. Positioned in such a manner, cells are able to communicate with one another and share their contents through special conduits.

3.3.2.5 Structure of Animal Cell Envelopes. Animal cell envelopes comprise of very elaborate membrane and cytoskeletal structures, but the basic foundation is the "fluid-mosaic lipid bilayer" model proposed by Singer and Nicolson (1972). Cytoskeletal proteins (e.g., spectrin, fodrin, actin, and synapsin-1) play key roles in altering and stabilizing the shape of many kinds of cells. The key feature from the perspective of cell disruption is the absence of a cell wall structure, which makes animal cells very easy to disrupt. In fact, most animal cells are acutely sensitive to "shear" and lyse very readily, releasing DNA and other colloidal foulants, which can cause serious problems during removing of cells from metabolite-containing extracts. Separation operations such as centrifugation and filtration can seriously damage mammalian cells (and spheroplasts of microbial cells).

3.3.3 Cell Disruption Methods

Even though the cell wall structure and composition only have been studied in details for a few organisms, it is clear that there is a great diversity. The shape and strength of cell walls depend on structural polymers, mainly polysaccharides, within the cell wall, and the degree of cross-linking between these polymers and other cell wall components. For cellular disruption, the major resistance to overcome is breaking of covalent bonds between these structural components. There are basically two ways for disrupting cell walls: mechanical and nonmechanical disruption, and their variability is illustrated in Figure 3.9.

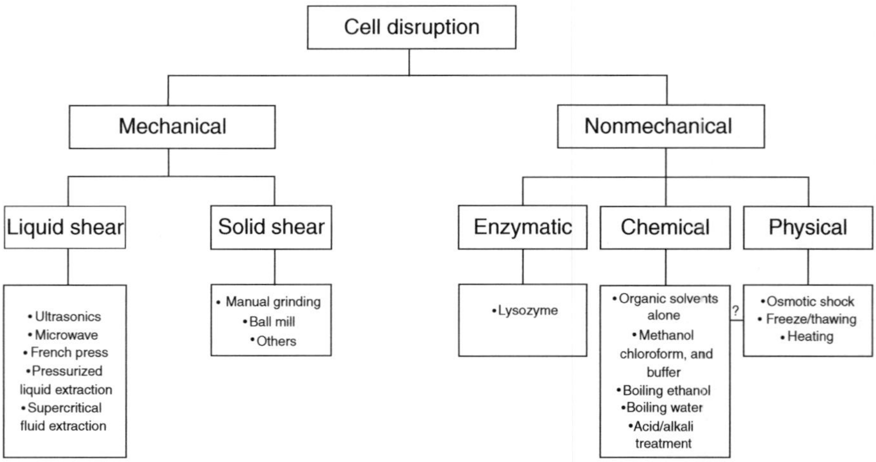

Figure 3.9 Tree diagram showing the range of the principal cell disruption methods available.

For mechanical disruption the important factors are (1) the size and shape of the cell, (2) the degree of cross-linking between the polymers, and (3) the polymer concentration in the cell wall. Although there is not much information available concerning the relative resistance of various organisms to mechanical disruption, the ease of disruption scale generally follows the order: animal cells > Gram-negative bacterial cells > Gram-positive bacterial cells > yeast cells > filamentous fungi > plant cells. A variety of methods are available that make use of mechanical forces to disrupt cellular walls and membranes resulting in the liberation of intracellular contents to a selected liquid solvent (Figure 3.9), but even though most of these have not been extensively applied for metabolome analysis, they are discussed in this section because of their great potential to enhance the "extraction" of intracellular metabolites, particularly, "extraction" of nonpolar compounds.

Nonmechanical disruption of cell envelopes, in contrast, comprises the most traditional techniques to extract intracellular metabolites from biological samples. These methods make use of chemical or physical agents to provoke sufficient permeabilization of cell envelopes to allow extraction of intracellular metabolites from the cytoplasmic medium. They can be differentiated into three different subgroups according to the nature of the disrupting agent: (i) enzymatic, (ii) chemical, and (iii) physical (Figure 3.9). Enzymatic and physical methods *per se* are not commonly applied in metabolome analysis, but sometimes they are combined with chemical methods to enhance the extraction process (especially physical methods). In contrast, chemical lysis of the cell envelopes includes the majority of procedures developed to extract intracellular metabolites from biological materials, and the available protocols will vary according to the structure and composition of cell walls.

3.3.4 Nonmechanical Disruption of Cell Envelopes

3.3.4.1 Enzymatic Lysis. Although not commonly applied in sample preparation for metabolome analysis, enzymatic lysis is attractive in terms of its delicacy and specificity for just the cell wall structure. If the wall is degraded under conditions where there is osmotic pressure, there will be lysis of cells and hence release of the intracellular metabolites into the extracellular matrix. Enzymatic methods have the advantages of having a high rate and yield in the extraction process, there is little metabolite degradation as it requires mild conditions of pH and temperature, and also they leave no fine debris that is difficult to remove from the sample. However, the enzymatic degradation of cell walls releases the monomers of cell wall polymers (mainly sugars, sugar derivatives, and amino acids) into the sample, adding artifacts to the pool of metabolites. In addition, lytic enzymes often require use of an aqueous medium and mild temperatures to degrade cell wall structures, and this may be incompatible with methods used to quench the metabolism and further biochemical activity in the samples.

The cell walls of different organisms are very diverse, thus, lytic enzymes are generally specific for particular groups of cells, and they have primarily been applied to disrupt microbial cells (Table 3.4). With few exceptions, one enzyme is not enough for degradation of cell walls and either a mixture of several enzymes

TABLE 3.4 Important Cell Wall Degrading Enzymes.

Organisms	Enzymes	Type of hydrolysed linkage
Bacteria	Glycosidases	β(1,4)-linkages between NAG and NAM residues in peptidoglycan
	Acetylmuramoyl-L-alanine amidases	Link between *N*-acetylmuramoyl residues and L-amino acid residues in certain glycopeptides
	Peptidases	peptide bonds (e.g., Gly-Gly, Ala-Gly)
Fungi, yeasts	β(1,3)-Glucanases	Random β(1,3)-linkages in glycans
	β(1,6)-Glucanases	Random β(1,6)-linkages in glycans
	Mannanases	(1,2)- or (1,3)- or (1,6)- β-D-mannosidic linkages
	Chitinases	β(1,4)-linkages of NAG polymers found in chitin and chitodextrins
	Proteases	Peptide bonds
Algae	Cellulases	β-(1,4)-linkages in cellulose

acting synergistically or a chemical pretreatment may be required. For bacterial cells, a single enzyme, such as lysozyme, can lyse the peptidoglycan of Gram-positive bacteria, but chemical destabilization of the outer membrane of Gram-negative bacteria is necessary to enable the enzyme to access the underlying peptidoglycan. More details on applications of lysozyme can be found in Box 3.3.

3.3.4.2 Physical Lysis. Physical lysis of cell walls as the sole mechanism has not found wide application in sample preparation for metabolome analysis, but it is very often combined with chemical or enzymatic methods. There are, however, three physical processes that are worth mentioning, even though they are usually

◊ Text box 3.3 Lysozyme.

Lysozyme is a relatively small enzyme that degrades the peptidoglycan of bacterial cell walls. It is a highly stable glycosidase that hydrolyses the glycosidic bond between C-1 of NAM and C-4 of NAG, but not between C-1 of NAG and C-4 of NAM. Chitin (poly NAG joined by β-1,4-linkages) is also a substrate for lysozyme. The main source of commercial lysozyme is hen egg white lysozyme (HEWL) and it is inexpensive. However, its use is limited because very few cells are susceptible to an efficient disruption. Although lysozyme has been mostly employed in the extraction of proteins and genetic material (Kheirolomoom et al., 2001; Santiago-Santos et al., 2004; van Hee et al., 2004; and others) with very few reports on using this enzyme for extraction of intracellular metabolites (Tondo et al., 1998; Michalke et al., 2002), the potential for its application on extraction of intracellular metabolites of bacteria exists, but methodology should be adapted to a metabolomics scale approach.

combined with chemical extractions: (i) cold osmotic shock, (ii) freeze-thawing, and (iii) heating.

(i) Cold osmotic shock: Osmotic shock, induced by a rapid change in the salt concentration of the medium, is effective in disrupting animal and specially red blood cells. Plant and microorganisms, having tough cell walls in addition to a membrane, are less susceptible to such treatment. Nevertheless, a limited effect can be observed with *E. coli* and other Gram-negative bacteria, where great part of the intracellular pool of amino acids leak from the cells under hyposmotic conditions (e.g., distillated water), although hyperosmotic shock has little effect.

(ii) Freeze-thawing: Water molecules are polar and, triangular, in shape, and, therefore, their charge distribution is asymmetric. Furthermore, water molecules are highly cohesive and link to each other via hydrogen bonds. In its liquid state, water has a partially ordered structure with an average of 3.4 H-bonded neighbors. Normal low pressure ice exists as "type I (or ice-Ih)" with four H-bonded neighbors. Since the ice structure forms more H-bonds, its volume expands compared to the volume of liquid water, disrupting or damaging the cell envelopes. Therefore, freeze-thawing cycles have the ability to make the cells permeable, easily releasing the intracellular metabolites to a liquid solvent. Freeze-thawing is very often an indirect consequence of sample storage at −20/−80°C, and hence precedes many other extraction methods, but its effects are mostly beneficial in terms of adding to the extraction process.

(iii) Heating: Heating increases the permeability of cell envelopes by denaturating cell wall related proteins and hereby decreasing the viscosity of the cytoplasmic membrane resulting in leakage of intracellular metabolites. However, heating is used to enhance the extraction efficiency of some chemical agents and these methods will be discussed later in this section. Nonetheless, several metabolites are very sensitive to high temperatures, which result in great losses of these thermo-labile compounds during hot extraction methods.

3.3.4.3 Chemical Lysis. In metabolome analysis, the intracellular metabolites are usually extracted using chemical agents to lyse the cells and extract the intracellular compounds. Table 3.5 presents a summary of the most popular extraction methods using chemical lysis. All methods make use of the same basic set of concepts to concentrate the metabolites in one phase. Any metabolite will be distributed between two phases according to the partitioning coefficient, solubility, temperature, and the relative volumes of the phases. However, the extraction rates are based on the migration kinetics and hence are governed by temperature and diffusion rates in the two phases, in addition to solvent access to the intracellular compounds, and hence it is directly related to the degree of cell permeabilization. There are a variety of chemical agents and extraction conditions that can be applied to different class of cells. Some chemical extraction methods will dissolve selectively a targeted group of metabolites (e.g., lipids or polar compounds), while others will be able to dissolve

TABLE 3.5 Summary of the Main Chemical Extraction Methods.

Method	For extraction of	*Applied for	Ideal Conditions	Advantages	Disadvantages	References
Buffered methanol–water–chloroform	Polar (methanol–water phase) and nonpolar (chloroform phase) compounds	Plant tissues Animal tissues Yeast cells Bacterial cells Filamentous fungi cells	Low temperatures (−40 to −20°C) Vigorous shaking (~300 g for 45 min)	Denaturation of enzymes by chloroform avoiding further reactions Possibility to separate polar from nonpolar compounds Good recovery of phosphorylated metabolites and thermolabile compounds Good reproducibility	Tedious and time consuming Toxic effects of chloroform Presence of buffer may pose problems for many analytical techniques	De Koning and van Dam, 1992 Cremin et al., 1995 Smits et al., 1998 Le Belle et al., 2002 Maharjan and Ferenci, 2003 Villas-Bôas et al., 2005a,b
Boiling ethanol	Polar thermostable metabolites	Yeast cells Bacterial cells Filamentous fungi cells	High temperatures (>80°C) Evaporation of ethanol–water mixture and resuspension of pellet in water	Simple and fast Denaturation of enzymes by hot ethanol Enhanced cell disruption by heating Good reproducibility	A number of metabolites are not stable at high temperatures for extraction Possible oxidation of reduced metabolites	Gonzalez et al., 1997 Hans et al., 2001 Castrillo et al., 2003 Maharjan and Ferenci, 2003 Villas-Bôas et al., 2005a

Cold methanol	Polar and mid-polar metabolites	Plant tissues Animal tissues Bacterial cells Yeast cells	Freeze-thawing cycle previous to extraction Low temperatures (< −20°C). Wash the cells with cold methanol once or twice after extraction to enhance recovery	Simple and fast Easy removal of solvent after extraction Excellent recovery of metabolites Excellent reproducibility Broad range of metabolites extractable	Not complete denaturation of enzymes Bad recovery of non-polar compounds	Shryock et al., 1986 Roessner et al., 2000 Maharjan and Ferenci, 2003 Villas-Bôas et al., 2005a,b
Acidic extraction	Polar and acid-stable metabolites	Plant tissues Animal tissues Bacterial cells Yeast cells Filamentous fungi cells	Low temperatures (0 to 4°C). Freeze-thawing cycle during the extraction. Neutralization of the sample pH after extraction	Simple Excellent recovery of amines and polyamines Denaturation of enzymes by extreme low pH	Bad recovery of metabolites Oxidation of reduced compounds Hydrolysis of proteins and polymers	Shryock et al., 1986 Kopka et al., 1995 Hajjaj et al., 1998 Buziol et al., 2002 Villas-Bôas et al., 2005a
Alkaline extraction	Polar and alkali-stable metabolites	Yeast cells Filamentous fungi cells	Low temperatures (0 to 4°C). Freeze-thawing cycle during the extraction Neutralization of the sample pH after extraction	Simple Excellent disruption of cell walls Denaturation of enzymes by extreme high pH	Bad recovery of metabolites Hydrolysis of proteins and polymers Saponification of lipids	Hajjaj et al., 1998 Villas-Bôas et al., 2005a

a broader range of metabolite classes. However, discrimination of certain groups of metabolites will always be observed, which will call for the use of multiple extraction agents in combination or not with some physical or mechanical process to enhance cell permeability and extraction efficiency.

Organic solvents are widely used for extraction of intracellular metabolites. Frequently, more than one solvent is used in the extraction procedure: polar solvents like methanol, methanol-water mixtures, or ethanol to extract polar metabolites, and nonpolar solvents like chloroform, ethyl acetate, or hexane to extract lipophilic compounds. The organic solvents destabilize the cell wall and cell membrane proteins and lipids forming pores on the cell envelopes from where the intracellular metabolites are eluted and solubilized by the extracting solvent.

Classical protocols make use of exhaustive extraction in a Soxhlet system in which the solvent is continuously recycled through the sample for many hours. The analytes must be stable in the refluxing boiling solvent and many primary metabolites are not. These classical procedures can be interesting for targeted analysis of secondary metabolites of plants, where cell permeabilization is difficult due to the very rigid cell wall that poses severe problems in the extraction of certain group of metabolites. However, these processes are often quite slow and require the use of significant amounts of sample and large volumes of organic solvents to ensure complete extraction. The subsequent workup employ solvent evaporation and concentration of the sample is slow and manually laborious and any impurities in the extraction solvent is also concentrated.

In contrast, the aims of most recent methods used for the extraction of intracellular metabolites within the metabolomics context have been to reduce the amount of solvent and sample, reduce the time required for extraction, and enhance the broadness (extraction of several different groups of metabolites simultaneously). Most cell envelopes can be made permeable by just being in contact with organic solvents for a certain period of time and an efficient extraction can be achieved by simply stirring the samples vigorously or submitting the sample to a previous freeze-thawing cycle before extraction. However, plant materials and, at some extension, also filamentous fungi mycelia require some previous mechanical disruption or cell envelopes such as grinding the frozen biomass using a mortar and pestle or applying microwave or sonic wave to enhance cell disruption (mechanical assisted methods will be discussed later).

Although there are a vast number of different protocols and method adaptations using organic solvents for extraction of intracellular metabolites, we are going to discuss, in the following, the most popular protocols that have been applied in metabolomics field using organic solvents.

3.3.4.3a Buffered Methanol–Chloroform–Water. De Koning and van Dam (1992) adapted a methodology, originally designed for extraction of total lipids from animal tissues (Folch et al., 1957), based on a buffered methanol–water mixture and chloroform at low temperatures (−40 to −20°C), to extract polar metabolites in a yeast-cell suspension. This method is widely used for extraction of intracellular metabolites of bacteria, yeasts, animal tissues, and filamentous fungi.

This method has the advantage of extracting two large groups of metabolites (polar and nonpolar) simultaneously and selectively into two solvent phases (chloroform and methanol/water, respectively) under very mild conditions (low temperatures). In addition, chloroform has a great ability in denaturating proteins, which prevents any biochemical reaction to take place in the sample during the extraction process. Excellent recoveries of amino and non-amino organic acids, sugar phosphates, and sugar alcohols have been reported for this method (Smits et al., 1998; Jensen et al., 1999; Villas-Bôas et al., 2005a), but nucleotides do not seem to be extracted very efficiently, and this method is considered tedious and time-consuming besides the use of chloroform being undesirable due to its toxic and carcinogenic effects.

3.3.4.3b Boiling Ethanol. Extraction at elevated temperatures with boiling solvents is another very popular extraction method. This method was proposed by Gonzales et al. (1997) for extraction of polar metabolites from yeasts and was based on the use of boiling ethanol as first described by Entian et al. (1977). The samples containing quenched cells free of extracellular medium are boiled at 80°C for a few minutes in a buffered ethanol solution 75% (v/v). The heating enhances the extraction efficiency of ethanol solution and its protein-denaturating power, deactivating all the enzymes in the sample. The solvent is evaporated after extraction and the water-soluble metabolites are resuspended in water for analysis.

This method has been mainly used for extraction of intracellular metabolites of microbial cells, but not all metabolites are stable at the high temperature applied during extraction and particularly poor recovery of phosphorylated metabolites, nucleotides and tricarboxylic acids has been observed using this method (Maharjan and Ferenci, 2003; Villas-Bôas et al., 2005a).

3.3.4.3c Cold Methanol. Methanol is a very powerful organic solvent used for extraction of intracellular metabolites from a wide range of cells. It has been used alone or mixed with water for extraction of intracellular metabolites of animal cells (Shryock et al., 1986), but only recently it has been recognized as an efficient extracting agent for intracellular metabolites of bacteria (Maharjan and Ferenci, 2003) and yeast cells (Villas-Bôas et al., 2005a). This method makes use of a single organic solvent that is not as toxic as chloroform and can be easily removed from the sample by solvent evaporation. It is important, however, that the extraction process is done at low temperatures (< -20°C) to avoid further biochemical reactions and degradation of thermo-labile compounds. Usually, a freeze-thawing cycle is included in the procedure to enhance cell permeability. It is a quick and very simple method and presents excellent reproducibility and recovery of polar and mid-polar metabolites. Plant cell envelopes are usually disrupted mechanically before extraction with methanol, and, although there is no report on using this procedure for extraction of intracellular metabolites of filamentous fungi, this method has a great potential to be adapted to all biological systems.

3.3.4.3d Acidic and Alkaline Extraction. Acidic and alkaline extractions are classical methods for the extraction of intracellular metabolites. These methods have been widely used for extraction of metabolites from animal and plant tissues, filamentous fungi, and microorganisms.

Perchloric acid (PCA), trichloroacetic acid (TCA), hydrochloric acid (HCl), potassium hydroxide (KOH), and sodium hydroxide (NaOH) are the most common acids and alkalis used for extraction of intracellular metabolites. The extraction is performed in aqueous medium and the concentration of acid or alkali varies according to the easy to disrupt property of the cells. The procedures are always performed under low temperatures (0–4°C) to avoid degradation of thermo-labile compounds, and freeze-thawing cycle is sometimes included in the process to enhance cell disruption. After extraction, the cell debris is removed from liquid medium and the pH is neutralized. A huge amount of salts are precipitated during pH neutralization, which are removed usually by centrifugation. It is possible, however, that co-precipitation of metabolites takes place during this process.

Acidic and alkaline extractions are the fastest nonmechanical cell disruption methods, acting immediately and reaching completion in a matter of minutes, depending on the concentration and temperature employed. Acids and alkalis added to a cell suspension react with the cell walls in numerous ways, i.e., hydrolysis of macromolecular polymer networks, saponification of lipids in cell envelopes, and denature most proteins avoiding further biochemical reactions. But these extractions at extreme pH are very harsh and several metabolites are not stable at these conditions. Great losses of nucleotides and many other primary metabolites have been demonstrated by using these methods (Hajjaj et al., 1998; Maharjan and Ferenci, 2003; Villas-Bôas et al., 2005a).

3.3.5 Mechanical Disruption of Cell Envelopes

As mentioned previously, mechanical disruption is not often used in metabolome analysis and they have been more widely applied for extraction of proteins or targeted analysis of secondary metabolites. These methods are based mainly on the use of mechanical forces to disrupt cell envelopes, releasing the intracellular contents into a liquid medium. The guidelines for the use of mechanical extraction methods are as follows: (i) choose a compatible liquid medium or solvent that is able to dissolve the group of metabolites of interest and avoid further biochemical reactions in the sample and (ii) be sure that the metabolites to be extracted are stable during the applied mechanical force. The mechanical extraction methods can be classified as *liquid shear*, where the cell disruption takes place in a liquid medium and the metabolites are extracted simultaneously with the cell disruption, or *solid shear*, where the cells are disrupted in absence of any solvent or liquid medium and the metabolites are dissolved later after the cell envelopes had been disrupted (Figure 3.9).

3.3.5.1 Liquid Shear Methods

3.3.5.1a Ultrasonics. Ultrasonication is one of the most widely used and efficient mechanical extraction methods in the laboratory. An *ac* output from an oscillator and amplifier is converted into mechanical waves by a transducer. The output from the transducer is coupled to the treated suspension by a metal probe, which oscillates at the required frequency. The wave amplitude generated is inversely proportional to the probe tip diameter, and the choice of probe diameter is governed by the volume of cell suspension being treated.

Ultrasonic disintegrators generally operate at frequencies of 15–25 kHz. Small cavitation bubbles generated at the tip of an ultrasonic probe immersed in a liquid expand, collapse, and move, causing free radical formation, shock wave propagation, and streaming off the liquid around the bubbles. The probe is mounted just bellow the liquid surface and heats up rapidly, and consequently intermittent use is recommended. During disruption, the cell suspension is cooled by ice or coolant passing through a jacketed cup and the probe is cooled with ice water between cycles. Successful breakage is proportional to the sound intensity and to some extent this can be judged by the ear ("white noise" is created and so wearing ear protectors is strongly recommended). Disruption efficiency can be affected by several operation parameters that include the amplitude of vibrations, surface tension, vessel characteristics, flow rate (if applicable), and use of additives.

Implosion of cavitation bubbles produces shock waves and viscous dissipative eddies that shear and "wear out" (or "fatigue") the cell walls. In general, microorganisms are more readily broken by ultrasound than by other methods. Sonication can cause significant denaturation of enzymes by a combination of cavitation and heating effects, but the use of an enzyme-denaturating solvent is recommended to avoid further biochemical reactions in the samples. Small ballotini beads (glass or steel) or diatomaceous earths can act as triggers for cavitation, and will also exert an additional grinding action, the net effect being increased cell breakage. Free radical formation occurs at high frequencies and while it has no effect on cell breakage, it can adversely affect the integrity of metabolites. Free radical accumulation can be alleviated by addition of free radical scavengers such as cysteine or glutathione (if it will not interfere in the posterior metabolite analysis).

3.3.5.1b Microwave-Assisted Extractions. Microwaves have been employed to assist and enhance chemical extractions of metabolites from diverse biological materials (Table 3.6). The microwaves irradiated on the samples produce rapid agitation of the molecules enhancing the penetration of the extracting agent into the cells, resulting in a more efficient extraction than simple boiling solvents. The advantages are that multiple samples can be extracted simultaneously and it is a very quick procedure. However, similar to extractions using boiling solvents, degradation of thermo-labile compounds is likely to occur.

3.3.5.1c French Press. The French press was developed in 1950 and is still a frequently used and effective apparatus for laboratory scale cell disruption. In its simplest form, it consists of a steel cylinder with a small orifice and needle valve at its base and a piston with a pressure tight seal. Pressures of up to 210 MPa are applied to the sample contained in the cylinder by means of a tight-fitting piston driven by a hydraulic press.

TABLE 3.6 Summary of the Main Mechanical Extraction Methods.

Method	For extraction of	*Applied for	Ideal Conditions	Advantages	Disadvantages	References
Ultrasonics	Free radical-resistant metabolites (the group of metabolites extracted will depend on the polarity of the solvent used) Specially applied for extraction of lipids	Plant tissues Animal tissues (Potentially applicable to other matrices)	15–25 kHz Low temperatures (<0°C) Use of enzyme denaturating solvent Addition of free radical scavengers (e.g., cysteine, glutathione)	Good for extraction of lipids e nonpolar compounds Multiple samples can be extracted simultaneously	Production of free radicals that can react with metabolites	Sargenti and Vichnewski, 2000 Goulas et al., 2000 Pernet and Tremblay, 2003 Yegles et al., 2004 Waksmundzka-Hajnos et al., 2004 Shah et al., 2005 Smedsgaard, 1997
Microwave	Thermostable metabolites (the group of metabolites extracted will depend on the polarity of the solvent used)	Plant tissues Yeast cells Bacterial cells Filamentous fungi cells	Use of enzyme denaturating solvent Fast cooling the samples after extraction to minimise degradation	Simple and fast Enhanced cell disruption by fast heating Multiple samples can be extracted simultaneously	A number of metabolites may be not stable during the process	Stout et al., 1996 Castro et al., 1999 Namieśnik and Górecki, 2000 Smith, 2003
French press	All class of compounds, which can be selected dissolved with different solvents after cell disruption	Plant tissues Bacterial cells (Potentially applicable to other matrices)	Use of compressed CO_2 or precooled nitrogen for cooling the needle valves, to prevent thermo degradation of metabolites	Simple and fast Broad range of metabolites extractable	Not complete deactivation of enzymes Tedious work specially when multiple samples have to be processed	Koutsovelkidis et al., 1999 Yi and Hackett, 2000 Bellevik et al., 2002 Strauss, 2003

Pressurised liquid extraction (PLE)	Mainly secondary metabolites	Plant tissues Yeast cells (Potentially applicable to other matrices)	Scarce information applied to metabolite extraction on literature	Fast Small sample sizes Very concentrated extracts Suitable for high-throughput screening	Possible degradation of thermo-labile compounds	Bethin et al., 1999 Namieśnik and Górecki, 2000 Smith, 2003 Gomez-Ariza et al., 2004 Alonso-Salces et al., 2005
Supercritical fluid extraction (SFE)	Nonpolar to mid-polar compounds	Plant tissues Animal tissues Bacterial cells Yeast cells Filamentous fungi cells	Low temperatures Addition of modifiers, such as methanol, to the carbon dioxide enables more polar compounds to be extracted	Fast Reduced amount of solvents Small sample sizes Easy automation Possibility of on-line coupling to GC/LC-MS Easy sample concentration	Optimization is strictly related to sample source Difficult to extract polar compounds Decomposition under high pressure may be observed for some labile compounds	Abdullah et al., 1994 Gharaibeh and Voorhees, 1996 Murga et al., 2000 Namieśnik and Górecki, 2000 Beek, 2002 Lim et al., 2002 Stolker et al., 2002 Smith, 2003
Grinding	All class of compounds, which can be selected dissolved with different solvents after cell disruption	Specially applied for: Plant tissues Filamentous fungi cells	Very low temperatures (under liquid N_2)	Effective breakage of hard cell walls Enhance any chemical extraction	Tedious work specially when multiple samples have to be processed	Kopka et al., 1995 Roessner-Tunali et al., 2003

During operation, the press is cooled to 0°C (~273 K) and is then filled with the cell suspension. Air must be forced out of the open needle valve, which is then closed before pressure is applied. At the selected pressure, the valve is cautiously opened and the sample is bled through the needle valve, while keeping the pressure constant. Various modifications to the original design exist, notably is the use of compressed CO_2 or precooled nitrogen for cooling the needle valves, to prevent thermo degradation of metabolites (e.g., a modern laboratory apparatus is the "SLM Aminco French Pressure Cell Press").

3.3.5.1d Pressurized Liquid Extraction (PLE). Conventional organic solvents can be maintained liquid at elevated temperatures above their atmospheric boiling points by employing a closed flow-though system. This method, known as pressurized liquid extraction (PLE), is commercially available in an automated or manual version known as accelerated solvent extraction (ASE) and consists' in principle, in a combination of physical chemical extraction method enhanced by a mechanical force (high pressure). Pressurized solvents at elevated temperatures have an enhanced power to dissolve chemicals, a lower viscosity and higher diffusion rates, resulting in an increased extraction rate.

PLE is a highly optimized alternative for exhaustive extraction in a Soxhlet system, reducing the time required for extraction from hours to minutes, using a smaller sample and requiring a small fraction of the original solvent volume. This method is easy to automate and has the ability to carry out multiple extractions. The extracts obtained from this method are generally much more concentrated than from conventional extractions, reducing the time spent in sample concentration. This method has been often applied for extraction of secondary metabolites of plant materials (Smith, 2003), but potentially it can be useful for extraction of other biological matrices. However, degradation of thermo-labile metabolites is expected to take place using this technique.

3.3.5.1e Supercritical Fluid Extraction (SFE). Supercritical fluid extraction is a long established method that has been used industrially for many years. However, only recently it started to be recognized as an extraction technique for metabolite analysis (for detailed information, see Westwood, 1993; Luque de Castro et al., 1994; McHugh and Krukonis, 1994).

Carbon dioxide is the most employed supercritical fluid for extraction of metabolites. There are other alternatives such as nitrous oxide and xenon, but the first has a strong oxidizing power that damage and modify several metabolites and the latter is considered too expensive. Carbon dioxide combines low viscosity and high diffusion rate with a high volatility, making it an ideal solvent. Its ability to dissolve metabolites can be increased by increasing the pressure and extractions can be carried out at relatively low temperatures, which is very beneficial for recovering thermo-labile compounds. Because of the high volatility of CO_2, the samples can be readily concentrated by simply reducing the pressure and allowing the supercritical fluid to evaporate.

Nevertheless, carbon dioxide has a very low polarity, which is the ideal solvent for extraction of nonpolar compounds such as lipids and fats, but unsuitable for most primary metabolites. The addition of modifiers, such as methanol, to the carbon dioxide enables more polar compounds to be extracted and increases the application of the method. It is increasingly being used for extraction of intracellular metabolites from plant cells (Table 3.6), whereas there are only few examples of applying SFE to other matrices.

3.3.5.2 Solid Shear Methods. Due to the absence of liquid solvents, the procedures using solid shear methods must be done under very low temperatures to ensure inactivation of any enzymatic activity in the samples. There are three solid shear methods that are relevant for metabolome analysis: manual grinding, ball mill, and Ultra-Turrax.

3.3.5.2a Manual Grinding. By using mortar and pestle, frozen cells can be grounded manually in liquid nitrogen. This very ancient method for enhanced extraction of biological compounds from solid matrices is still extremely useful for disrupting cell envelopes, mainly those cells with hard cell wall structures such as filamentous fungi and plant tissues. The samples are grinded under very low temperatures and the metabolites are dissolved in a selected solvent(s) after the grinding process. Although efficient, this process is laborious and can be very time consuming depending on the number of samples to be processed.

3.3.5.2b Ball Mill. Cell disruption in ball mills is regarded as an optimized alternative for the classic mortar and pestle. Various designs of ball mills have been used for cell disruption, and these consist of either vertical or horizontal cylindrical chamber, with a motor-driven central shaft supporting a collection of off-centered discs or other agitating elements. The cylindrical grinding tank is usually surrounded by a cooling chamber, and the temperature can be controlled. The grinding process can be enhanced by adding beads such as ballotini glass beads or steel beads into the samples. Similarly to manual grinding, the metabolites are dissolved in a selected solvent(s) after the grinding process.

3.3.5.2c Ultra-Turrax. The Ultra-Turrax homogenizers-dispenser has long been a laboratory favorite devise to grind and homogenize quenched plant or animal tissues. It is a round-shape knife that rotates rapidly like an automatic hole saw. Using this equipment, frozen plant and animal tissues can be easily homogenized at low temperatures, but it tends to work better for harder tissues than soft ones. Special care must be taken to ensure that all tissue peaces are grinded homogenously and ears protection is always recommended due to the high noise generated by this device.

3.4 METABOLITES IN THE EXTRACELLULAR MEDIUM

Metabolites in the extracellular medium are usually of great interest for metabolome analysis because they are more accessible and easy to handle, and recent approaches on metabolic footprinting analysis (Allen et al., 2003; Villas-Bôas et al., 2005b,

2006) have demonstrated how useful phenotypic information can be obtained by analyzing these compounds. There are two main groups of extracellular metabolites concerning sample preparation procedures: (i) metabolites in solution and (ii) metabolites in the gas phase.

3.4.1 Metabolites in Solution

Typical samples containing extracellular metabolites in solution are spent microbial/cell culture media or body fluids such as plasma, urine, milk, root exudates, apolastic, and others. After handling these samples, according to the guidelines presented in Box 3.2, they are ready to be analyzed. However, very often the sample composition poses problems for the analytical technique that will be used, i.e., high level of salts, proteins or lipids, or even presence of water. To minimize these problems, the metabolites of interest can be extracted from the liquid samples either by partitioning into an immiscible solvent, trapping the metabolites onto a column or solid-phase matrix, or simply evaporating the samples to dryness followed by selectively dissolving the compounds in an appropriate solvent.

Partitioning the metabolites into an immiscible solvent is very laborious and, therefore, has not found extensive applicability in metabolome analysis. Trapping the metabolites in a solid-phase matrix, on the contrary, gained great popularity in analysis of metabolites, and two methods specifically is worth mentioning in further details: (i) solid-phase extraction (SPE), and solid-phase microextraction (SPME). Simply evaporation of the samples to dryness and selectively dissolving the compounds is also applied extensively and will therefore be discussed in details in Section 3.4.

3.4.1.1 Solid-phase Extraction (SPE). SPE is an extraction method that uses a solid phase and a liquid phase to isolate one or one type of analyte from a solution. It is usually used to clean up a sample before using a chromatographic or other analytical method to quantify the amount of analyte(s) in the sample. The general procedure is to load a solution onto the SPE phase, wash away undesired components, and then wash off the desired analyte(s) with another solvent into a collection tube.

The concept of passing a liquid sample through a solid matrix (usually a short hand-packed column) has been employed for many years for cleaning samples before analysis. However, the introduction of disposable prepackaged SPE cartridge offered two important advantages: (1) standardization resulting in better reproducibility and (2) a more diverse range of solid-phases resulting in an increased applicability of the method.

Solid-phase extractions use the same type of stationary phases as used in liquid chromatography columns. The stationary phase is contained in a glass or plastic column above a frit or glass wool (Figure 3.10a). The column might have a frit on top of the stationary phase and might also have a stopcock to control the flow of solvent through the column. Commercial SPE cartridges generally have 1–10 mL capacities and are discarded after use. Figure 3.10b shows an SPE cartridge on a vacuum manifold, which increases the solvent flow rate through the cartridge. A collection tube

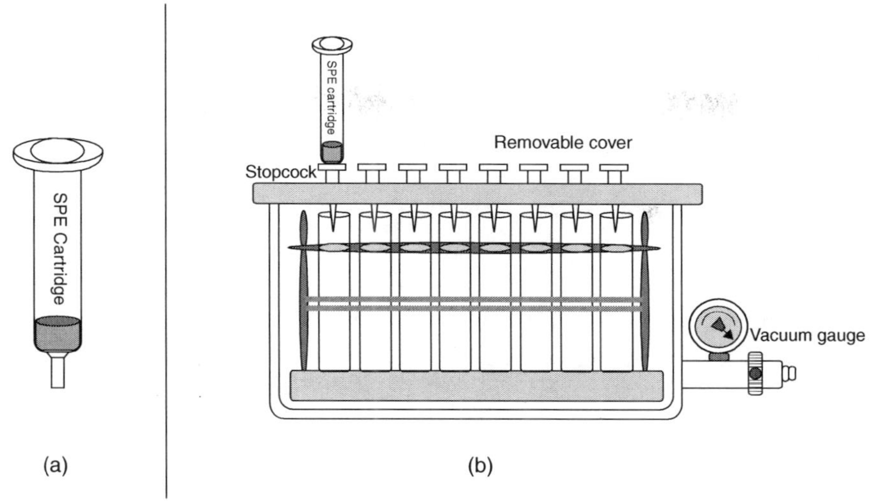

Figure 3.10 Schematic illustration of a solid-phase extraction (SPE) machinery. (a) SPE column cartridge, which are usually disposable. (b) SPE cartridge on a vacuum manifold device, which increases the solvent flow rate through the cartridge.

is placed beneath the SPE cartridge (inside the vacuum manifold for the example in Figure 3.10b) to collect the liquid that passes through the column.

Although, in some occasions, the impurities of the sample are trapped and the metabolites of interest pass thorough the cartridge, the metabolites are in most cases trapped in the solid matrix and can thereafter be released into a small volume of an extraction solvent by altering the polarity, pH, or ionic strength of the mobile phase. Usually the SPE cartridge is washed with the sample solvent to activate the solid matrix and then the sample is loaded. The cartridge containing the analyte(s) trapped in the solid phase is washed with a weak solvent to elute weaker components that were trapped together with the analyte(s). Then, the solid-phase is washed with a small volume of a stronger solvent to elute the analyte(s). A final washing step with an even stronger solvent is usually added to the protocol to elute strongly adsorbed components in order to clean up the SPE cartridge. This basic general protocol is adapted to any specific SPE phase and their main differences are summarized in Box 3.4. When a large number of samples need to be processed simultaneously, the process can easily be automated using robotic or automation devices, commercialized by different manufacturers, eliminating almost completely the sample handling and leading to a high reproducibility.

SPE has a considerable scope for analysis of metabolites, principally applied for extraction of metabolites from body fluids (Conneely et al., 2002; Kabbaj and Varin, 2003; Smith, 2003). The disposable cartridges reduce the handling of body fluids, such as urine and blood, and consequently the biohazard to the analyst is minimized. A wide range of cartridge material, eluents, and sample matrices are described on manufacturers' websites and in the literature. The great limitation of SPE, however,

◊ Text box 3.4 General elution protocols for different SPE phases.

Normal phase

1. Condition the cartridge with six to ten hold-up volumes of nonpolar solvent, usually the sample solvent
2. Load the sample into the cartridge
3. Elute unwanted components with a nonpolar solvent
4. Elute the first component(s) of interest with a polar solvent
5. Elute remaining components of interest with progressively more polar solvents
6. When recovered all components of interest, discard the used cartridge in a appropriate manner.

Reversed phase

1. Solvate the bonded phase with six to ten cartridge hold-up volumes of methanol or acetonitrile
2. Flush the cartridge with six to ten hold-up volumes of water or buffer (do not allow the cartridge to dry out)
3. Load the sample dissolved in strongly polar solvent
4. Elute unwanted components with strongly polar solvent
5. Elute weakly held components of interest with a less polar solvent
6. Elute more tightly bound components with progressively more non-polar solvents
7. When recovered all components of interest, discard the used cartridge in an appropriate manner.

Ion-exchange phase

1. Condition the cartridge with six to ten hold-up volumes of deionized water or weak buffer
2. Load the sample dissolved in a solution of deionized water or buffer
3. Elute unwanted weakly bound components with a weak buffer
4. Elute the first component(s) of interest with a stronger buffer (change the pH or ionic strength)
5. Elute other components of interest with progressively stronger buffers
6. When recovered all components of interest, discard the used cartridge in an appropriate manner.

Some important troubleshooting tips

- Poor analyte retention = dilute the samples with weaker solvent, use stronger sorbent, use larger cartridges
- Matrix variability = buffer samples to constant pH, ionic strength
- Volume overload = decrease load volume, use larger cartridge
- Mass overload = decrease load volume, use larger cartridge.

is its selectivity that is ideal for targeted analysis but unsuitable for broad metabolite profiling, where different class of metabolites should be analyzed together. The cartridge material and elution condition tend to be very selective for a specific group of metabolites, which is due to ensure the good reproducibility offered by SPE.

3.4.1.2 Solid-Phase Microextraction (SPME). Pawliszyn and co-workers (Chen and Pawliszyn, 1995; Lord and Pawliszyn, 2000) invented the ingenious SPME method to improve the throughput of SPE by eliminating the necessity of eluting the analytes of interest from the solid phase before injection into a separation/analytical method. SPME is based on the use of a fiber coated with a stationary phase as an extraction medium. After carrying out an extraction from a sample solution, the fiber is placed in the injection port of a gas chromatograph so that the analytes are thermally desorbed directed into the carrier gas stream. Although nonvolatile analytes can be extracted directly into the eluent stream of a liquid chromatograph system (Chen and Pawliszyn, 1995) or even be on-fibre derivatized prior to analysis (Lord and Pawliszyn, 2000), the SPME methods gained popularity mainly for the analysis of volatile compounds by GC/GC–MS.

The principle of SPME is that the objective of this technique is never an exhaustive extraction of the analyte(s) from the sample solution but to obtain a representative sample of the analyte(s) of interest trapped on the coated-fibre matrix that can be compared with the extraction of a standard solution. In SPME, a small amount of extracting phase associated with a solid support is placed in contact with the sample matrix for a predetermined amount of time. If the time is long enough, an equilibrium is established between the sample matrix and the extraction phase. When equilibrium conditions are reached, the fibre does not accumulate more analyte(s). The phase distribution and the amount extracted depend on the partition coefficient between the sample solution and the fibre.

The main advantages of SPME system are that no solvent is required to elute the sample from the fibre and unless the sample is very complex and rich in nonvolatile compounds that can be bound to the fibre, the fibre can be reused several times as the thermal elution step also cleans up the fibre. However, the coated-fibre is relatively expensive and fragile, and nonvolatile compounds can easily be bound on it and are difficult to be removed. In addition, the extraction process can be relatively slow because good reproducibility requires that an equilibrium is established. The SPME technique can also be used to assay the headspace above the sample (see the following section) and this method is preferred for volatile metabolites as the fibre avoids contact with the matrix solution. Similar to SPE, SPME is ideal for targeted analysis of metabolites because the equilibrium is dependent on the analyte and it will be favored depending on the fibre matrix being used, which is unsuitable for a broad metabolite profiling.

3.4.2 Metabolites in the Gas Phase

Most biological matrices contain volatile metabolites that are usually lost to the environment and that represent valuable information on the phenotype. Gas samples are volatile and they can therefore be analyzed directly by gas chromatography leaving no residues. However, several volatile metabolites are present at very low concentration near to the detection limit and the integrity of a gas sample is very difficult to maintain from the collection point to the analyzer due to the high diffusion rates of gases. There has, therefore, been considerable interest in concentrating and trapping relevant metabolites to increase the sensitivity.

A series of methods have been developed to trap and concentrate components from gases. Some of the more efficient methods rely on passing of the gas over a cold adsorption tube packet with a form of GC stationary phase, including adsorptive materials, such as porous carbon, or sorptive polymers, such as Tenax, polystyrene-divinyl benzene or PDMS (e.g., Larsen and Frisvad, 1995; Demyttenaere et al., 2003). The gas may be pumped for a specific time or can be allowed to diffuse into the trap in long-term exposure studies. The trapped metabolites are usually desorbed thermally and transferred directly into a gas chromatograph for separation and quantification.

3.4.2.1 Headspace Analysis. Metabolites in the gas phase of a cultivation flask are usually analyzed by determining their levels in the headspace gas above the culture either by taking a direct gaseous sample with a syringe or by trapping the volatile compounds on a SPME fibre. Alternatively, liquid samples can be harvested and heated to increase the vapor phase concentration in the headspace phase, and both manual and automated systems are available, the latter giving higher reproducibility.

The analysis of volatile metabolites in the headspace of a sample or cultivation flask is rarely a quantitative approach and commonly, the sampling conditions are established and fixed and the profile of volatile compounds obtained from different cultures are then compared. Rather than directly sampling the gases from the headspace of a cultivation flask or bioreactor, the metabolites in the headspace can be trapped on a SPME (Nilsson et al., 1996; Mills and Walker, 2000; Demyttenaere et al., 2003). It is important, however, to be aware that the distribution is between the fibre and the matrix. Thus, raising the temperature reduces the deposition onto the fibre even though it increases the concentration of metabolites in the headspace, because it increases the vapor concentration above the fibre as well as above the sample. Therefore, SPME can give very distinct profiles compared to direct headspace analysis. The headspace will favor the high volatile metabolites, while the fibre will favor the less volatile ones.

3.5 IMPROVING DETECTION VIA SAMPLE CONCENTRATION

The samples obtained during extraction of intracellular metabolites and even some samples from extracellular metabolites are characteristically diluted. Thus, prior to sample analysis, the solvent(s) must be partially or totally removed from the samples. Freeze-drying, or lyophilization, is commonly used to remove water from aqueous samples in order to avoid thermal degradation. The process of freeze-drying consists of freezing the sample and subsequently removing the frozen solvent by sublimation. This method combines the advantage of both deep-freezing and dehydration. The metabolites are stabilized by a nonaggressive technology, avoiding heat.

However, freeze-drying is also a relatively time-consuming process. The mechanisms are complex by which freeze-drying of a particular sample is achieved. In general, larger surfaces are preferred rather than thick ice layers to obtain a fast drying. In storage of the dry material, care has to be taken to avoid degradation by oxygen and light. Indeed, in some instances, interactions with oxygen can prove to be very deleterious to some organic compounds by provoking molecular oxidation and undesirable free radicals. It is recommended to break the vacuum with a dry inert gas (nitrogen or argon) and the samples should be stored under oxygen-free conditions or even under high vacuum at low temperatures.

The freeze-dry method has given rise to an intensive development of new instruments. From manually operated to fully automated devices are commercially available nowadays. In classical setups, the frozen samples are dried at room temperature that accelerates the sublimation process. However, the metabolites are exposed to room temperature after finishing the drying process, which can be damaging to those thermal sensitive metabolites. Modern designs enable the drying process to be performed at very low temperatures (i.e., −56°C) consisting in a great advantage in analysis of metabolites. However, freeze-drying process can be significantly affected by several other variables such as the concentration of organic solvents in the solution, the pH of the solution, additives (e.g., sugars, buffering substances), and others.

Organic solvent solutions cannot be frozen even under the low temperatures and pressures reached by the newer freeze-dryer devices. Since most of the extraction procedures make use of organic solvents, these samples can be freeze-dried merely adding extra volume of deionized water in order to increase the water:solvent ratio and thus, allowing the mixture been kept frozen during the process. However, the sample volume will increase resulting in a longer freeze-drying process. Aqueous samples containing high concentrations of sugars (e.g., 100 g/L of glucose) present extremely low drying rate, being practically impossible to finish the drying process and ending with a highly viscous product. For this particular case, the differences in the final volume of the sample after resuspension must be taken into account when quantitative analysis is aimed.

Furthermore, losses of metabolites during lyophilization are often observed and the losses are certainly related to discrimination during resuspension. The different metabolites present different solubilities in the solvent used for resuspension, and, therefore, discrimination during dissolving these solutes in a very small volume of solvent are likely to happen. In addition, the recovery of the resuspended solution from the lyophilization flask is another important source of losses. Considering that for most extraction procedure we end up with large volume of extracts, we are forced to use considered large flasks for lyophilization. To dissolve the remaining salts after the concentration process by adding a small volume of solvent is definitely a challenge and, hence, can explain some of the general losses observed.

Alternatively, nonaqueous extracts can be concentrated by solvent evaporation using several different commercial devices designed for this proposal. Organic solvent evaporation seems to be a very reliable method for concentration of samples containing primary metabolites (Villas-Bôas et al., 2005a). It is fast enough to minimize losses by thermo-degradation. However, this technique is dependent on the type of extraction procedure used, since this procedure is not well suited for aqueous sample extracts as water takes long to dry under vacuum and it is often necessary to heat the samples. Nonetheless, solvent evaporation has several advantages over the lyophilization because it is faster, less aggressive, and less discriminative.

REFERENCES

Abdullah MI, Young JC, Games DE. 1994. Supercritical fluid extraction of carboxylic and fatty acids from *Agaricus* SPP. mushrooms. J Agric Food Chem 42:718–722.

Allen J, Davej HM, Broadhurst D, Heald JK, Rowland JJ, Oliver SG, Kell DB. 2003. High-throughput classification of yeast mutants for functional genomics using metabolic footprinting. Nat Biotechnol 21:692–696.

Alonso-Salces RM, Barranco A, Corta E, Berrueta LA, Gallo B, Vicente F. 2005. A validated solid-liquid extraction method for the HPLC determination of polyphenols in apple tissues—Comparison with pressurized liquid extraction. Talanta 65:654–662.

Bethin B, Danz H, Hamburger M. 1999. Pressurized liquid extraction of medical plants. J Chromatogr A 837:211–219.

Britten RJ, McClure Y. 1962. The amino acid pool in *Escherichia coli*. Bacterial Rev 26:292–335.

Buziol S, Bashir I, Baumeister A, Classβen W, Noisommit-Rizzi N, Mailinger W, Reuss M. 2002. New bioreactor-coupled rapid stopped-flow sampling technique for measurements of metabolite dynamics on a subsecond time scale. Biotechnol Bioeng 80:632–636.

Beek TA. 2002. Chemical analysis of *Ginkgo biloba* leaves and extracts. J Chromatogr A 967:21–55.

Bellevik S, Summerer S, Meijer J. 2002. Overexpression of *Arabidopsis thaliana* soluble epoxide hydrolase 1 in *Pichia pastoris* and characterization of the recombinant enzyme. Protein Expres Purif 26:65–70.

Castrillo JI, Hayes A, Mohammed S, Gaskell SJ, Oliver SG. 2003. An optimized protocol for metabolome analysis in yeasts using direct infusion electrospray mass spectrometry. Phytochem 62:929–937.

Castro MDL, Jiménez-Carmona MM, Fernández-Pérez V. 1999. Towards more rational techniques for the isolation of valuable essential oils from plants. Trends Anal Chem 18:708–716.

Chen J, Pawliszyn JB. 1995. Solid phase microextraction coupled to high-performance liquid chromatography. Anal Chem 67:2530–2533.

Conneely A, Nugent A, O'Keeffe M. 2002. Use of solid phase extraction for the isolation and clean-up of a derivatized furazolidone metabolite from animal tissues. Analyst 127:705–709.

Cook AM, Urban E, Schlegel HG. 1976. Measuring the concentrations of metabolites in bacteria. Anal Biochem 72:191–201.

Cremin P, Donnelly DMX, Wolfender JL, Hostettmann K. 1995. Liquid chromatography-thermospray mass spectrometric analysis of sesquiterpenes of *Armillaria* (Eumycota: Basidiomycotina) species. J Chromatogr A 710:273–285.

Demyttenaere JCR, Moriña RM, Sandra P. 2003. Monitoring and fast detection of mycotoxin-producing fungi based on headspace solid-phase microextraction and headspace sorptive extraction of the volatile metabolites.

De Koning W, van Dam K. 1992. A method for the determination of changes of glycolytic metabolites in yeast on a subsecond time scale using extraction at neutral pH. Anal Biochem 204:118–123.

Entian KD, Zimmermann FK, Scheel I. 1977. A partial defect in carbon catabolite repression mutants of *Saccharomyces cerevisiae* with reduced hexose phosphorylation. Mol Gen Genet 156:99–105.

Fiehn O. 2002. Metabolomics–the link between genotypes and phenotypes. Plant Mol Biol 48:155–171.

Folch J, Lees M, Stanley GH. 1957. A simple method for the isolation and purification of total lipids from animal tissue. Biol Chem 226:497–509.

Gharaibeh AA, Voorhees KJ. 1996. Characterization of lipid fatty acids in whole-cell microorganisms using *in situ* supercritical fluid derivatization/extraction and gas chromatography/mass spectrometry. Anal Chem 68:2805–2810.

Gomez-Ariza JL, de la Torre MAC, Giraldez I, Morales E. 2004. Speciation analysis of selenium compounds in yeasts using pressurized liquid extraction and liquid chromatography-microwave-assisted digestion-hydride generation-atomic fluorescence spectrometry. Anal Chim Acta 524:305–314.

Gonzalez B, Fronçois J, Renaud M. 1997. A rapid and reliable method for metabolite extraction in yeast using boiling buffered ethanol. Yeast 13:1347–1356.

Goulas A, Papakonstantinou E, Karakiulakis G, Mirtsou-Fidani V, Kalinderis A, Hatzichristou DG. 2000. Tissue structure-specific distribution of glycosaminoglycans in the human penis. Int J Biochem Cell Biol 32:975–982.

Hajjaj H, Blanc PJ, Goma G, François J. 1998. Sampling techniques and comparative extraction procedures for quantitative determination of intra- and extracellular metabolites in filamentous fungi. FEMS Microbiol Lett 164:195–200.

Hans MA, Heinzle E, Wittmann C. 2001. Quantification of intracellular amino acids in batch cultures of *Saccharomyces cerevisiae*. Appl Microbiol Biotechnol 56:776–779.

Jensen NBS, Jokumsen KV, Villadsen J. 1999. Determination of the phosphorylated sugars of the Embden-Meyerhoff-Parnas pathway in *Lactococcus lactis* using a fast sampling technique and solid phase extraction. Biotechnol Bioeng 63:356–362.

Kabbaj M, Varin F. 2003. Simultaneous solid-phase extraction combined with liquid chromatography with ultraviolet absorbance detection for the determination of remifentanil and its metabolite in dog plasma. J Chromatogr B 783:103–111.

Kopka J, Ohlrogge JB, Jaworski JG. 1995. Analysis of *in vivo* levels of acylthioesters with gas chromatography/mass spectrometry of the butylamide derivative. Anal Biochem 224:51–60.

Koutsovelkidis I, Neopikhanov V, Soderman C, Lorenz A, Uribe A. 1999. Butyrate inhibits and *Escherichia coli* derived mitogen(s) stimulate DNA synthesis in human hepatocytes *in vitro*. Prep Biochem Biotechnol 29:121–138.

Larsen TO, Frisvad JC. 1995. Characterization of volatile metabolites from 47 *Pinicillium* taxa. Mycol Res 99:1153–1166.

Larsson G, Törnkvist M. 1996. Rapid sampling cell inactivation and evaluation of low extracellular glucose concentrations during fed-batch cultivation. J Biotechnol 49:69–82.

Le Belle JE, Harris NG, Williams SR, Bhakoo KK. 2002. A comparison of cell and tissue extraction techniques using high-resolution 1H-NMR spectrometry. NRM Biomed 15:37–44.

Leder IG. 1972. Interrelated effects of cold shock and osmotic pressure on permeability of the *Escherichia coli* membrane to permease accumulated substrates. J Bacteriol 111:211–219.

Letisse F, Lindley ND. 2000. An intracellular metabolite quantification technique applicable to polysaccharide-producing bacteria. Biotechnol Let 22:1673–1677.

Lim GB, Lee SY, Lee EK, Haam SJ, Kim WS. 2002. Separation of astaxanthin from red yeast *Phaffia rhodozyma* by supercritical carbon dioxide extraction. Biochem Eng J 11:181–187.

Lord H, Pawliszyn J. 2000. Evolution of solid-phase microextraction technology. J Chromatogr A 885:153–193.

Luque de Castro MD, Valcácel M, Tena MT. 1994. *Analytical Supercritical Fluid Extraction*, Springer, Berlin.

Maharjan RP, Ferenci T. 2003. Global metabolite analysis: the influence of extraction methodology on metabolome profiles of *Escherichia coli*. Anal Biochem 313:145–154.

Marshall S, Nadeau O, Yamasaki K. 2004. Dynamic actions of glucose and glucosamine on hexosamine biosynthesis in isolated adipocytes. J Biol Chem 34:35313–35319.

Mashego MR, van Gulik WM, Vinke JL, Heijnen JJ. 2003. Critical evaluation of sampling techniques for residual glucose determination in carbon-limited chemostat culture of *Saccharomyces cerevisiae*. Biotechnol Bioeng 83:395–399.

McHugh MA, Krukonis VJ. 1994. *Supercritical Fluid Extraction: Principles and Practice* (2nd edition), Butterworths, London.

Michalke B, Witte H, Schramel P. 2002. Effect of different extraction procedures on the yield and pattern of Se-species in bacterial samples. Anal Bional Chem 372:444–447.

Mills GA, Walker V. 2000. Headspace solid-phase microextraction procedures for gas chromatography analysis of biological fluids and materials. J Chromatogr A 902:267–287.

Murga R, Ruiz R, Beltráan S, Cabezas JL. 2000. Extraction of natural complex phenols and tannins from grape seeds by using supercritical mixtures of carbon dioxide and alcohol. J Agric Food Chem 48:3408–3412.

Namieśnik J, Górecki T. 2000. Sample preparation for chromatographic analysis of plant material. J Planar Chromatogr 13:404–413.

Nilsson T, Larsen TO, Montanarella L, Madsen JØ. 1996. Application of headspace solid-phase microextraction for the analysis of volatile metabolites emitted by *Penicillium* species. J Microbiol Met 28:113–122.

Orth HCJ, Rentel C, Schmidt PC. 1999. Isolation, purity analysis and stability of hyperforin as a standard material from *Hypericum perforatum* L. J Pharm Pharmcol 51:193–200.

Pernet F, Tremblay R. 2003. Effect of ultrasonication and grinding on the determination of lipid class content of microalgae harvested on filters. Lipids 38:1191–1195.

Rizzi M, Baltes M, Theobald U, Reuss M. 1997. *In vivo* analysis of metabolic dynamics in *Saccharomyces cerevisiae*: II. Mathematical model. Biotechnol Bioeng 55:592–608.

Roessner-Tunali U, Hegemann B, Lytovchenko A, Carrari F, Bruedigam C, Granot D, Fernie AR. 2003. Metabolic profiling of transgenic tomato plants overexpressing hexokinase reveals that the influence of hexose phosphorylation diminishes during fruit development. Plant Physiol 133:84–99.

Sargenti SR, Vichnewski W. 2000. Sonication and liquid chromatography as a rapid technique for extraction and fractionation of plant material. Phytochem Anal 11:69–73.

Schaefer U, Boos W, Takors R, Weuster-Botz D. 1999. Automated sampling device for monitoring intracellular metabolite dynamics. Anal Biochem 270:88–96.

Shah S, Sharma A, Gupta MN. 2005. Extraction of oil from *Jatropha curcas* L. seed kernels by combination of ultrasonication and aqueous enzymatic oil extraction. Biores Technol 96:121–123.

Shryock JC, Rubio R, Berne RM. 1986. Extraction of adenine nucleotides from cultured endothelial cells. Anal Biochem 159:73–81.

Singer SJ, Nicolson GL. 1972. The fluid mosaic model of the structure of cell membranes—cell membranes are viewed as 2 dimensional solutions of oriented globular proteins and lipids. Science 175:720–731.

Smedsgaard J. 1997. Micro-scale extraction procedure for standardized screening of fungal metabolite production in cultures. J Chromatogr A 760:264–270.

Smeaton JR, Elliott WH. 1967. Selective release of ribonuclease-inhibitor from *Bacillus subtilis*. Biochem Biophys Res Com 26:75–81.

Smith RM. 2003. Before the injection—modern methods of sample preparation for separation techniques. J Chromatogr A 1000:3–27.

Smits HP, Cohen A, Buttler T, Nielsen J, Olsson L. 1998. Cleanup and analysis of sugar phosphates in biological extracts by using solid-phase extraction and anion-exchange chromatography with pulsed amperometric detection. Anal Biochem 261:36–42.

Stout SJ, daCunha AR, Picard GL, Safarpour MM. 1996. Microwave-assisted extraction coupled with liquid chromatography/electrospray ionization mass spectrometry for the simplified determination of imidazolinone herbicides and their metabolites in plant tissues. J Agric Food Chem 44:3548–3553.

Tondo EC, Andretta CWS, Souza CFV, Monteiro AL, Henriques JAP, Ayub MAZ. 1998. High biodegradation levels of 4,5,6-trichloroguaiacol by *Bacillus* SP. isolated from cellulose pulp mill effluent. Rev Microbiol 29:265–271.

Villas-Bôas SG, Højer-Pedersen J, Åkesson M, Smedsgaard J, Nielsen J. 2005a. Global metabolite analysis of yeast: Evaluation of sample preparation methods. Yeast 22:1155–1169.

Villas-Bôas SG, Moxley JF, Åkesson M, Stephanopoulos G, Nielsen J. 2005b. High-throughput metabolic state analysis: The missing link in integrated functional genomics of yeasts. Biochem J 388:669–677.

Villas-Bôas SG, Noel S, Lane GA, Attwood G, Cookson A. 2006. Extracellular metabolomics: A metabolic footprinting approach to assess fiber degradation in complex media. Anal Biochem 349:297–305.

Waksmundzka-Hajnos M, Petruczynik A, Dragan A, Wianowska D, Dawidowicz AL. 2004. Effect of extraction method on the yield of furanocoumarins from fruits of *Archangelica officialis* Hoffm. Phytochem Anal 15:313–319.

Westwood SA. 1993. *Supercritical Fluid Extraction and its Use in Chromatographic Sample Preparation*, Blackie, London.

Weuster-Botz D. 1997. Sampling tube device for monitoring intracellular metabolite dynamics. Anal Biochem 246:225–233.

Wittmann C, Krömer JO, Kiefer P, Binz T, Heinzle E. 2004. Impact of the cold shock phenomenon on quantification of intracellular metabolites in bacteria. Anal Biochem 327:135–139.

Yegles M, Labarthe A, Auwärter V, Hartwig S, Vater H, Wennig R, Pragst F. 2004. Comparison of ethyl glucuronide and fatty acid ethyl ester concentrations in hair of alcoholics, social drinkers, and teetotallers. Forensic Sci Int 145:167–173.

Yi EC, Hackett M. 2000. Rapid isolation method for lipopolysaccharide and lipid A from Gram-negative bacteria. Analyst 125:651–656.

4

ANALYTICAL TOOLS

BY JØRN SMEDSGAARD

This chapter will present in a short but concise form the principles of the key techniques of chromatography (GC and LC) and mass spectrometry (MS) (used alone or in combination with chromatography) as needed for metabolite profiling of biological samples. The focus will be on the small biomolecules in complex samples, and it is intended to guide the reader to select and optimize a methodology. The techniques: GC-injection, EI ion source, ESI-source, Quadrupole analyzer, tof analyzer, iontrap analyzer, and MS detection will be introduced, and the advantages and limitations of each technique will be highlighted and related to the different metabolite classes described previously in Chapter 2, and the text will guide the reader into the differences in target analysis, metabolite profiling, and fingerprinting, all analytical approaches important for metabolomics studies.

4.1 INTRODUCTION

The complexity of the metabolome is very large as discussed in the previous chapters, in terms of both chemical diversity and quantities of each metabolite. Therefore, metabolome analysis presents a serious challenge for any analytical chemist. Adding to the challenge is the requirement to determine all these metabolites in a large number of small samples—and possibly even to quantify the amount of each of them. With current analytical technologies, it is not possible to detect the complete metabolome (*all the smaller metabolites*) in one single analysis, not even from the simplest organisms. On the contrary, the advances in analytical methodologies combined with new data processing techniques (chemometrics and other multivariate techniques as

Metabolome Analysis: An Introduction, by Silas G. Villas-Bôas, Ute Roessner, Michael A. E. Hansen, Jorn Smedsgaard and Jens Nielsen

discussed in Chapter 5) have so far been the major driving force behind development of metabolomics. Of these analytical technologies, MS and chromatography in particular, are the core analytical technologies behind metabolome analysis.

This chapter aims to introduce these key analytical techniques from a practical perspective to give the reader the basics to understand and select techniques for metabolome analysis. The understandings of the analytical principles are included whereever needed to evaluate the quality of the data. However, the reader is referred to specialized textbooks for an in-depth theoretical and practical discussion of analytical methodologies like MS and chromatography. Reference to a few textbooks will be given at the end of the chapter.

4.2 CHOOSING A METHODOLOGY

Choosing a suitable analytical strategy requires a clear formulation, the problem to which we want some answers. In metabolome analysis, it can be difficult to formulate problems in such a way that it can be solved by one or a few analytical methods. An example is often found in functional genomics studies: gene functions are studied by producing knock-out mutants leading to the question: *I deleted this gene—how did that affect the metabolite pattern?* This may seem as a very simple question, but it can be very difficult to answer. Many metabolites take part in many different pathways; there may be unknown intermediates, other secondary changes, and so forth, and deletion of a single gene may, therefore, result in numerous changes. On the contrary, the expression of the changes might be insignificant, given the cultivation conditions. Also, some of the changes might not be detectable by the analytical procedure commonly used for the wild-type profiles. The result is that we have to deal with a number of changes or minute changes and may be even with completely new or unknown metabolites. Also, extracting information from may be 100 chromatograms, each with hundreds of peaks, also possesses a serious challenge for data processing as discussed in Chapter 5. Á-priori knowledge can greatly simplify the problems and may enable us to split the problems into subproblems allowing a more sensible analytical or targeted strategy to be planned.

Planning an efficient strategy for metabolome analyses requires consideration of the following questions: what kind of information is needed? what kind of chemistry is expected? and what are the analytical facilities available?

In general, the approaches used for metabolome analysis are often divided into three different strategies:

Fingerprinting In this strategy, a chemical fingerprint or picture is made by a direct analysis of crude sample extracts, typically by MS, nuclear magnetic resonance spectrometry (NMR), or infrared spectrometry. These fingerprints can be an efficient tool to compare and classify samples but do not always give information about occurrence of specific metabolites (whether they are known or unknown). A derivation of fingerprinting is *footprinting* where the cell-free spend media is analyzed for left metabolites (sometime also called the exometabolome).

Profiling It aims to detect as many metabolites as possible, whether these are known or unknown. However, the metabolites detected by profiling must be recognized consistently and should be also quantified. Profiling is typically done by chromatography in combination with MS or by capillary electrophoresis (CE) combined with MS.

Target Target analysis aims to detect and quantify specific metabolites. A multitude of different analytical methods might be used for this purpose, each being able to detect one or more metabolites.

Although there is an overlap between these strategies, they can give not only quite different but also complementary results. These strategies share some common methodologies and analytical approaches but are typically implemented quite differently. It is crucial to remember that no single technique can give a complete "picture" of all metabolites present in an organism and can even less enable quantification of them. Therefore, no matter what methodology is used, the chosen method will bias the results. This is particularly the case for fingerprinting and profiling analyses that cannot be compared without taking the analytical procedure into account. Although fingerprinting analyses are mostly based on direct spectrometric measurement of more or less crude samples (see Chapter 3) by, e.g., ultraviolet-visual spectrophotometers (UV), NMR, or mass spectrometers (MS), profiling and target analyses require, in general, a separation of the compounds by, e.g., gas or liquid chromatography (GC or LC) or CE prior to the spectrometric detection by, e.g., UV, NMR, or MS. The combinations GC–MS and LC–MS are so far, the most important; however, analyses by CE coupled with MS have shown impressive results. Both fingerprinting and profiling can be somewhat misleading as two quite different samples may show the same fingerprint or metabolite profile using one analytical approach whereas another analytical strategy may reveal important metabolic differences. Both terms are much older than metabolomics and are frequently found in the analytical literature (e.g., in flavor and fragrance analyses, profiling and fingerprinting have been used for more than 30 years for analytical strategies, not too different from that of metabolomics). There seems to be a general consensus that fingerprinting is a crude spectroscopic measurement whereas metabolite profiling requires some compound separation as described above. However, neither approach can be used without a careful check of the analytical strategy and assessing the analytical limitations. The use of these terminologies for metabolomics is being still debated and no clear consensus has been reached yet.

The nature of the metabolome chemistry, as discussed in Chapter 2, is very complex, and no single methodology can detect the complete metabolome in one procedure. The following key parameters have to be evaluated to select an analytical procedure:

Chemistry	polarity (polar, nonpolar) pK_a: acidic, alkaline, neutral concentration (sensibility of detectors) detectability (chromophors, ionizability, or others) volatility
Concentration	trace or massive amount (ppb range or percent range)
Matrix	interference from coextracted substrate or may be from major components in the sample

In the following chapters, the different methodologies are discussed in terms of their application range and their usability.

On the contrary, one should keep an eye open for information that can be collected for free, information that may not necessarily be needed immediately to address the question posed, but that might be useful at a later point (also see the discussion in the introduction), e.g., collecting full spectra rather than measuring single wavelength or masses.

4.3 STARTING POINT—SAMPLES

No analysis is better than the quality of the samples analyzed, and it is therefore of outmost importance to ensure that the samples are prepared in such a way that they are a true representation of the original samples, and that they are compatible with the planned analytical approach. Sample extraction was discussed in the previous chapter and in one of the case stories; however, it may be necessary to do further sample work-up before continuing with the instrumental analysis. Metabolome analyses are often based on specialized sampling and sample preparation procedure; therefore, the procedure must be developed together with the instrumental methods to avoid many problems. However, one should be aware that anything that comes into contact with the sample or any sample experience (light, temperature, and so forth) can influence the results. Also, are often biological samples too complex to be analyzed directly or may contain impurities that hamper detection of target metabolites. In these cases, some kind of extended sample preparation are needed, e.g., solid-phase extraction, ion-exchange purification, or other similar techniques may have to be applied. Although elaborate sample preparation techniques may improve the quality of, e.g., target analyses, these procedures will reduce sample throughput. Selecting or developing an analytical protocol is very much a balance between the effort put into sample preparation, performance of the instrumental analysis, and the requirement of the data. Whether the effort is best spent on sample preparation as discussed in the previous chapter, on the instrumental analysis, or on data analysis depends very much on the problem to be solved. A few illustrations of the different approaches can be found in the examples at the end of this book. In any event, development of an extraction procedure should always be done in conjunction with the instrumental analysis planned to ensure that the two protocols will match each other.

4.4 PRINCIPLES OF CHROMATOGRAPHY

Chromatography is a very efficient separation technique where compounds are separated by using small differences in their distribution in two-phase systems, typically using gas – liquid or liquid – liquid systems (or similarly adsorption coefficient in gas/liquid – solid systems). In practice, one of the phases (the stationary phase) is not really a liquid phase, but rather a film chemically bound to a surface behaving like a liquid. Although chromatography has been around for about a century, it developed dramatically between the 1960s and the 1990s mostly because of the improvements of columns, detectors, and electronics. Today, nearly all types of chemical components can be separated by chromatographic techniques, often even when they are found in complex mixtures. Metabolomics, where many small metabolites have to be separated, is nearly always based on high-performance chromatographic separation with either a gas or a liquid as the mobile phase.

All chromatographic techniques utilize small differences in distribution coefficient (and their temperature dependence) to separate compounds in a two-phase system, e.g., liquid – liquid or gas – liquid systems. Similar rationales exist for separations based on adsorption (e.g., liquid/gas – solid systems), using ion exchange as well as other physical principles. As adsorption chromatography is rarely used for metabolome analysis, the reader is referred to chromatographic textbooks for further information.

4.4.1 Basics of Chromatography

The principle of chromatography is illustrated in Figure 4.1 where two compounds at a specific time-point are distributed in the two phases as given by the distribution

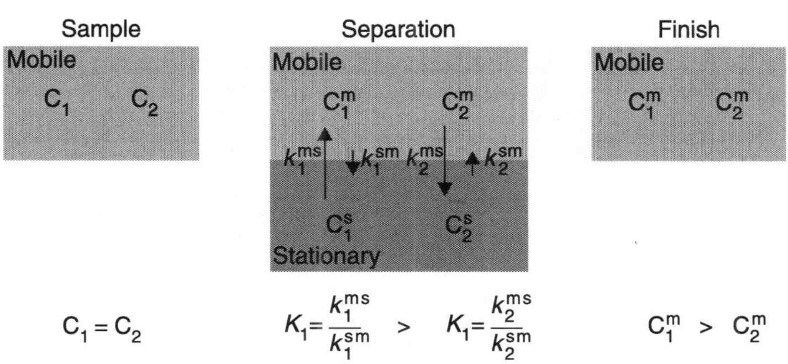

Figure 4.1 The chromatographic separation used in metabolome analysis is normally based on distribution between two phases. In these systems one phase is a stationary phase behaving as a liquid and a mobile that can be either a gas or a liquid (liquid–liquid chromatography or gas–liquid chromatography). The compounds C_1 and C_2 are separated due to small differences in their distributions K_1 and K_2.

coefficient K. One of the two phases is chemically bound to a surface and fixed in a column but acts as a liquid phase (designated as the stationary phase). The other phase is usually a liquid or gas which can be exchanged (designated as the mobile phase). Figure 4.1 illustrates one step of the separation: A sample with equal amount of two compounds is placed in contact with the stationary phase. When equilibrium has been reached, the two compounds are distributed as given by their distribution coefficient. If K_1 is greater than K_2 , more of C_2 will be in the stationary phase than C_1; hence, we have increased the amount of C_1 as compared with C_2 in the mobile phase. Moving the mobile phase to a new section of the stationary phase, more C_2 migrate into the stationary phase than C_1. Similarly, if we add clean mobile phase to the stationary phase with the two components, more C_1 migrate into the mobile phase than C_2. If we repeat this process many times and keep measuring the concentration of the two compounds in the mobile phase, we will find that we have separated C_1 from C_2. In practical chromatography, the stationary phase is held in a column (tube) where the mobile phase is constantly fed through the column. The whole separation process is initiated by placing a small sample in the mobile phase at the beginning of the stationary phase (column). The separation process is a dynamic process where small differences in distribution coefficients determine how much time the different compounds spend in the stationary phase: compound C_2 will spend more time in the stationary phase than C_1 as C_2 "favors" the stationary phases as compared with C_1. By continuously feeding fresh mobile phase to the column and assuming ideality (at a rate ensuring that equilibrium is a prevailing mechanism), we will dynamically separate the compounds until the end of the column is reached. If we continuously measure composition at the end of the column, we will obtain a relation between the amount of mobile phase passed through the column and composition/concentration (quite often the term is used instead of the mobile-phase volume particularly in GC). A plot of concentration vs. time is a *chromatogram* where compounds eluting are seen as peaks.

Several factors can deteriorate the chromatographic separation. These factors are jointly referred to as *dispersion* and consists of effects from the system (the gas or liquid chromatograph) and the separation process in the column. It is outside the scope of this book to go into details of these effects, but the major effects are illustrated in Figure 4.2 as they also illustrate key points required for understanding the fundamentals of chromatography: (1) *Eddy diffusion*: Not all compounds will

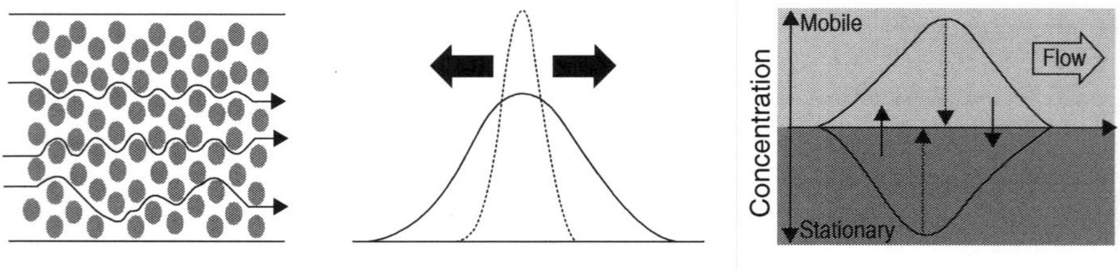

Figure 4.2 The three major dispersion effects that can deteriorate the separation in the chromatographic column resulting in derivations from ideality see the discussion in the text.

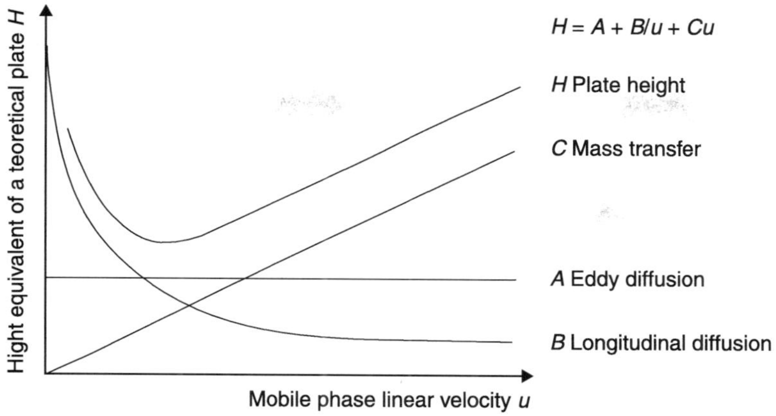

Figure 4.3 The van Deemter plot illustrates combined effects of the different dispersions shown in Figure 4.2 and can be used to find a flow optimum.

follow the same flow path in a packed column, (2) *Axial diffusion along the column,* (3) *Resistance to mass transfer in the mobile and stationary phase.* These effects depend on the flow rate of the mobile phase, often measured as the linear flow u as illustrated in the bottom graph in Figure 4.3. The eddy diffusion is independent of the flow rates and depends only on the column geometry—an open tubular column will have zero eddy diffusion, and a column with a more uniform packing will have smaller eddy diffusion. The axial diffusion depends reciprocally on the flow rate and is much more pronounced when the mobile phase is a gas rather than a liquid. A higher flow rate (higher linear velocity) will reduce the effect of axial diffusion. Finally, the resistance to mass transfer is actually made up of at least two terms: one for the liquid phase and one for the stationary phase.

In simple terms, the resistance to mass transfer is a measure for how well the equilibrium is reached at any point in time illustrated in Figure 4.2. If the resistance to mass transfer is high, equilibrium will not reach for a small length of column as illustrated in Figure 4.2; hence, the concentration profiles are different in the two phases. This effect depends on the two phases and on the analyte, and the effect increases with an increase in flow rate (not perfectly linear as indicated in Figure 4.3). These three effects can be combined to get a measure of the separation efficiency of the system, often referred to as the van Deemter curve as shown in Figure 4.3: H is the height equivalent of a theoretical plate, thus a measure of the system separation power (column length divided by the theoretical plate number), u is the linear mobile phase velocity (flow rate) and A, B, and C are parameters that are used to combine and quantify the effect of the column dispersion. A more detailed description and analysis of A, B, and C can be found in the chromatographic theory, see Jönsson (1987) and Giddings (2002). As it can be seen, there is an optimum u where we get the lowest H (most plates for a given column), thus the best separation power for a given chromatographic system. In a more practical context, it is important to note that there is a flow optimum, and that the performance deteriorates more dramatically by using lower flow rates than by using higher flow rates. This effect is most pounced in gas chromatography where it is, in general, an advantage to use a relatively higher linear flow rate, but other parts of the analytical system may limit the usable flow rates, e.g., back-pressure in HPLC and ion sources of mass spectrometers. See Section 4.5.

Other dispersion effects, most of which are related to the chromatographic system, can have serious influence on the performance of chromatographic systems. The most important of these are discussed in the following sections in conjunction with the relevant systems.

The reader is referred to the supplemental literature for an in-depth discussion of theory and dispersion in chromatography (see, e.g., Jönsson, 1987 and Giddings, 2002).

4.4.2 The Chromatogram and Terms in Chromatography

A chromatogram is basically a plot of a detector signal recorded at the end of the column vs. time usually starting at the time of injection. The analytes will start migrating through the column immediately after injection and hopefully be separated by the chromatogram. A simple chromatogram is shown in Figure 4.4 illustrating the most important parameters used to describe a chromatogram: *retention time, peak height, and peak width.* The shortest possible time from injection to the first nonretained metabolite elute is usually referred to as the dead-time.

An analyte is described by the retention time (time from injection to its elute), the peak width, the area under the peak, or the peak height (maximal signal). [The latter two parameters require that a sensible baseline should be established for the area and

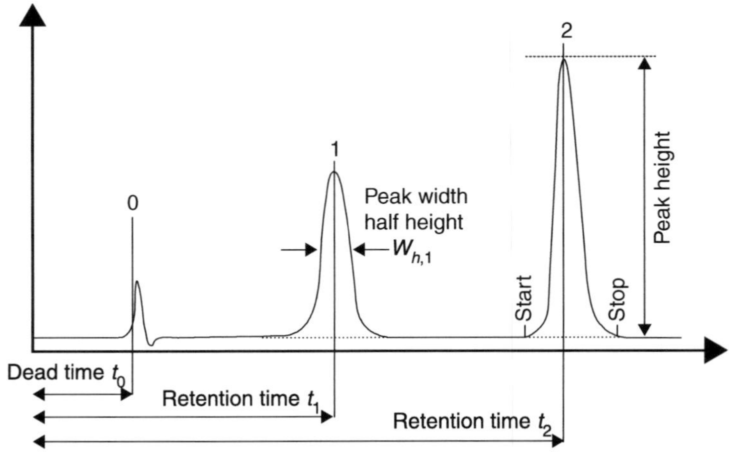

Figure 4.4 This simple chromatogram show the most important terms used to describe a chromatogram. Each of the two peaks 1 and 2 are characterized by their retention time, peak width, peak height, and peak area (determined as the area under the curve from the peak start to the peak stop). The dead time is the time it takes the solvent front to pass from injector to detector and is often seen as a baseline disturbance.

Capacity factor $k_r = \frac{t_r - t_0}{t_r}$ and $k = \frac{k_2 + k_1}{2}$ Selectivity $\alpha = \frac{k_2}{k_1}$

Resolution $R = \frac{t_2 + t_1}{1/2(\sigma_2 + \sigma_1)} \approx \frac{\alpha - 1}{\alpha} \frac{k}{1 + k} \frac{\sqrt{N}}{4}$

Plate number $N = \left(\frac{t}{\sigma}\right)^2 = 5.55 \times \left(\frac{t}{W_{1/2}}\right)^2$ Plate height $H = \frac{L}{N}$

Figure 4.5 By measuring the terms described in Figure 4.4 some simple key parameters can be calculated and used to evaluate and compare the performance of a chromatographic separation. Most interesting is the resolution *R* that describes how well separated two compounds are and the plate number that describe the overall performance (can also be used to do a van Deemter plot, see Figure 4.3).

also that the beginning and the end of the peak should be determined.] This is not always easy, but a multitude of different techniques are implemented in modern software that in most cases will give reliable peak areas. The process of finding peaks, peak areas, and other features is often referred to as integration. It is advisable to evaluate the performance of the integration; thus, peak detection—area determination manually on selected real data as the automated processes can be way off.

By calculating some of the simple parameters as shown in Figure 4.5, the basic performance of a chromatographic system can be assessed. The capacity factor *k* is one way to express retention of a compound in the column by calculating a fraction of the total retention time spent in the stationary phase (*k* has no unit). The selectivity is used to compare the behavior of a compound in two different columns or the behavior of two compounds in the same column. The selectivity expresses how much time one compound spends in the stationary phase compared with the other compound. Quite often, a chromatographic column will be described as having a higher selectivity for some types of compounds, which means that some compounds will spend more time in the stationary phase than others under the same conditions, i.e., these compounds will have higher *k*-values. The resolution *R* is a measurement of how well two peaks are separated; $k \approx 1.2$ corresponds to baseline separation. As resolution is a combination of retention (how much time each compound spends in the stationary phase) and the width of the peak, it can be improved by decreasing the peak width (e.g., narrow bore columns, smaller particles, or change of solvent systems) or by a longer retention (e.g., use of longer columns, slower gradients, or other solvents). The plate number N or the plate height *H* are used to describe the performance of a column; the more the plates (or lower plate height *H*) the better the separation power. However, the plate number depends on the compound and the mobile phase, but by using a test system, plate numbers can be used to compare the performance of columns. For a given system and sample, a van Deemter plot is calculated as shown in Figure 4.5, using measurement of the plate height as a function of the flow rate, thereby, to find an optimal mobile flow velocity (most useful in gas chromatography). Using the expression for resolution in Figure 4.5, it can be seen that the resolution is proportional to the square root of the plate number, and hence a doubling of the resolution requires four times as many plates, which in practice requires a column four times longer. However, the retention time increases linearly with column length, hence gives much longer analysis times. To improve resolution between two compounds, it is often advisable to choose another chromatographic system (e.g., change either the mobile phase or the column phase) rather than just using a longer column of the same type and with the same mobile phase.

Optimizing a separation is almost always a matter of increasing the selectivity, thus increasing α by changing one (or both) of the two phases. One may select a column with different characteristics even under the same conditions, or in case of HPLC, one may use different solvents. Examples of this can be found in the example section. In general, separations are almost always optimized to give a sufficient separation of all relevant metabolites (or as many as possible in metabolomics) in the shortest possible time.

In real life, very few chromatograms are as simple as the one shown in Figure 4.4. Particularly, in the case of metabolomics, where highly complex samples are studied, peaks that are not, or poorly, separated will be encountered as illustrated in Figure 4.6. Although the shoulder-separated peaks can be recognized in many cases, the separated peaks can of course not be identified in any simple way. Therefore, while analyzing complex samples, one should be aware that two or more compounds might be present in each chromatographic peak. Having spectral data (particular mass spectra) helps to determine whether more compounds are found in each peak as described later.

In metabolomics, compounds of quite different chemical nature and varying concentrations are the most likely to be encountered as discussed in Chapter 2. A chromatographic system will, in general, perform better for some classes of compounds than for others. We will therefore often see peak shapes as illustrated in Figure 4.7 whereas other compounds produce perfect sharp peaks.

Overloading occurs when we saturate the stationary phase by injecting so much of the compound that equilibrium cannot be reached, hence the samples are spread over a long section of the column. In severe cases, the compound is spread all the

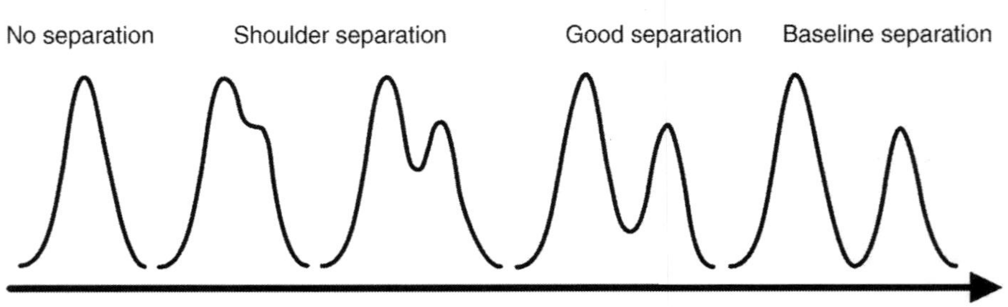

Figure 4.6 Analyzing complex samples it is not always possible to get an ideal baseline separation as shown to the right. In most cases all situation from no separation at the left to a perfect baseline separation at the right will be encountered. In very complex samples each peak can very well be the result of several overlapping compounds.

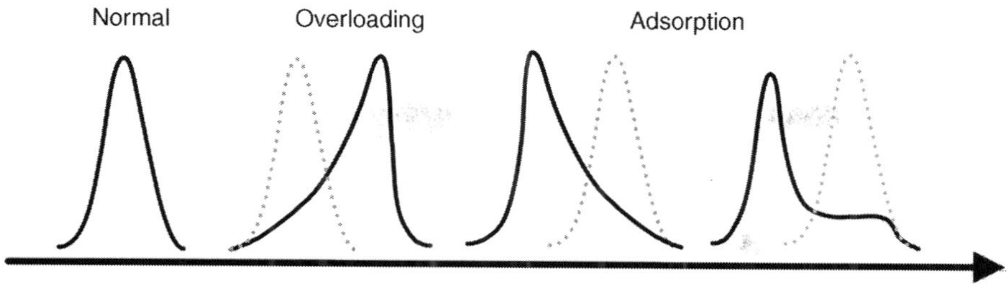

Figure 4.7 Chromatography, neither using gas nor liquid as a mobile phase, will be the result of just one separation mechanism or at done equilibrium. The result is skewed peaks as illustrated either as a result of overloading where the stationary phase is saturated (or equilibrium cannot be reach) or as a mixed mechanism where compounds are adsorbed on the silica surface and released at another rate than the distribution. The perfect peak shape to the left is only obtained for well-behaved compounds.

way from injector to detector looking like a high background. Only, the front part of the "peak" follows the chromatographic principle as described in the previous section, whereas the tail part is just passing through the column with the eluent. Adsorption is often causing errors in chromatography, and here compounds are retained in the column by a mixed mechanism: distribution as described previously and adsorption to the column surface (typically the silica is used as a carrier material in most columns). The distribution coefficients and adsorption coefficients are normally very different for a given mobile-phase composition, the latter often being larger; the result is a tail on the peaks: the front forms nice peak shape as expected from distribution, but the adsorbed molecules are released slower giving a long tail on the peak. Again, this can be quite severe giving peak tails that are several minutes long. Finally, these mechanisms are often combined, thus some compounds give a relatively nicer peak shape if injected at a low concentration, but showing serious tailing if injected at a higher concentration. Typical examples in HPLC are organic acids separated on standard C-18 column under acidic conditions or alkaloids separated under neutral-to-alkaline conditions—in both cases, the adsorption is due to the formation of hydrogen bonds in uncovered silanol groups on the column carrier material. Similar problems are common in GC when apolar phases are used.

4.5 CHROMATOGRAPHIC SYSTEMS

As described in the previous sections, the principles and theories of gas and liquid chromatography are quite similar, and so are the analytical systems. In both cases, they consist of a supply of the mobile phase, an injection system, the column, and a detector—and, of course, some electronics (and computers) to control the system as well as to collect and process the data. However, these components are of a quite different design for gas and liquid chromatography and are therefore described separately in the following sections.

4.5.1 Gas Chromatography

Gas chromatography is a remarkably simple but capable analytical system with an amazing separation power, where up to thousands of compounds can be separated within an hour. Although the theory and most of the core technologies have been fully developed for more that 20 years, technical developments are still improving the performance of GC. The key elements of a gas chromatograph are illustrated in Figure 4.8, and these are discussed in more details in the following sections.

4.5.1.1 Gas Supply and Mobile Phase. The mobile phase, typically helium, is delivered from a compressed gas supply and the flow is controlled by pressure and flow regulators. GC analysis can be done using constant flow, constant pressure, or a flow program—the latter as a result of more recent technical developments. The gas supply system is a critical component of a gas chromatograph; however, most modern GC systems have very stable and precise flow and pressure controls, and if well maintained, these are rarely a source of errors (see also the injector discussion below). However, the quality of the gas used can give rise to errors in the form of ghost peaks due to impurities in the gas or the gas supply system. Therefore, it is important to use a high-purity carrier gas and, often in combination with gas purifiers, to remove the minute amount of oxygen and water still present in the gas. The gas purity is often specified in percentage, e.g., as 99.9995% pure, often written as N55 or 5N5, meaning five 9s followed by a 5 (similarly N57 is five 9s followed by a 7, thus 99.9997%). The purer the better; however, it is important to check what

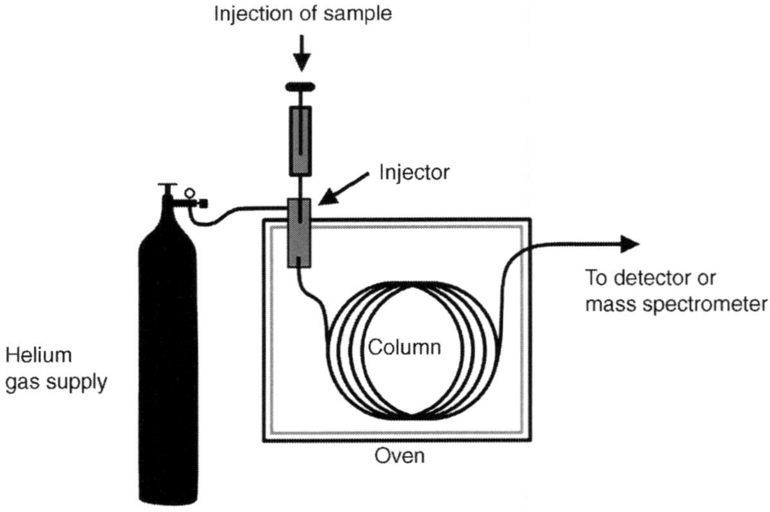

Figure 4.8 The key element of a gas chromatograph: a gas supply, (typically helium), pressure and flow regulators, an injector to transfer the sample into the mobile gas phase, a column placed in an oven where the temperature can be controlled and program, and finally connected to a detection system, typically a mass spectrometer.

impurities are left in the gas, in particular, oxygen and water can ruin columns (particularly polar substances are most sensitive to oxygen) and hydrocarbons give a high background.

4.5.1.2 Columns and Oven in Gas Chromatography. Separation of the evaporated compounds from the sample is done in a column, which in modern gas chromatography is almost always a long open tubular, narrow bore fused silica tube where a stationary phase is bound to the inner surface. These quart tubes are produced using the same technology as is used to produce optical fibers with a diameter ranging from 50 to more than 500 μm and with a length ranging from 10 to 100 m. The outside of the column is coated with a polymer (typically a polyimide), which makes it very durable as long as the surface is not scratched. The inside of the column is coated with a stationary phase often of a lipophilic nature. Figure 4.9 shows examples of the chemical structure of some of the most common stationary phases.

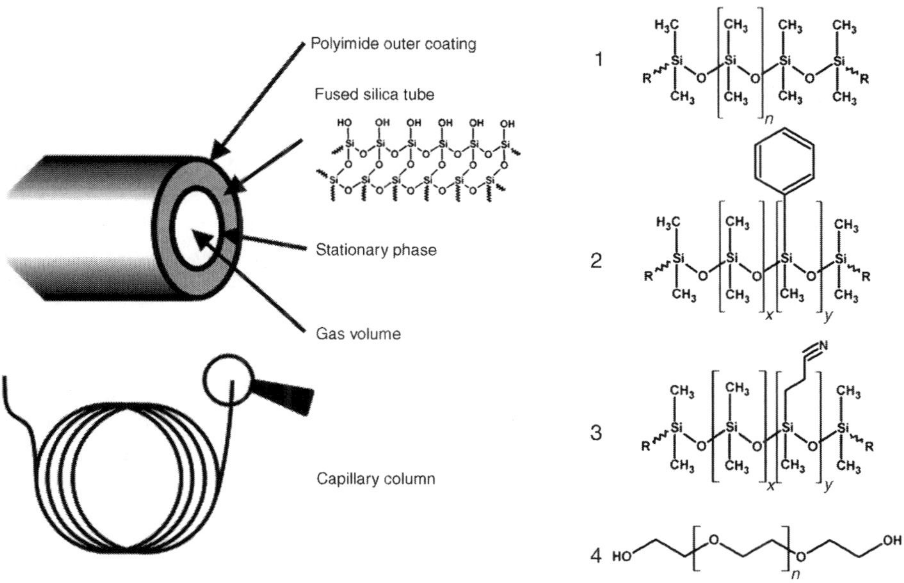

Figure 4.9 Most modern GC columns are made from fused silica made in much the same way as optical fibers. Purified quartz tube is pulled to a capillary typical up to 100 m long and with inner diameters from 50 to 530 μm. The outer surface is coated with a polyimide polymer giving an impressive strength. The inner surface is coated with the stationary phase, were the most popular are based on silicone polymers: (1) methyl-silicone, (2) methyl-silicone where some phenyl groups replace the methyl groups, 5 or 50% are common, (3) methyl-silicone where some cyano-propyl groups replace methyl groups, 17% is common, and (4) cabowax, a polar polyethylene glycol polymer. The phases are normally chemically bound to the silica surface and also cross-linked to increase stability. The residual silanol groups are covered by deactivation, typical methylation. The phase thickness is carefully controlled between 0.1 and 5–8 μm.

So far the most popular general-purpose stationary phases are the apolar methyl-silicone phases, the more polar methyl-silicone phases with 5% phenyl groups, the even more polar cyano-propyl methyl silicone phases, and the very polar carbowax phases. These phases are nowadays always chemically bound to the wall and are often also cross-linked to increase the stability; however, there are temperature limits for all types of columns, which in general are lower for the more polar columns. A key parameter for retention is the ratio between the two phases, thus how much gas phase and how much stationary phases is found in a section of the column as discussed earlier in this chapter. This ratio is often called β and is determined by dividing the gas phase volume by the stationary phase volume both of which are easily calculated from the column diameter and the phase thickness. This is a central parameter for selection of a column, lower phase ratio gives more retentions (corresponds to more stationary phase in the column) but fewer plates. Therefore, a thick-phase column (low β) is typically selected for volatile compounds with low retention, whereas thin-film columns (high β) are used for less-volatile compounds eluting at high temperature.

Column length is also important in relation to the number of theoretical plates as discussed earlier in this chapter, but remember as illustrated by equations in Figure 4.5 that retention time is proportional to the time spent in the stationary phase which again is proportional to the column length, but the longer the time in the column the wider the peaks get because of band broadening effects. Therefore, the separation power (theoretical plate number N) is proportional to the square root of the retention time, hence a column four times longer is required to double the chromatographic resolution.

The distribution between the phases depends strongly on the temperature in gas chromatography; therefore, controlling the temperature is critical in gas chromatography. This is done by placing the column in an oven where the temperature is controlled carefully. The distribution coefficient depends strongly on the temperature; therefore, changing the temperature can be used to improve the separation during analysis. This is called temperature programming where the oven is set at a low temperature during injection and at the beginning of the analysis, and then the temperature is increased at a specific rate to a maximal temperature. Temperature programming is also used to optimize analysis time.

4.5.1.3 Injection in Gas Chromatography. The most critical part of gas chromatography is the sample injection—that is, to transfer the typical liquid sample to the gaseous mobile phase and focus it at the beginning of the column. Volatile metabolites are quite unfair and are often not considered as a part of the metabolome; therefore, injection of gaseous samples is described here, and the reader is referred to the extensive literature on flavor analysis. Liquid samples encountered in metabolomics contain a broad range of more or less volatile analytes and matrix components in a large volume of solvent. These samples can give serious problems in gas chromatography if the injection technique is not well adapted, and injection problems are so far major source of problems in gas chromatography. The problems arise from the slow and incomplete evaporation and transfer of the sample to the column in a time that is insignificant compared with the peak width. Therefore, the widely used split/splitless injection is discussed in some details in the following sections focusing on some of the key problems. All practitioners of gas chromatography should consult the very comprehensive textbooks written by Konrad Grob (2001), a pioneer in modern gas chromatography.

Split/splitless injection is based on rapid evaporation of the samples in a small heated chamber and the transfer of the vapors onto the column by the carrier gas and is the single most difficult part of gas chromatography. In the days of packed columns, operated at high gas flow rates (30–50 ml/min), it was easy to get a rapid and efficient transfer of the sample to the column. The introduction of capillary columns that are operated at low flow rates (typical 1–2 ml/min) required adaptation of the injection technique from the previously used techniques. Initially, this was done by venting a part of the sample out of the injector maintaining the high flow rate through the injector but with a significant loss of sample (sensitivity)—the split injection. A later development was closing the split-vent during injection and circumventing the long transfer time by focusing the analytes on the column—the splitless injection. Figure 4.10 illustrates a typical design of a modern split/splitless injector. The injector contains the following elements: gas flow regulation (column flow and split operation), evaporation

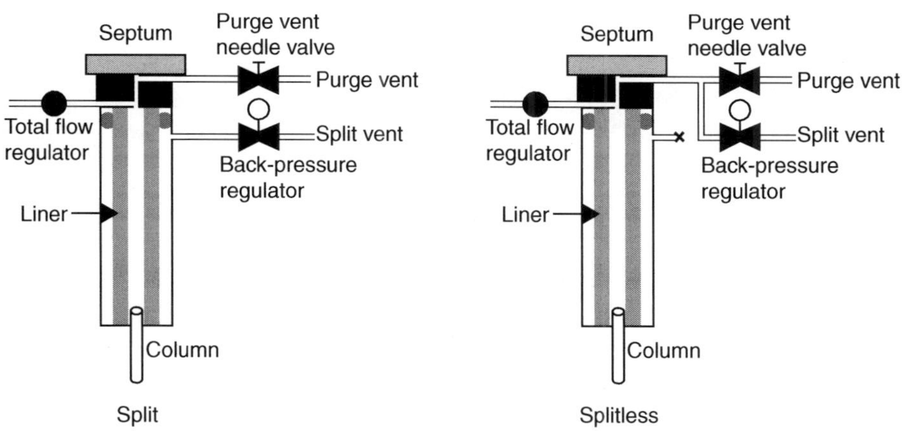

Figure 4.10 A typical split/splitless injector with flow control and back-pressure control. In both split and splitless mode a total flow is delivered to the injector camber. The pressure in the injector governs the flow through the column (determined by column dimension and temperature typical from 1 to 5 ml/min). At the stop of the injector is a septum purge vent that vents a small stream of carrier gas (few millimeters per minute) from beneath the septum to prevent leakage and evaporated septum compound to enter the column. A back-pressure regulator vents gas from the injector to maintain a constant pressure in the injector. In split mode (to the left) this is done from the bottom of the liner, thereby venting a part of the sample. In splitless mode (to the right) the gas is vented from the top through the septum purge vent thereby preventing injector overload to go back into the gas line. The injector is heated and a replaceable glass liner is used as an evaporation chamber.

chamber—the gas liner, a septum, and a heated block. These elements are described in the following sections.

The carrier gas flow is regulated either by a constant column head pressure or by a constant flow rate through the injector. As the viscosity of the mobile phase (normally helium) depends on the temperature, the flow rate will change with the temperature if the pressure is kept constant. With the design illustrated in Figure 4.10, a constant gas flow is maintained through the injector while the column head pressure is kept constant by a back-pressure regulator venting a part of the carrier gas through a split vent and a septum purge vent. The septum purge vent will continuously vent a small stream of gas, typically a few milliliters per minute, from the top of the injector (beneath the septum) to prevent contaminated evaporation from the septum to enter the column, to remove oxygen leaking through the septum after many penetrations, to prevent overloading of the injector to get into the gas supply system, and finally to vent excess carrier gas during the splitless period, as shown later.

The liner, typically a glass tube, serves as an evaporation chamber where the sample is evaporated. These come in many designs with and without packing materials, various deactivation, insertions, and sizes. A large volume (wide bore) liner is normally used for splitless injection and a smaller volume (narrow bore) liner for split injection. The inner diameter is typically around 2–4 mm and typical length is around 8–10 cm; a wide bore liner has a volume around 1 ml, which is important to remember. The column entrance is typically positioned 1–2 cm toward the bottom of the liner but should be optimized together with the needle length for each injector design and injection technique used; for further details see the books by Grob (1987 and 2001).

The liner is placed in a temperature-controlled heated block. In some modern injectors, the temperature can be programmed with very steep temperature gradients where the temperature can be raised from ambient to, e.g., 250°C in a few seconds (the programmed temperature vaporizer, PTV injector). It is important that the injector should have sufficient heating capacity to evaporate the sample without a large temperature drop.

The first step of a typical injection process is illustrated in Figure 4.11 where the goal is an instant and complete transfer of the sample to the gas phase. The injection begins when the syringe penetrates the septum/seal at the top of the injector. When the plunger is pushed down, the sample is injected (sprayed) into the hot glass liner where solvents and analytes are ideally flash evaporated. The evaporation is a rather complex process that can result in many types of problems. The major problems arise from incomplete evaporation, from dirt (involatile matrix), and heat stress. As illustrated in Figure 4.11, droplets and involatile materials may hit the wall of the liner where they are deposited and are slowly released by thermal degradation. Another situation is when either the gas flow through the liner is so high that the droplets are transported past the column entrance before they are completely evaporated, or when they simply shoot past the column entrance before they are evaporated (e.g., if the needle is too close to the column entrance). Also, the sample may start evaporating out of the needle even before the plunger is pushed down. Finally, an often overlooked problem is overfilling the injector: One microliter solvent will

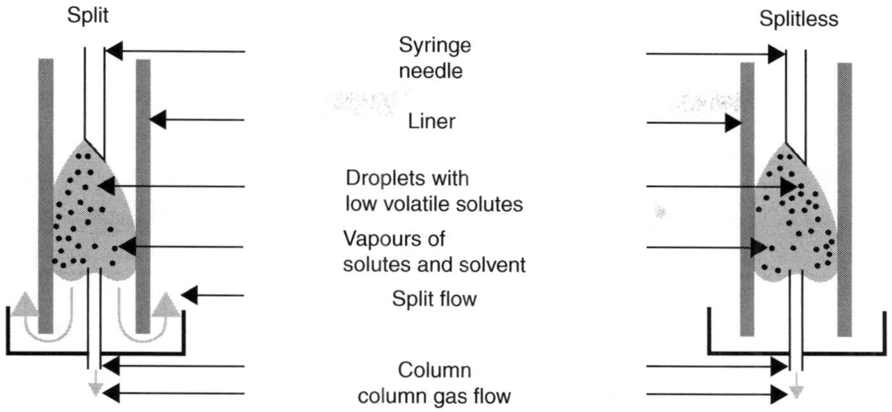

Figure 4.11 The injection starts by a syringe needle penetrates the septum and injects the sample into the hot glass liner. The goal is instant evaporation of solvent and sample, however this is not always the case and sample and nonvolatile matrix components may end on the hot liner wall. Deposited sample and matrix components on the liner wall can serious deteriorate the performance and can result in "ghost peaks." In split mode where a significant part of the sample is vented from the bottom of the injector, the amount is determined by the ratio between total flow (minus the septum purge flow) going into the injector and the column flow. Ratio between 1:10 and 1:100 is common. In splitless mode all gas going through the liner will enter the column; hence most of the sample will be transferred to the column. After a specific time the split-vent is opened to vent the remaining sample from the liner (40–90 s).

give 0.5–1 ml gas, thus completely filling a normal wide bore liner. If the gas flow through the injector is high, the evaporated solvent is rapidly removed, tolerating larger volume injections, but in case of splitless injection where the flow rate through the injector is low, overfilling the liner is a common source of injection problems (e.g., cross-contaminations, high variability, and high back ground). The complete injection—evaporation—process will take seconds; however, transferring the evaporated samples to the column depends, of course, on the flow rate through the liner. The key parameters in this process are geometry of the injector (column and needle), liner type, temperature, gas flow rate, and syringe/injection technique used.

In *split injection*, a large portion of the flow through the injector liner is vented from the bottom of the liner, see Figure 4.11. In the injector design illustrated in Figure 4.10, a constant flow is fed to the injector where a constant pressure is maintained by venting a portion of the gas from the bottom of the injector. This will give a constant column-head pressure used to adjust a suitable column flow, e.g., 1 ml/min. The total flow lead into the injector is then used to adjust the flow that needs to be vented from the bottom of the liner (and through the septum purge vent). Venting, e.g., 30 ml/min will give a split ratio of 1:30. A longer distance between the needle and the column entrance/bottom of the injector often allows more time for sample evaporation when using high flow rates. At the same time, a narrow bore liner is often used to give an efficient heat transfer and to ensure that the sample vapors are as concentrated as possible. Although split injection gives very good injections with sharp peaks, a significant portion of the sample is lost (approximately 97% in the above example), resulting in decreased sensitivity. If sensitivity is not an issue, split injection should be the first choice. Also, split injections can be done at any column temperature, as shown below.

Splitless injection is used to increase the amount of sample transferred to the column by closing the split vent during the injection. Hence, all the gas flowing through the liner is going onto the column but only at the column flow rate which is in the range of a few milliliters per minute (the excess total flow going into the injector is vented through the septum purge vent at the top of the injector). Therefore, transfer of the sample to the column will take quite a while, typically in the range 30–90 s; hence, measures must be taken to focus the sample at the beginning of the column to obtain a good chromatographic separation. In simple terms, the injection time has to be short compared with the peak width in the chromatograms. By using conditions that allow recondensation of the solvent in the column, a section with very high retention is created. In this section, the recondensed solvent will effectively trap the analytes and at the same time minimize the migration into the column. This recondensation is crucial to splitless injection to get a narrow injection profile best obtained in a *retention gab* as described below. In case of compounds eluting at a high temperature, one may get away by keeping the column at sufficiently low temperature to minimize migration during injection. It is important to remember that evaporation of 1 μl solvent corresponds to 0.5–1 ml gas at 250°C; therefore, it takes quite a while to transfer the sample to the column at a few milliliters per minute, which is around the maximum that can be maintained in a standard wide bore liner. Overloading results in uncontrolled sample loss through the septum purge vent or even pushing the sample back into the carrier supply gas lines giving a high background in the following samples. Large-volume injection is not described in this book but can be done by on-column injection or PTV injectors, described in detailed in the literature listed below. After transfer of the sample to the column, the split vent is opened (after 30–90 s) to vent the remaining sample from the injector.

Condensation of the sample solvent on a *retention gab* mounted at the beginning of the column is a very efficient way to focus the sample at the beginning of the separation column—also called solvent effect. The retention gab is a piece of fused silica column, which is deactivated, but without stationary phase. Two to five meters of the same dimension is normally mounted in the beginning of the column. Solvent effect is obtained by keeping the retention gab around 20° below the boiling point of the solvent during injection. This results in condensation of solvent on the retention-gab wall as illustrated in Figure 4.12, spreading over may be 10–30 cm retention gab. The sample is equally spread in the condensed solvent, which now acts as a stationary phase with a very strong retention of the sample molecules. As the solvent evaporates, the sample molecules will be trapped in a still smaller section of the retention gab. When all the solvent is evaporated, the sample molecules will move with the carrier gas through the remaining retention gab as a narrow band. When the sample molecules reach the stationary phases in the separation column, they will be retained again and will now be focused as a narrow injection band.

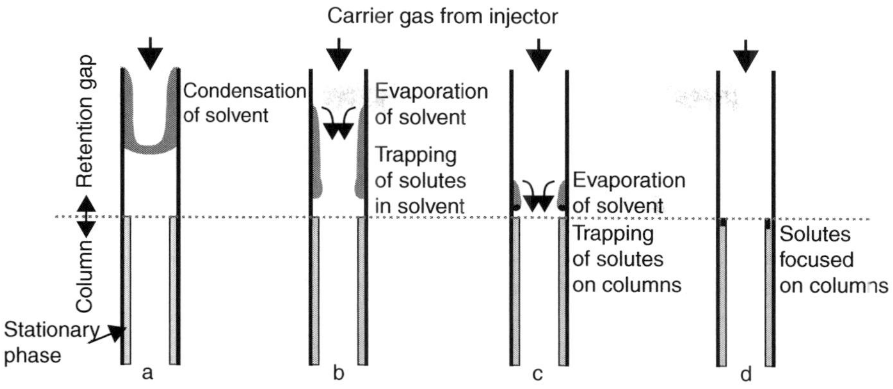

Figure 4.12 As the transfer of sample is slow in splitless injection using solvent effect is efficient tool to focus the analytes at the beginning of the column. By keeping the column at temperature low (typical 20 degrees below the boiling point of the solvent) the solvent is recondensed in the first part of the column (or rather a precolumn or retention gab which is an empty piece of fused silica with a deactivated surface). The recondensed solvent will then act as a stationary phase with very high retention power, retaining the analytes until all the solvent is evaporated. A retention gab is crucial for efficient use of solvent effect to avoid a mixed mechanism from both the solvent and the stationary phase.

As in the case for split injection, the injector parameters are quite important. In general, sample is evaporated closer to the column entrance in splitless injection than in split infection, and a larger liner is used. However, the same parameters need to be optimized. Furthermore, splitless injection requires efficient use of solvent effects, and the oven temperature during injection is therefore not only important for separation but also for obtaining a good injection profile. A wellperformed splitless injection can give extremely narrow peaks and very good separation where more than 90% of the sample is transferred to the column. On the contrary, by *not* paying attention to the problems in splitless injection, it is possible to completely ruin any separation giving ghost peaks, peak splitting, and many other errors.

4.5.1.4 Derivatization for GC. Gas chromatography requires that the sample is sufficiently volatile to be evaporated in the injector. This is easily achieved for small molecules with low boiling points (below 200–300°C) whereas nonvolatiles need to be made more volatile by chemical derivatization before they can be analyzed by gas chromatography. Of interest in metabolomics are the amino acids, sugars, small organic acids, and other polar metabolites along with other larger apolar metabolites like fatty acids and sterols. Most of these metabolites are in their normal nonvolatile form, but they can be made volatile by derivatization by "covering", e.g., the carboxylic, hydroxylic, and amino groups with an apolar functionality, thereby making them more volatile so that they can be analyzed by gas chromatography. The derivatization is often done by methylation or silylation; however, numerous chemical procedures are available. It is outside the scope of this book to go into details of the different chemical reactions usable for deritivazations in gas chromatography, but examples can be found in the second part of this book and in, e.g., Drozd (1981) and Toyo'oka (1999).

However, it is important to remember that derivatization will also produce artifacts in the sample and the sample may also contain surplus reagents. These reagents can seriously disturb the split/splitless injection as they, in general, are involatile and hence may be deposited in the injector.

4.5.2 HPLC Systems

Liquid chromatography is based on a liquid mobile phase delivered to the separation column by a pumping system. Compared with gas chromatography, a very wide selection of mobile phases can be used in liquid chromatography, together with a huge selection of columns and stationary phases. Therefore, nearly all types of compounds that can be dissolved in a mobile phase can be separated from apolar (lipid) to ionic, small to very large, and acidic to alkaline. The separation efficiency (total plate number) is often lower in liquid chromatography than in gas chromatography because of the shorter columns; however, the per meter plate column can be much higher in liquid chromatography. Although an HPLC system, as shown in Figure 4.13, is technically more complex than a gas chromatograph, it is quite simple to operate and, in general, gives relatively fewer problems. On the contrary, as both the stationary phase *and* the mobile phase can be used in optimization of the separation process, we have almost infinite number of ways to ensure separation of compounds of interest making optimization of liquid chromatography far more complex than gas chromatography, as shown below.

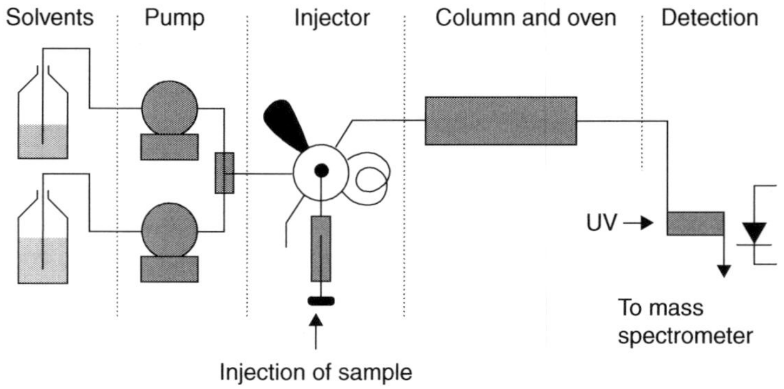

Figure 4.13 The key parts of a high performance liquid chromatograph. The liquid mobile phase is delivered from the solvent reservoirs by a pumping system, where the flow and composition can be controlled precisely. The sample is filled into a loop—a length of tube—and placed inline with the solvent flow. From the injector the sample and flow is lead to the column, see Figure 4.14. The column may be placed in a thermostat to control the temperature. From the column the flow with the separated analytes is lead to a detector, e.g., a flow-cell in a UV spectrophotometer and a mass spectrometer.

4.5.2.1 The Liquid Chromatograph. The key components of a simple liquid chromatograph are shown in Figure 4.13. Generally, a liquid chromatograph comprises solvent reservoirs, pumps, injector, column, and one or more detectors.

4.5.2.1a LC Pumps. From the solvent reservoirs, the mobile phase needs to be supplied to the column by a high-performance pump(s). The pump has to deliver a constant and pulse-free flow at a rate suitable for the separation column, often against a high back-pressure. In normal analytical chromatography, the flow rate used is between 0.1 and 1 ml/min, and in micro- and nano-flow HPLC, flow rates as low as a few nanoliters per minute are used. At the same time, the pump has to be able to mix two or more solvents, where the composition can be programmed as a function of time. Keeping the flow rate constant, the amount of each solvent is changed over time—the composition is typically given as a percentage of each: % solvent A, % solvent B, % solvent C, and so forth where the total, of course, is 100%. If only two solvents are used, it is quite common to only state the percentage of solvent B, thus 15% B means that 85% of the flow is solvent A and 15% is solvent B (if the flow rate is 1 ml/min, we have 0.85 ml/min solvent A and 0.15 ml/min solvent B). By changing the composition of the mobile phase, the selectivity is changed and hence the performance of the separation (see Figures 4.1 and 4.5). This corresponds to the temperature gradient in gas chromatography but is much more powerful as the number of possibilities is much higher. In general the solvent with the lowest eluting power is labeled as solvent A and the strongest eluting solvent as B.

To ensure a stable and pulse-free flow, most modern pumps incorporate a degasser system to remove dissolved air from the solvents. Any bubbles in the solvent lines will act as small springs giving a highly unstable flow. The solvent mixing may be done either on the low-pressure side by controlling the solvent delivery to the pump or on the high-pressure side by using multiple pumps controlling the flow from the individual pumps. Both systems can deliver very reproducible flow and gradients in the normal range, but a detailed description of the advantages and disadvantages of the two types of pumps is outside the scope of this book.

Beside pulsations, as mentioned above, the major problems with HPLC pumps are delay volume and errors in gradients near the 0 or 100% composition. Many pumps have a significant volume within the pump-head, mixer, pressure gauge, and so forth; and the significant volume, therefore, needs to be pumped through the system before the specified composition is actually delivered to the column. The precision of the gradient also deteriorates near the end, where very small amounts of one of the eluents cannot be delivered accurately. [This happens when both low and high pressure mixing are used]. Therefore, the best gradient reproducibility and retention stability will be with solvent composition (given as percentage of solvent A) in the range of 5–95%.

It is very important that the solvents are very pure. Any impurity in the solvents will have an effect on the separation and even more as background in the detection. As very narrow bore tubing and small particulate columns are used, it is also important that solvents are free from any particulate materials. Therefore, solvents are typically filtered through filters with a pore size of 0.45 μm or less.

4.5.2.1b LC Injection. An injector is needed to inject the sample into the solvent stream. In most systems, injections are done by a rather simple loop injector where a small piece of tube is filled with the sample, which is then moved into the mobile phase stream by a rotating valve. Modern HPLC injectors are rarely a source of problems in liquid chromatography and if properly designed and maintained, will give nearly perfect injections, but, as in gas chromatography, the time it takes to transfer the sample to the column should be small compared with the elution peak width, unless a trapping technique is applied. As a rule of thumb, the volume injected should be transferred to the column in less than 1 s (e.g., 1000/60 µl thus more than 20 µl at 1 ml/min). On-column trapping can be done by dissolving the sample in a solvent with low elution power in an eluent with low elution power.

4.5.2.1c LC Columns. The columns used in liquid chromatography are normally short steel tubes packed with a particulate material, which are often spherical porous silica particles, polymer particles, or in some modern columns, a monolithic structure. The stationary phase is chemically bound to the surface of these particles or the material as such serves as the stationary phase. Today, a huge number of different columns are available for general and very specialized analysis. The most common type used for metabolomics are columns based on silica particles onto which a stationary phase is chemically bound. A typical example is shown in Figure 4.14. These columns normally have an apolar phase, and therefore a solvent gradient going from a polar solvent (water) to a more polar solvent (e.g., acetonitrile or methanol) is used. For historical reasons, this is called reversed-phase chromatography whereas chromatography on the bare silica is called normal-phase chromatography. As mentioned above, reversed-phase chromatography is commonly applied in metabolome analysis and a very popular phase is octyldecyl chains bound to the silica surface, which are normally referred to as C-18 columns, see Figure 4.14. These C-18 columns are found in many variations, which can behave quite differently. Even with the same type of phase bound to the particles, there can be differences in particle size, particle shape (perfect spheres are better as they can be packed more densely in the column), pore diameter (thus surface area), degree of coating, deactivation of uncoated silica, chemistry of the silica, and so forth. As in the case for columns, for gas chromatography, the dimension of the HPLC column also affects the separation efficiency.

The column length will increase the number of plates (also see Figure 4.5) in the same way as in gas chromatography, and also a smaller diameter will give a better resolution. As columns for liquid chromatography contain particles, the Eddy diffusion plays a role in the deterioration of the separation efficiency (also see Figure 4.2). Therefore, smaller particles will, in general, give a better separation efficiency.

Combining all these, the best column will be a long, narrow bore column with small particles. However, such columns will give very high back-pressures and are difficult to make. In practice, today a general-purpose column is around 100 mm long, has an inner diameter of 2 mm and is packed with 3 µm particles.

Many specialized columns, where the selectivity of the stationary phase has been optimized for certain type of compounds, can be found in the catalogs from different manufactures. These columns include stereospecific phases, carbohydrate phases,

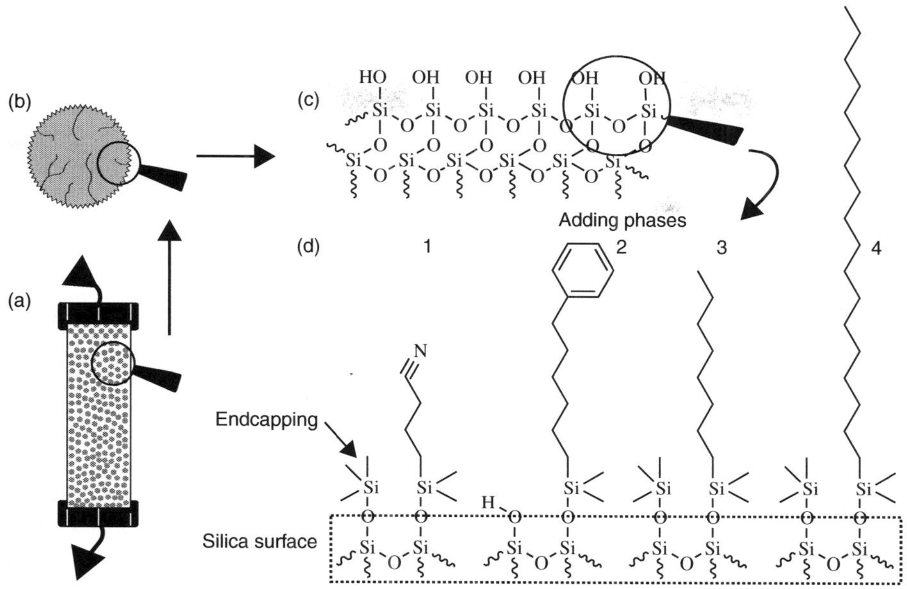

Figure 4.14 (a) HPLC columns are typically steel tube packed with silica particles. The particles are held in place of steel frits in each end and end caps with connectors for capillary tubes. (b) The silica particles is mostly spherical porous particle a few micrometers in diameter (3–5 μm, and around 1.5 μm for UPLC columns) with a considerable pore volume and a pore diameter in the 80–200 Å range. The pore volume significantly increases to surface area hence the area that can be used for chromatography. Smaller particles will give better separation and also higher back-pressure, thereby limiting the flow rates that can be used. (c) The bare silica surface is covered with silanol groups, which in reversed phase chromatography be covered with stationary phase, or used directly in normal phase chromatography. (d) Common stationary phases bound to the surface fro use in HPLC are: (1) cyano-propyl chains, (2) phenyl-hexyl chains, (3) *n*-octyl (or C-8) chains, and (4) octyldecyl (C-18) chains. The carbon load, hence the amount of surface is a key factor determining the performance of a column. The uncovered silanol-groups are normally end-capped to reduce adsorption effects either by methyllation or by using other functional groups to give the column specific properties.

and so forth. See, e.g., Neue (1997). The reader is advised to consult catalogs from the different manufactures to get an up-to-date picture of what is available.

4.5.2.1d LC Detection by Spectroscopy. The eluent from the columns can easily be passed through a flow cell in a spectrometer for nondestructive detection of all compounds that possess spectrometric features, e.g., a chromophore or a fluorphore. This requires that the eluent in itself does not have absorption in the range of interest. Also, a flow cell with a sufficient small volume is needed for matching the elution volume for the chromatographic peaks to retain the separation obtained in the column. In general, UV and fluorescence spectrometers are very versatile detectors in HPLC with several usable features: These detectors have a very large linear response range (3–5 orders of magnitude) with very good performance for quantitative analysis, they can give information about the bond structure in the molecules (aka chromophores), and are nondestructive, therefore can be combined with other detectors like mass spectrometers as described in Section 4.5. The limitation in the use of UV and fluorescence spectrometry in HPLC detection is the availability of a chromophore and/or a fluorophore in the molecules that the eluents need for being transparent, and that particularly for UV detection the sensitivity is limited. In metabolomics, many important metabolites do not have chromophores and/or fluorophores; spectrometry is therefore of limited usability as a general technique.

4.5.2.1e LC, Other Hardware Components. Pluming the solvent lines in an HPLC is not trivial. It is important to ensure that the flows in all the solvent lines are laminar and that there is no "dead-volume", that is, small volumes where samples can be withheld and hence mixed. These dead-volumes are particularly critical at low flow rates. The longitudinal diffusion (see Figure 4.2) in the tubing connecting the different parts of the HPLC also plays a role, and the tube diameter should therefore be matched to the flow rate to ensure a true laminar flow. Even at the higher flow rates, around 0.5–1 ml/min used with 4 mm internal diameter columns, wide bore tubing between the injector and the column and between the column and the detector can deteriorate the separation efficiency (e.g., using tubing with an internal diameter of 0.5 mm rather than 0.12 mm as required).

In general, an HPLC is rather easy to operate, but it can be challenging to optimize. The most common problems are (i) unstable flows due to air in the solvents or tubing, (ii) blocking of tube fittings or of columns because of particulate material in the solvent or sample, (iii) crystallization/precipitation of sample components in the column, and (iv) leakage from poor connections. It is crucial that high-quality solvents are used and these are free from air and particulate material, and care is taken to ensure that samples are free from particulate material (by filtration or high-speed centrifugation) and that they are truly dissolvable in the eluent at starting conditions.

Although leakage is, in general, easy to find at higher flow rates, it can be very difficult to find at lower flow rates as the solvent evaporates faster than it leaks. Also, in some cases solvent tends to "creep" out around seals (e.g., in the pump and injector) or connections, and if the eluents contain nonvolatile modifiers (salts), a buildup can be seen. It is important that these are removed by washing to avoid a buildup that may cause problems with stable operation of the system, and, in particular, deteriorate the pump seals.

4.6 MASS SPECTROMETRY

The mass spectrometer is both an analytical instrument in its own right by which very complex samples can be analyzed and a very versatile detector for chromatography providing very high sensitivity, and at the same time providing chemical or structural information. Development of modern biological MS has more or less been the driving force behind the development of metabolomics and MS today is probably one of the most important analytical methodologies in biotechnology. Nearly all analytical problems in biotechnology can be addressed by MS, ranging from the analysis of small volatile molecules, complex natural products, and proteins to intact viruses.

The core principle in MS is the determination of the mass to charge ratio, m/z, of charged compounds: molecules, clusters of molecules, complexes or fragments, and any combination of these. In principle, it is possible to determine the mass-to-charge ratio of anything with a charge on it (or which can be charged) and which can be transferred into the gas phase of the mass spectrometer. The developments during the last decades have dramatically expanded the range of molecules that can be determined by MS and also increased the sensitivity significantly. At the same time, MS has become much cheaper and the instruments have become easier to operate. This section only addressed the basics of MS with relevance to metabolome analysis by first introducing the instruments followed by a short discussion of the kind of results that are typically obtained. The reader can find a more in-depth description of MS in many recent reviews and textbooks listed at the end of the chapter.

4.6.1 The Mass Spectrometer—An Overview

The mass spectrometer is an instrument that performs all the required processes for mass spectrometric analysis starting from a sample in either a gas or a liquid phase: ionization/transfer of sample to the gas phase and transfer to vacuum, separation according to mass-to-charge ratio (m/z), detection of ions and processing, and presenting the data in a usable format. An overview of an instrument is shown in Figure 4.15, and a more detailed description of selected parts is given in the following sections.

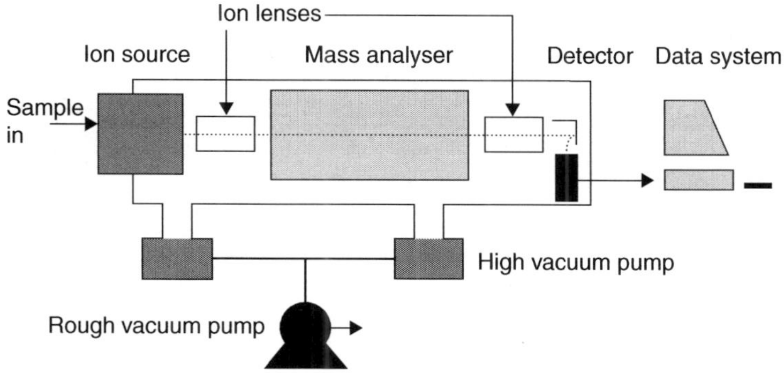

Figure 4.15 The mass spectrometer consist of a relative few elements: the ion source where the analytes are ionized and transferred to the high vacuum of the mass spectrometer, a mass filter where ions are separated according to mass to charge ratio, a detector to measure the ion current, a data system for control, and finally vacuum pumps to maintain high vacuum. Ion lenses are used to focus the ion bean so that ions will follow a narrow path through the instruments.

4.6.1.1 The Ion Source. The samples can be introduced into the ion source directly either as a gaseous sample from a gas chromatograph (where it is already in the gas phase), as a liquid sample into the instrument, or eluting from a liquid chromatograph dissolved in the mobile phase. The key processes in the ion source are transfer of the sample to the gas phase, ionization, and transfer to vacuum. Depending on the sample type (gas/liquid) and ionization method, these processes can be done in reverse order, i.e., ionization in the solvent followed by transfer of the ions into the gas phase. So the far, most common ionization techniques are electron impact ionization (EI) used with gas chromatography and electrospray ionization (ESI) used either with direct sample infusion or combined with liquid chromatography. These techniques are discussed in more details below. In general, the ion source is a part of the mass spectrometer that requires most attention in terms of both operation and maintenance. Many ionization parameters play a significant role for the results obtained, particularly, the solvent used for sample introduction as the solvent composition is a core part in the ionization process.

4.6.1.2 The Mass Analyzer. Determination of mass-to-charge ratio is done using a combination of electric and/or magnetic fields and several types of mass analyzers are in the market today. Some of the most popular mass analyzers are described in some details below. All mass analyzers have to be operated in high vacuum to ensure that ions do not collide with uncharged molecules, e.g., air or with each other. Mass analyzers are often grouped according to their performance: nominal mass analyzers where the mass resolution is unit mass separation, i.e., resolution around 1:1000–2000 and presenting integer mass accuracy; and high resolution mass analyzers, where the resolution is more than 1:7000 reaching as high as 1:100,000, presenting mass accuracy below 1 ppm. The latter type of mass analyzer will be able to separate all formulas and isotopic compositions with relevance to metabolomics approximately below 1000 Da.

4.6.1.3 The Detector. The detector will measure the current (amount of ions) or the number (by counting) as a function of time. As the m/z transmission of the mass analyzer is changed over time, the detector will measure mass as a function of m/z. Detection is, of course, crucial for the quality of the data obtained. Very sensitive high-speed amplifiers and analog to digital conversions are very important integrated parts of all detector systems. These electronic parts obviously depend on the detector design, which are described in some more details in Section 4.6.7.

4.6.1.4 The Data System. All modern mass spectrometers are designed around a data system that not only controls the instrument but also plays a significant role in data processing. Therefore, the data system should be considered as the fourth leg of the mass spectrometer and it is as important as the other parts. However, more advanced processing, e.g., chemometrics as described in Chapter 5, is normally done using separate systems and programs.

4.6.1.5 Other Hardware. Besides the above-mentioned elements, a mass spectrometer consists of a pumping system to maintain the required vacuum for the mass analyzer and a significant amount of control electronics and power supplies. High-vacuum systems based on two pumping stages are normally used to reach the pressure required in the range between 10–5 and 10–7 hPa, where high-resolution mass analyzer requires the lowest pressure. The first stage is normally a rotary oil pump backing one or more turbomolecular pumps capable of reaching these low pressures. In general, these vacuum systems are reliable but require some care and attention. The second important hardware is the high-voltage power supplies. All mass spectrometers use high voltages in the range of 1 kV to more than 20 kV depending on the ionization technique and mass-analyzer design. Particularly, the stability and control of the high-voltage power supplies for the mass analyzers can have a significant influence on the quality of the mass resolution, accuracy, and sensitivity. The problem is that although these high-voltage power supplies, in general, are very good, they do change over the years as do high voltage wires and connectors, and therefore they occasionally require attention.

4.6.2 GC-MS—the EI Ion Source

In many ways, the electron impact (EI) ion source and GC–MS represent the classical mass spectrometer configuration that has been around more or less since the invention of MS. This is due to the perfect match of a gaseous mobile phase to the vacuum in the mass spectrometer. Modern GC–MS systems are therefore highly developed, representing a mature technology with high performance, easy to operate, and delivering highly reproducible results. Furthermore, the theory and mechanisms are well developed, and extensive reference materials and databases are available.

Figure 4.16 shows a simplified view of an electron impact source used for GC–MS. The source consists of a small-heated volume, across a few centimeters, where a beam of energetic electrons ionizes the compounds eluting from the GC-column by impact. The electrons are emitted from a heated filament and accelerated to typically 70 eV before they are led through the source volume. Two small magnets are normally used to ensure a narrow beam of electrons through the source volume, and a trap plate on the opposite side is used to control the electron flux (current) through the source. The capillary column enters the source and terminates close to the electron beam. This ensures that the eluting compounds of a peak are kept together and as many molecules as possible reach the electron beam. As the source is operated in high vacuum ($<5 \times 10^{-5}$ hPa), the gas and eluting compounds will expand violently out of the column with the effect that the mean distances between molecules are increased dramatically. Thereby, molecule–molecule collisions and reactions are prevented, and from an analytical point of view, sample molecules are removed rapidly from the source, giving a very rapid response (in the low mile-second range or below). The ions formed by impact of the electrons (see below) are dragged out of the source by an electrical acceleration potential—in case of positive ions, by applying a higher potential (positive) to the source with respect to an acceleration plate outside the source. The acceleration voltage depends on the type of mass analyzer in use, and may be in the range from a few hundred volts in quadrupoles up to 10 kV in sector instruments. Finally, a repeller plate within the source is used to control

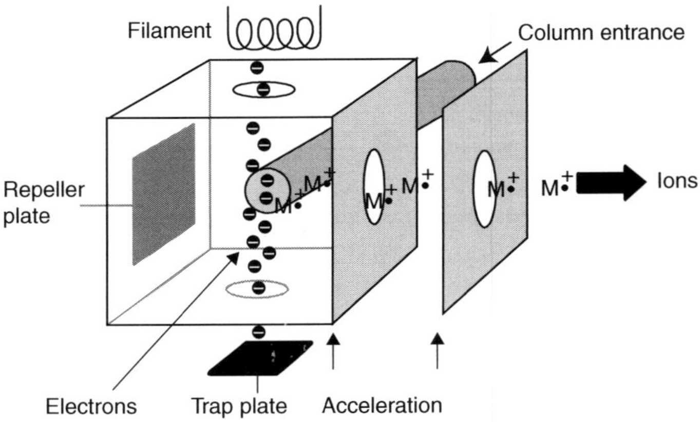

Figure 4.16 In case of gas chromatography with a gaseous mobile phase the electron impact source is very efficient to produce ions from the analytes in high vacuum. Modern mass spectrometers can easily deal with the typical flow from capillary columns hence the column ends as close to the electron beam used for ionization as possible. The electrons are emitted from a heated tungsten filament and accelerated to 70 eV before they enter the source volume. On impact with analyte compounds these are ionized, see Figure 4.17. The electron current can be controlled by measuring the current reaching a trap plate. A repeller electrode is used to control the electric fields in the source and an acceleration lens pulls the ions out of the source and accelerates them to a specific energy.

the electric fields in the source. The source is heated to prevent condensation and the high vacuum is used to remove nonionized compounds and carrier gas.

The ionization mechanism is illustrated in Figure 4.17 where a high-energy electron hits one of the electrons in the molecule. An electron energy of 70 eV is commonly used and is far more than what is required to break the strongest bond in organic molecules (the bond energy is typical in the range from a few electron volts

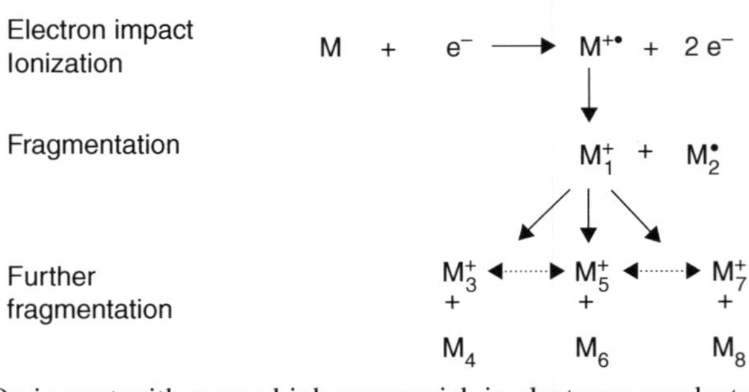

Figure 4.17 On impact with a very high energy rich in electrons, an electron is kicked out of the compound. This produce a positive-charged radical ion. As the electron energy is very high, excess energy is often transfer to the compound and this energy disperse through the molecule and will in most cases lead to bond breakage – fragmentation. This fragmentation is to some extent compound specific and can be used to deduce the structure.

to maybe 10 eV). These high-energy electrons will, with impact with an organic molecule, produce a radical ion by "shooting of" a bonding electron. The resulting radical ion will have the same mass as the original ion (except for the mass of an electron) and is called the *molecular ion*. Owing to the use of very energetic electrons, excess energy may be present in the molecule after the impact. This energy is dispersed through the molecule and may lead to further bond breakage and fragmentation. Thus the molecular ion may undergo fragmentation to the ion M_1 by the loss of a neutral radical, which again may fragment further. The molecule may also undergo internal rearrangement and reactions to disperse energy and form stable ions. The complete fragmentation and rearrangement pattern is highly compound specific in terms of both masses seen and their ratio and is therefore a powerful tool for identification of unknown compounds. Comprehensive discussion about fragmentation in EI ionization can be found in McLafferty (1993) and should be consulted by all practitioners of GC–MS. The very compound-specific fragmentation in EI ionization has also led to collection of very large libraries of spectra that can be a great assistance for identification of unknown compounds. The use of these libraries does, however, require a critical evaluation of the results, as the search results can be way off.

4.6.3 LC–MS—the ESI Ion Source

The main obstacle for LC–MS-based techniques has been the incompatibility of the liquid eluent coming from the column and the vacuum of the mass spectrometer. Initially direct liquid introduction of the solvent (at very low flow rates) into the EI source was tried, but even very powerful vacuum pumps performed rather poorly. Techniques based on separation of analytes from solvents have been used prior to ionization by EI but with rather poor performance. Development of atmospheric ionization techniques in the mid-1980s, particularly electrospray ionization (ESI), LC–MS revolutionized analytical chemistry, and today it is one of the most important analytical techniques in biotechnology.

ESI mass spectrometry is so far the most used ionization technique in biological MS, but other techniques, such as atmospheric chemical ionization, are used for specific application. In many cases, the combined ion sources allow the user to switch between the different techniques. ESI is the predominant technique in metabolome analysis and is therefore described here in more detail.

The principle of ESI is illustrated in Figure 4.18; for simplicity it is shown in positive mode for the detection of positive ions. The eluent from the column is pumped through a narrow steel capillary tube into an open source chamber held at atmospheric pressure. The outer diameter of this steel tube is typically in the range of 0.2–0.3 mm and is often referred to as the spray needle. If a voltage above a certain threshold is applied to the needle, a so-called Taylor cone is formed at the end of the capillary, which is stretched into a highly charged thin filament. When this solvent filament reaches a certain diameter, the Rayleigh limit (where the number of charges exceeds the number that can be held together by the surface tension forces and hence results in an instability of the filament), a series of fine droplets are expelled, and

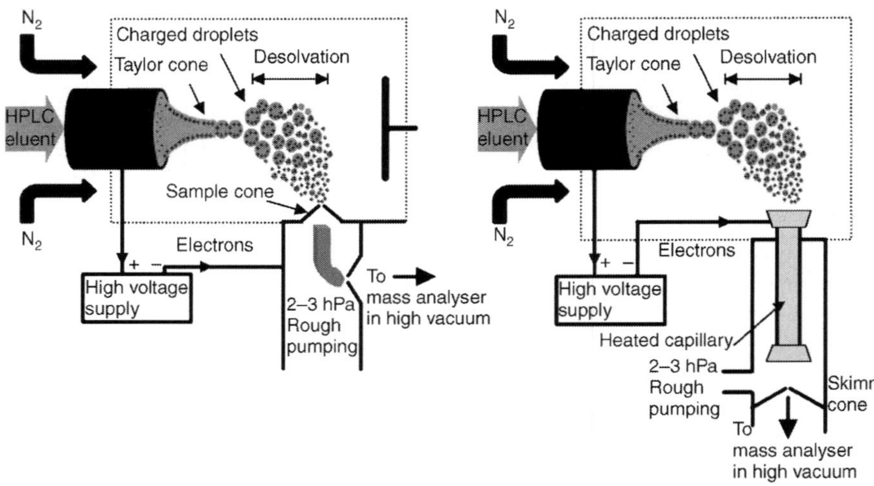

Figure 4.18 In the electrospray source the eluent coming from the HPLC is sprayed through a narrow bore steel capillary (about 0.2 mm OD) at atmospheric pressure. When a high voltage is applied to the capillary, a Taylor cone will form and a spray of fine highly charged droplet will be emitted. To facilitate evaporation of solvent from the droplets, a stream of heated nitrogen is blow through the source. The ions are sampled through a small orifice in a sample cone of a heated capillary into vacuum.

a spray of highly charged fine droplets are formed—the so-called *electrospray.* A flow of heated nitrogen gas is used to evaporate solvent from the charged droplets. As the solvent evaporates the Rayleigh limit is reached and a series of smaller droplets are expelled from the initially formed droplets. This process continues until the droplets are capable of carrying the remaining charge. However, the physical details of the electrospray process are not fully understood and other mechanisms may also play a role: *ion evaporation*, where ions evaporate directly from the droplets or *coulomb explosion*, where the droplets explode to a multitude of small droplets when the Rayleigh limit is reached. The electrospray mechanism illustrated in Figure 4.19 shows a hypothetical desolvation pattern from a 1-µm droplet formed by electrospray from classic steel capillary spray tube around 0.2 mm diameter. A series of small droplets are ejected from the parent droplet as the solvent evaporates until the Rayleigh limit is reached. This process continues until no more solvent can be evaporated and the remaining molecules can accommodate the remaining charge. The goal is to end with a charged molecule in the gas phase. The overall process is governed by several factors including droplet size, surface tension of the solvent, surface activity of ionizable compounds, the ion strength of the solvent, pH, counter ions, and temperature (which, by the way, will always be below or at the boiling point of the solvent).

As illustrated in Figure 4.19, smaller parent droplet will produce more ions because of two facts: There are less droplet fragmentation steps before we have the ion in the gas phase and we have a much larger surface area, thus more molecules are

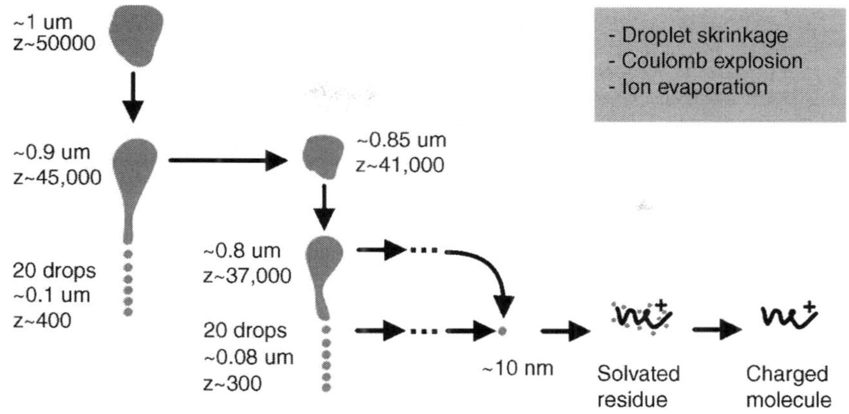

Figure 4.19 Although the mechanism of the electrospray process is still a matter for some debate the key points can be summarized to the following: from the Taylor cone formed at the spray needle a series of highly charged droplets around 1 μm in diameter is formed. As the solvent evaporates from these droplets to a point where the surface strength cannot overcome the coulomb repulsion, a Taylor cone is formed from the droplet emitting a series of smaller droplets (nm size droplets). The process is repeated from the new droplets as the solvent evaporates and at the end we have charged molecules. Alternatively, there is some evidence that ions may be emitted directly from the droplets to reduce the number of surface charges. As the solvent only evaporates completely from the small droplet, it is an advantage to produce the smallest droplets from the initial spray. The process is governed by the surface strength of the solvent, surface activity of the analytes and additives, ion strength, nature of counter ions, size of droplets (needle size and flow rate), concentration, evaporation rate, and several other factors.

exposed to the surface for ionization. In other words, the smaller droplets give much higher ionization efficiency. The size of the droplets is predominantly governed by the diameter of the spray needle and surface tension of the solvent. To increase the efficiency of ionization, nanoelectrosprays have been developed using spray nozzles with diameters in the low micrometer range (or even lower) producing droplets in the low nanometer range. The overall result is an amazing increase in sensitivity.

The ionized residues are transferred into the vacuum of the mass spectrometer through sampling orifices as illustrated in Figure 4.18, e.g., sampling cone or narrow bore capillaries using multiple pumping stages. The ions are guided by electrical potentials and the supersonic gas jet created by the pressure drop across the sampling orifice. The potential between the sampling orifices in pumping stages can be used to induce fragmentation by acceleration of the ions so that they collide with the gases in the intermediate pumping stage, a technique called *in-source collision induced dissociation* (in-source CID). Modern electrospray interfaces, as illustrated, can accommodate the flow-rates used in normal analytical HPLC up to around 1 ml/min; however, most interfaces work better at lower flow rate from below 0.1 to 0.3 ml/min. In nanoelectrospray, the flow-rate is typically below 50 nl/min; therefore, either a splitting device is needed for HPLC, or capillary HPLC columns are used.

Besides the physical design of the ion source, the composition of the solvent and selection of source parameters is crucial for the ionization efficiency. Obviously, ions are required in the solvent, but too high ion strength can completely ruin the electrospray, and it has been shown that an optimal ionization is obtained between 10^{-5} and 10^{-2} M. It is nearly impossible to get a stable electrospray from an apolar organic solvent both because of a low surface tension and because of the presence of very few ions. Normally, volatile acids or bases are added to the solvent used in electrospray to facilitate more efficient ionization. Other modifiers can also be used to enhance ionization, e.g., various salts at lower concentrations.

The ESI is very soft and will (in positive mode) predominantly produce protonated $M + H^+$ ions and depending on conditions also produce sodiated $M + Na^+$ ions; clusters with solvent molecules can also be seen. Fragments and compound-specific spectra as seen in EI ionization are not found in ESI, and ESI mass spectrometry can therefore not be used for compound identification to the same extent as EI–MS, unless fragmentation techniques are applied either in the source or by MS–MS. Furthermore, the fragmentation process is governed more by gas phase chemistry and is not as specific as in EI ionization. On the contrary, producing only one or very few ions from each compound enhances the sensitivity and hence the usability of the mass spectrometer as a selective detector. Limited fragmentation can also be used to analyze complex samples without prior separation as described in one of the case studies in the second part of this book (Chapter 9). A more detailed discussion about ions seen from ESI can be found in Section 4.8.

The major issue encountered in ESI is what has become known as *matrix effects*. Matrix effects, in general, result in loss of sensitivity and discrimination so that ion intensity observed from some compounds is much lower or completely missing in the presence of other compounds, e.g., from the sample matrix. It can be seen as these compounds "steal" more than their part of the charges because they are better at carrying charge or having better surface properties. This is a common problem in positive ESI if, e.g., TWEEN or PEG (poly-ethylene-glycol polymers) is present in the sample as the signals from sample compounds can be completely hidden or lost in the numerous peaks from these compounds. If the ion strength is too high (e.g., because of buffers or salts), the ion source may "short circuit" and quench the electrospray process completely. Finally, by analyzing complex mixtures directly, one of the components may be much more efficiently ionized than other compounds in the sample, thereby stealing more charges than expected by its concentration, and resulting in suppression of other compounds.

Not all compounds can be protonated by positive electrospray MS. In these cases, the voltages can be reversed, thus producing a negatively charged spray. The ionization mechanism in negative electrospray is not as well studied but it predominantly leads to the formation of deprotonated ions thus $M - H^-$. It is not always easy to á priori determine whether a compound will ionize better by positive or negative ESI and under what conditions will they do so. For some compounds, a better ionization can be obtained by spraying an acidic solvent, e.g., containing formic acid. Many sugars can only be ionized by negative ESI, whereas it is easy to find rather strong carboxylic acids that are much more efficiently ionized by protonation in positive electrospray. An advantage of negative ESI is that very few clusters are seen and often there are fewer matrix problems. On the contrary, it is, in general, more difficult to get a stable electrospray in negative mode.

A detailed discussion about the mechanism and optimization of ESI is outside the scope of this book but a few more general recommendations in relation to metabolomics can be found in Section 4.6.2 and in the suggestions for further reading. Also, some of the case stories in the second part of the book illustrate the use of ESI mass spectrometry in metabolomics. Besides being a very versatile analytical tool in metabolomics, electrospray MS has become one of the most important tools in protein and peptide analysis and is widely used for sequencing, study protein of modifications, and so forth.

Other LC–MS techniques available are all based on the basic design of the electrospray source. The techniques that are most frequently used is atmospheric pressure chemical ionization (APCI) and atmospheric pressure photo ionization (APPI). None of these techniques are generally used for metabolome analysis; however, these techniques have advantages for target analysis. The reader is referred to a more specialized analytical literature for details of these techniques.

4.6.4 Mass Analyzer—the Quadrupole

The quadrupole mass analyzer is one of the simplest and most versatile mass analyzers and is widely used particularly for GC–MS (see Figure 4.20.) The key characteristics of a typical quadrupole mass analyzer is a mass resolution around 1:1500 nominal mass accuracy, and a mass range from 2 Da/e up to about 3000 or 4000 Da/e. The quadrupole mass analyzer consists of four parallel metal rods where an RF voltage supply is connected to adjacent rods creating an alternating electric field between the rods. The charged molecules enter the quadrupole axially after they have been accelerated to a required linear energy. Once inside the quadrupole, they start spinning within an imaginary cylinder created by the RF voltages. The diameter of the imaginary cylinder depends on the mass-to-charge

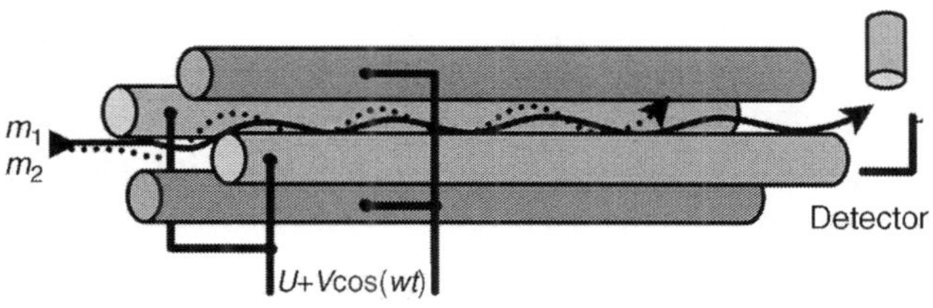

Figure 4.20 The Quadrupole mass analyzer is a simple and efficient mass analyzer. It consist of four metal rods place parallel few centimeters apart. If an RF-voltage is applied to adjacent rods, an ion injected along the axis will start spinning in an imaginary cylinder. Depending on the voltage and frequency the ion will pass through the quadrupole. If the imaginary cylinder is offset by a small direct current voltage only ions within a narrow mass to charge range will survive through the quadrupole. By selecting different voltages as illustrated in Figure 4.21, a wide range of ions can be separated.

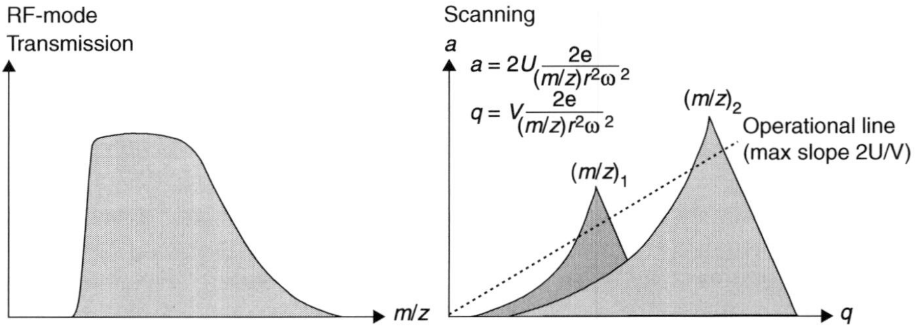

Figure 4.21 Transmission of ion in the quadrupole mass analyzer in RF only mode can be seen to the left. As shown, ion within a wide range on mass to charge ratio will pass through the quadrupole. If a DC voltage is applied on top of the RF voltage, the imaginary cylinder is offset and only ions within a certain range can pass through the quadrupole. Or in another way, an ion with a specific mass to charge ratio can pass through the quadrupole for all values below the curves as illustrated to the right. Here it can be seen it is possible to select values for a and q that allow separation of the two ions as illustrated. If the voltages are scanned at a fixed ration, ions are separated at a resolution determined by this ratio.

ratio (m/z) of the ion and the RF voltage. Only ions within a certain m/z range will survive all the way through the quadrupole.

If we apply only an RF voltage to the quadrupole, ions with a wide range of m/z values will pass the quadrupole, where heavy ions will spin in a narrow circle and light ions in a wider circle. There will be a rather sharp cut-off at the low-mass end where the low-mass ions hit the rods whereas in the high-mass end there will be a slow trailing off because of the lower transmission efficiency of heavy ions. In RF-only mode, the quadrupole (or hexa- or octa-poles) is called a wide pass filter and is commonly used for focusing ion beams and collision cells in MS–MS. This is illustrated in Figure 4.21a—in RF-only mode there is a high transmission of ions within a wide m/z range. If DC voltage is applied on top of the RF voltage, the m/z range transmitted is narrowed down and a mass separation is obtained. The DC voltage will offset the imaginary cylinder in which the ions spin, and only ions within a narrow m/z interval will survive to the end of the quadrupole. This is illustrated in Figure 4.21b where the effect of changing the DC voltage and RF-amplitude is illustrated. These voltages depend on the frequency ω and radius of the quadrupole; both are kept constant for a given instrument and often the actual voltages are replaced by the parameters a and q that are both proportional to the AC and DC voltages. As illustrated, the ion $(m/z)_1$ will survive through the analyzer with all combinations of DC voltage (U or a) and RF amplitude (V or q) in the dark grey area under the curve. Similarly, $(m/z)_2$ will survive for all combinations in the light grey area under the other curve. In the overlapping area, both ions will be allowed to pass through the quadrupole, corresponding to the situation illustrated in Figure 4.21a. By selecting a suitable combination of a and q, i.e., the DC voltage and the RF-amplitude only, a narrow m/z range will pass through the quadrupole. As quadrupoles are operated at a fixed frequency, scanning a quadrupole to allow different *m/z* values to pass is done by changing *a* (the DC voltage, *U*) and *q* (the RF amplitude, *V*) at a fixed ratio. The optimal ratio is obtained during the tuning of the instrument, and the calibration procedure establishes the relation between the *a/q* ratio and *m/z* passing through the quadrupole. Changing (scanning) the values of *a* and *q* (thus *U* and *V*) at a fixed ratio along the dotted lines shown in Figure 4.21b, also called the operational line, will give better than unit resolution if $(m/z)_1$ and $(m/z)_2$ are 1 Da apart.

The *advantage* of the quadrupole mass analyzer is that it is easy to build, easy to operate, and is very reliable. In general, it has a high sensitivity, thus a high ion transmission, but the transmission decreases with mass. This is because of the fact the quadrupole operates optimally within a certain ion velocity window (time the ion spends between the rods) that in general is a compromise set to favor the lower mass. Higher *m/z* requires higher acceleration in the source to get a good sensitivity, but the result is loss of low mass resolution (lower masses are just too fast to be separated).

A quadrupole allows only one *m/z* to pass at any one time; therefore, ions with other *m/z* are lost during that time. For example, scanning a quadrupole from *m/z* 50 to *m/z* 550 thus 500 Da in 1 s allows transmission of each *m/z* for 2 ms and the ions are lost for the rest of the time. If we reduce the mass range to 250 Da, we will have 4 ms per *m/z* value, thus, we may get a twofold increase in sensitivity. This is often used for selective high-sensitivity analysis where only a few selected *m/z* values are allowed, giving much more time to measure each *m/z*. This is called selective ion recording SIR (or SIM for selected ion monitoring), and it results in a dramatic increase in the sensitivity but with the loss of a diagnostic mass spectra that can be used for identification. Therefore, SIR mass spectrometry is only used for target analysis where it is very efficient, whereas a full scan mode is normally used for profiling purpose and when dealing with unknown metabolites.

4.6.5 Mass Analyzer—the Ion-Trap

The ion-trap (more correctly called a quadrupole ion-trap) is in family with the quadrupole mass analyzer as described above but instead of continuously transmitting ions through the quadrupole, the ion-trap can store ions and eject these when required. A classical ion-trap consists of two bowl-shaped end-caps placed on either side of a doughnut-shaped ring electrode as illustrated in Figure 4.22. Ions are injected into the ion-trap through one of the end-caps and trapped in the small volume within the ion-trap by applying an RF-voltage and a DC voltage to the ring electrode and end-caps. The ions will be trapped in a complex motion pattern within the trap and can be held for some time (μs to ms). To control the ion motions and cool the ions (lowering their energy), a damping gas, usually helium, is let into the trap at a pressure of about 0.01 Pa. By changing the amplitude of the RF-voltage and the DC potentials on one of the end-caps, ions with specific *m/z* values can be ejected from the ion-trap, and hence can separate the ions. The normal duty cycle is to trap ions with all *m/z*, close the inlet, and then eject ions according to their *m/z* values. However, there is a limit to the number of ions that can be stored in the small volume within

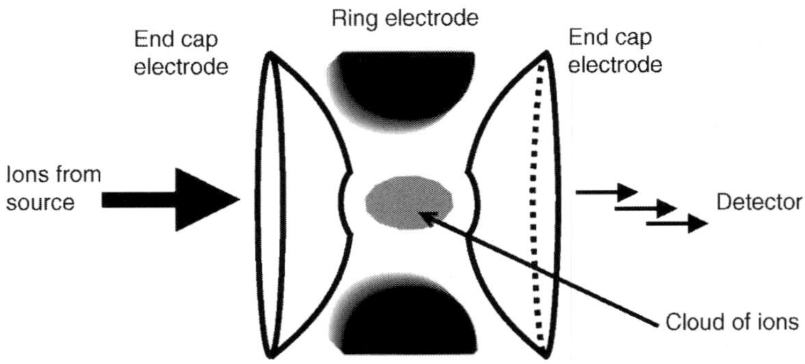

Figure 4.22 The ion trap mass analyzer consists of two cone-shape end-cap electrodes place on each side of a ring electrode. An RF voltage is applied to the end-cap and the ion beam enters through a hole in one of the end caps. Due to the RF-voltage the ions will be trapped between the two end caps forming a cloud of ions in the center of the trap. A gas (helium) is normally feed to the trap to cool the ions. By applying a DC voltage on top of the RF-voltage ions at specific mass to charge ratio will be emitted through one of the end cap electrodes. It is possible to emit all but one m/z value, which then can be fragmented by collision with gas in the trap to produce a second fragment spectrum, MS-MS.

the ion-trap before ion–ion interaction will start to reduce performance. Therefore, most ion-trap instruments include a gain controls that controls the number of ions collected in each duty cycle often to less than a few hundred. However, even with gain control, ion–ion reactions can be seen in the ion-trap often resulting in formation of unexpected ions and adducts seen in the spectra. Most noticeable is ion–ion reactions leading to protonation of molecular ions in GC–MS where radical ions are expected as describe above. This is particularly pronounced in GC–MS in analyses of samples with a wide concentration range and good chromatographic separation giving sharp peaks.

An ion-trap is not scanned like the quadrupole mass analyzer, but it collects ions and then the selective ejection of ions is used to measure a mass spectrum. Therefore, there is no gain of sensitivity by using selected ion monitoring, and therefore this is rarely used on ion-traps.

The major advantage of the ion-trap mass spectrometer is that besides providing full mass spectra, a selected ion can be kept in the ion-trap while all other ions are ejected. The energy of the selected ion can then be increased and lead to fragments by collision with the gas in the ion-trap. The fragments can then be ejected systematically to get fragment mass spectrum or a daughter spectrum of the selected ion, a technique normally referred to as tandem MS, or MS–MS. This process can be repeated keeping one of the fragment ions trapped and fragment it further. These fragment spectra provide useful structural information about the molecule and it is particularly useful in connection with ESI mass spectrometry as described above, because only very few diagnostic ions are formed in the ion source. This multistep MS–MS–MS is often referred to as MS^n. Besides being an efficient tool for structure elucidation, MS–MS techniques can also be used selectively by measuring a specific

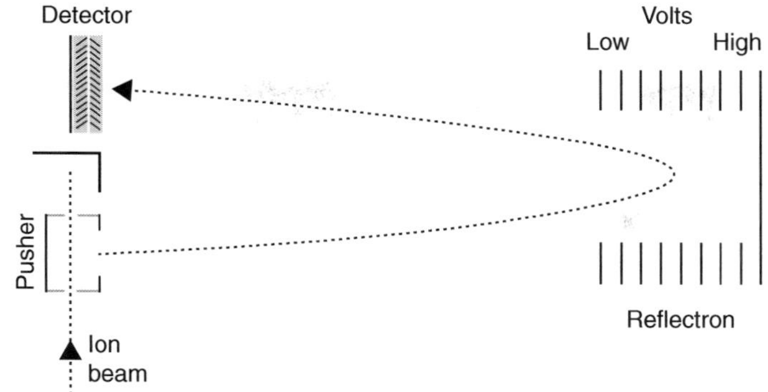

Figure 4.23 In the time-of-flight mass analyzer, ions enters a pusher region where at time zero, they are accelerated to a specific kinetic energy by a short electric pulse. At the same time a very precise timer is started. The ions drift through a flight tube, and in this case, the flight direction is reversed by an electric mirror (reflectron). The advantage of the reflectron is that the flight path becomes longer and that small differences in kinetic energy are even out thereby increasing the mass resolution and accuracy. When ions reach the detector, a time mark is noted for each ion and stored in the spectrum. Many push events are summarized to a spectrum.

ion that is transformed into another specific ion, combining a specific transformation with retention time, resulting in a highly selective analysis.

Ion-traps potentially have the possibility to provide very high resolution and also rather good mass accuracy within a limited mass range, but it is usually used at nominal resolution over wide mass ranges. The latest generation of ion-traps, the linear ion-trap, can store many more ions and provide higher resolution over a wider mass range. The reader is referred to dedicated textbooks for more details on ion-traps.

4.6.6 Mass Analyzer—the Time-of-Flight

The time-of-flight (TOF) mass spectrometer is in many ways one of the simplest mass analyzers as illustrated in Figure 4.23 where the mass-to-charge ratio is determined by giving the ions a push to the same kinetic energy and then measuring the time they take to fly a specific length. From the three simple relations from physics as shown in Figure 4.24, it can be deduced that the m/z is proportional to the squared

$$\left.\begin{array}{l} E = qU \\ E = \frac{1}{2}mv^2 \\ s = vt \end{array}\right] \blacktriangleright \quad m = \frac{2E}{v^2} = \frac{2qU}{v^2} = \frac{2zeU}{s^2}t^2$$

$$\text{In short} \quad m/z = kt^2$$

Figure 4.24 The relation between flying time and mass to charge ratio can be calculated from these simple equations where E is the kinetic energy, q is the charge on the mass m, accelerated by the potential U, flying the distance s by the speed v in the time t, and k is a constant determined by calibration. It is important to note that m/z is proportional to the flying time squared hence double mass to charge requires four time longer flying time.

flying time. In practice, the ions enter a so-called *pusher*, where a short electric pulse is used to accelerate the ions to the same kinetic energy and at the same time to start a timer. Great care is taken by designers to focus the ion beam ensuring a beam as narrow as possible that enters the pusher region as this minimizes spread in the kinetic energy (a major source of loss in resolution and accuracy). The ions then drift through a flying tube to the detector. In the TOF mass analyzer illustrated in Figure 4.23, an electric mirror is used to reverse the ion beam, which both lengthens the flying path and corrects the residual differences in kinetic energy from the pusher as not all ions started on exactly the same "starting line" when the pusher pulse was applied and the timer started. The electric mirror significantly increases the mass resolution and the mass accuracy that can be obtained. When an ion reaches the detector, a signal is generated and the arrival time of an ion is registered. The operation of a TOF mass analyzer requires lower pressure than the other mass analyzers, typically in the 10^{-7} hPa range to avoid any ion–ion or ion-gas molecule interactions.

As can be seen from the equations in Figure 4.24, low-mass ions will have a higher velocity than heavy ions and arrive first. In a typical reflectron TOF mass analyzer, the flying time for a 1000 Da/e ion is less than 50 μs, and therefore TOF analyzers are very fast, and up to 20,000 push events can be done per second. In general, spectra from many push events are summarized into one mass spectrum to improve ion statistics and reduce noise. It is obvious that accurate measurement of flying time is crucial for the TOF mass analyzer and, in general, requires very fast timers capable of measuring time in the nanosecond to picosecond (10^{-9} to 10^{-11} s) range. Just to illustrate: *If we assume that we want to measure a mass resolution of 10,000 (10^5) at mass 1000, we can separate mass 1000.0 Da/e from mass 1000.1 Da/e, and if mass 1000.0 Da/e has a flying time of 50 μs, then the flying time for mass 1000.1 Da/e will be 50.0025 μs or just 2.5 ns more (use the equations in Figure 4.24).* To accomplish measurement of 10,000 in resolution, a very fast and accurate timing and detection system is needed. In TOF-MS, two rather different approaches are used: ion counting in small time intervals (steps or bins) or measuring the ion current as a function of time. Although the second principle is quite similar to what is used with other mass analyzers, ion counting in time intervals is quite different. The detector system does influence the data obtained and is discussed in more detail in the next section.

The TOF mass analyzer is not scanned in a manner similar to the scanning of ion-trap, and does not store ions either. Ions of all masses are pushed into the flying tube at exactly the same time, and we will have to wait until all ions have reached the detector before the next group of ions is pushed. Therefore, there is no advantage in using selected ion monitoring (SIM or SIR) as the next push event cannot be done before all other ions have reached the detector whether we want to monitor these or not. However, the pusher rate has an impact on the sensitivity, and thus more the ions sent through the flight tube better the sensitivity, and many push events are normally summarized into one spectrum (*not a scan as the analyzer is not scanned*). Depending on the instrument, the requirement for resolution, accuracy, and sensitivity, many hundred spectra can be collected per second making TOF analyzer an ideal companion for high-speed GC–MS with deconvolution or the lasted generation of fast HPLC. Furthermore, with modern electronics, TOF analyzers can routinely give mass resolution more than 10,000 (full width half maximum) and mass accuracy below 5 ppm. However, for quantification, the TOF mass analyzers at present cannot match the quadrupole mass analyzer mainly because of limitation in the detection system which requires some attention to ensure a good performance (discussed in some more details in Section 4.5.7). Despite the poor quantification, the TOF-analyzer is becoming increasingly popular as the performance, sensitivity, and simplicity of operation is outstanding.

4.6.7 Detection and Computing in MS

When the ions have been separated in the mass analyzer, a detection system is used either to measure the ion current (flux) continuously as a function of the scan in progress (the voltages as illustrated in Figure 4.21) or to count the ions arriving in small time segments, so-called time bins. The ion current is normally measured by detectors based on a conversion dynode and *electron multiplier* commonly used in quadrupole and ion-trap instruments, whereas ion counting devices based on microchannel plate (MCP) detectors coupled to time-to-digital converter (TDC) are normally used in TOF instruments.

A conversion dynode—electron multiplier detector—is illustrated in Figure 4.25a. When an ion hits the conversion dynode, it leads to emission of one or more

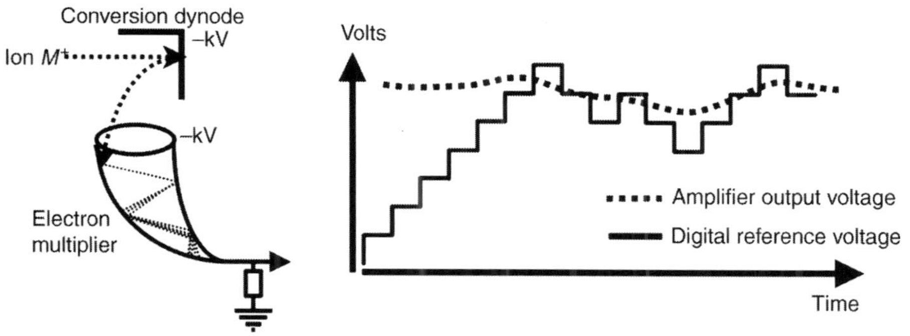

Figure 4.25 The most common detector in mass spectrometry is based on an electron multiplier as shown to the left. To avoid radiation directly from the source most detectors use a conversion dynode. Ion hit the dynode and secondary ions are emitted and these will hit the electron multiplier. An ion hitting the multiplier will start emission of a cascade of electrons, thereby amplifying the ion current up to 10^5 times. The output is further amplified before it is converted to digital number by an analog to digital converter (ADC). In the ADC the detector signal is compared to a small reference voltage, if the detector signal is larger, the voltage is step up by a specific amount. The output is the number of reference steps required to get closest to the detector voltage. The number of steps and the speed is crucial for the detector performance.

secondary ions. These ions will then hit the wall of an electron multiplier leading to the release of a cascade of electrons. One ion may lead to the release of more than 10^5 electrons that generate a current, which is further amplified and measured by an analog to digital converter (ADC). The ADC can be viewed as a counting device where the number of steps of a reference voltage has to be increased until it reaches the voltage received from the detector amplifier as illustrated in Figure 4.25b. There are two main issues that determine the performance of a detector: The *dynamic range*, thus the number of step it counts, and the *response time*. The dynamic range is determined by the total number of voltage steps the ADC can use to compare the reference voltage to the voltage received from the amplifier. This is typically given as the number of binary integers of the ADC outputs for further processing, e.g., as 12-bit, 16-bit, or even 24-bit words. A 16-bit output means that the ADC can count 2^{16} steps or 65,536 steps. In other words, the detector can assign 65,536 different values to the signal intensity. To enhance the dynamic range, the ADC may control the amplifier and turn the gain down if the maximum is reached (or up, if below a certain value). The response time of the electron multiplier itself is very fast, and the overall response time is determined by the ADC conversion rate. In general, a greater dynamic range or high resolution (many bits) will give a slower conversion.

The advantage of the electron multiplier detector is that it can measure the actual ion current coming through the mass analyzer continuously. Also, it has a large dynamic range covering several orders of magnitude. Therefore, electron multipliers are widely used in conjunction with nominal resolution mass analyzers or using slower scanning high resolution analyzers as the sampling rate is typically in the megahertz range which is more than adequate to get 10–30 data points per *m*/*z* value, as required for accurate peak determination. However, TOF analyzer requires very fast detection to precisely determine the arrival time, typically in the gigahertz range. This can be achieved by the latest generation of very fast electron multiplier detectors with 1 GHz ADC converters but only converting with 12-bit resolution (4096 steps). Compared with the MCP detectors, as described below, the electron multiplier detector has the potential to give superior quantification to TOF mass spectrometers.

The *MCP detector* consists of one or more thin plates with numerous small channels (in the 10 μm range) placed at an angle incident to the ion beam as illustrated in Figure 4.26a. An ion entering any of these channels will start a cascade of electrons similar to that of an electron multiplier, thereby generating a current. The advantage of the MCP detector is that it has a rather large surface area needed to detect the more scattered ions in TOF analyzers. This current is amplified and used to produce a stop signal to the timer in the TOF mass spectrometer. The timers used in conjunction with MCP are called *a time to digital converter* and is basically a single-start multiple-stop timers running at a very high frequency, typical in the range from 1 to 10 GHz. The pusher pulse starts the timer and whenever an ion generates a signal on the detector, the timer adds one to the current time step or bin. This is illustrated in Figure 4.26b showing the small time bins on the time scale. After the first push event in a spectrum, single ions will be counted in various time bins, as more push

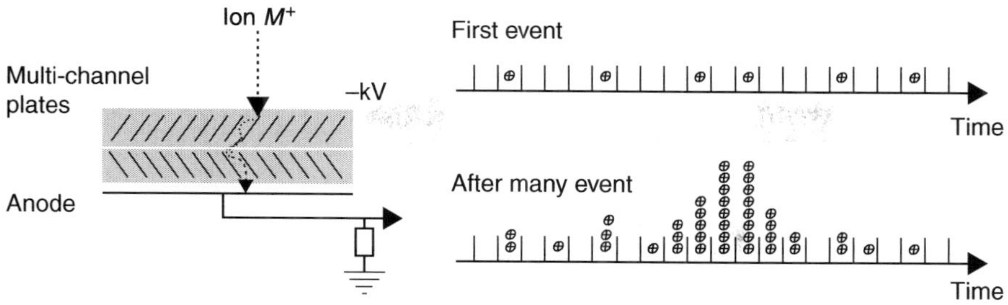

Figure 4.26 In most time-of-flight mass spectrometers the ions are detected by a multi-channel-plate detector (MCP) together with a time to digital converter (TDC). The MCP works as wide area electron multiplier with many hole each working as small electron multipliers as shown to the left. When an ion hit the MCP a cascade of electrons is generated in that hole and a small current is produced. This current will produce a stop signal to the TDC timer (which is a single start multiple stop timer) and 1 is added to that time bin, thus the smallest time step (to the right top). The next ion will generate a new signal and again 1 is added to that bin. Unfortunately, the MCP-TDC detector is blinded by the arrival of an ion corresponding to 2–4 time bins hence the ion current should be kept low so that only one ion arrive within this dead time period. A TOF spectrum is normally the result of many push events hence many ions may end in some of the time bins.

events have been done, more ions will be found in some bins while others are empty. When all push events requested for a spectrum have been carried out, the number of ions counted in each bin is transferred to the data system together with the bin time for further processing. The width of the time bins is very important for the resolution of the data that can be collected and it is on modern instruments in the range 0.2–0.5 ns. Two major issues require attention when working with MCP–TCD detector systems: *Only one ion can be detected at any one time*, thus if two ions arrive at the same time bin, they will be counted as only a single arrival and only one count is added to the bin. The second problem is that although they react very fast, the *detector system has a dead-time*; thus, it is blinded by an ion arrival for 1–2 ns which corresponds to several time bins, thus the detector cannot see if an ion arrives in that time span. The results of these two effects are that the ion current (flux) through the mass spectrometer has to be kept rather low to ensure that all ions are counted. If ions arrive at a very high rate, the detector goes into dead-time when the first ion is detected whereas the next few ions are therefore not seen. The consequence is that the ion profile is skewed to a shorter flying time, hence to a lower m/z as more of the first arriving ions are seen than later arriving ions. Also, dead-time problems will give a very low number of ions counted for each mass (m/z), and therefore give errors in isotopic patterns and a poor quantification. Today, advanced ion lens control and statistical data processing have given methods to reduce these problems in the MCP–TDC detector systems; however, optimal performance is best achieved avoiding dead-time in the detector.

When the data have been collected, they are transfered to a computer that links the detector signal to the scan or time information. The scan information or flying

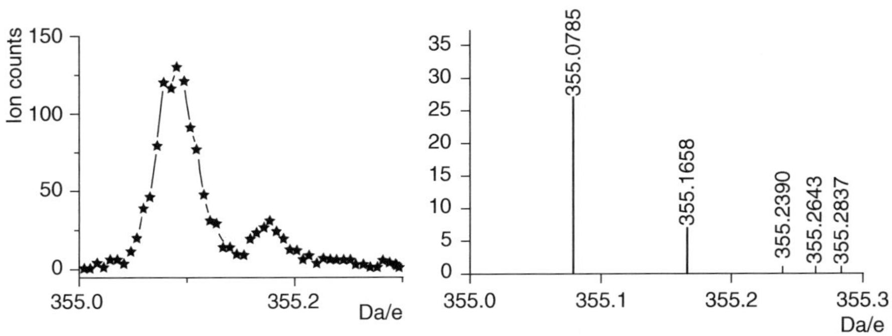

Figure 4.27 Structure of data from detectors as read from the detector is shown to the left. The sample rate (in this case number for bins) has to be sufficient for the resolution to get enough data point to precise peak detection. These raw data is commonly referred to as continuum data. In most case the continuum data is converted to centroid data on the fly by detecting the peak position (the centroid) and peak height or area. The result is a significant reduction of data file size as each mass peak is saved as two numbers rather than 10–20 data points.

time is converted into a mass-to-charge scale (normally just referred to a mass scale when dealing with small molecules) on the basis of a calibration table where the relation between, e.g., voltages or time and m/z is stored. These calibration tables are typically prepared by analyzing a known sample and calculating a relation between the measured m/z and the true monoisotopic mass. In most cases, a polynomial calibration curve is used to smoothen small errors. When the mass scale has been added to the data, we have what is called a raw mass spectrum or often called a continuum mass spectrum as shown in Figure 4.27a. Here the stars indicate the individual data points, as these data are from a TOF instrument with an MCP–TDC detector; they show how many ions arrived in each time bin. If they have been from an electron multiplier, they would have shown the ion current at each sampling point. In most cases, these continuum data are further processed, where the mass peaks are detected and the result is shown as a bar at the central m/z value and with a height corresponding to the ion count/current as shown in Figure 4.27b. These bar spectra are normally referred to as centroid spectra and are typically normalized to the highest peak in the spectrum. There is, of course, a considerable reduction of data file size in calculating centroid mass spectra with very little loss of information. In the example in Figure 4.27, about 110.000 data points were collected in the full continuum spectrum covering 900 mass units, whereas only around 700 ions were seen. If data are collected at a rate of one spectrum per second, continuum spectra can give very large data files, whereas centroid files are more manageable.

Beside collecting and preprocessing mass spectral data, the computer is generally used to control the instrument, perform data processing, and even library searches particularly for EI spectra. Data analysis is further discussed in Section 4.8 and data processing in Chapter 5.

4.7 THE ANALYTICAL WORK-FLOW

The driving force behind planning and carrying out chemical analysis can roughly be summarized as follows:

- the wish to determine a selection of known specific compounds in a series of samples
- to learn what a specific analytical methodology can tell about samples of interest.

Traditional chemical analyses are performed for determination of specific compounds normally driven by a hypothesis. With the widespread use of techniques like MS that can produce excessive information, it might be feasible to simply generate lots of data and subsequently mine the data for new information about the system studied. This represents a change toward data-drive research (see also discussion in Chapter 1).

When the samples are ready and the analytical protocol selected, then the analytical instruments and methodology have to be prepared and validated. A few of the choices and procedures used to get an analytical system ready are described in the following sections to give a rough idea about the typical work-flow used in metabolome analysis.

4.7.1 Separation by Chromatography

Chromatography is applied, as described earlier, if we need to separate compounds in the sample before detection. The first decision is to choose between gas or liquid chromatography:

Gas chromatography is chosen for volatile samples or when the expected compounds can be easily made volatile by derivatization, and high separation power is needed. Also, GC combined with EI–MS is well suited for compound identification and quantification.

Liquid chromatography is chosen for all other compounds, thus for nonvolatile compounds, complex extracts, where derivatization cannot be used and where a multitude of detectors will be an advantage (ESI–MS, UV, fluorescence, electrochemical, NMR, and so forth).

When the chromatographic principle has been selected, it is time to select a column and the analytical conditions needed. Sometimes these choices are driven by what is available, which, of course, is not optimal, and laboratories planning to do comprehensive metabolome analyses need to have a fairly wide selection of columns available. Although the overall strategy and goals are not that different when developing methods based on either GC or LC, there are significant differences in the practical implementation as illustrated below. In both cases the overall goal is that the compounds of interest are well separated in narrow sharp peaks in the shortest possible time (and, of course, the method should be reliable, simple, and stable).

In *gas chromatography*, the selection of a column is rather simple as only a few phases are used although there are differences in dimensions (diameter, length, and film thickness). Most of the problems in metabolomics are solved on weak to moderately polar columns, e.g., the 5% methyl-silicone phase or the 17% cyanopropyl-methyl-silicone phase both of which come in many variations in terms of cross-linking and deactivation. Specialty phases, e.g., chiral phases based on cyclodextrins may be an advantage in some cases. As discussed previously, injection into GC is nearly always on the basis of split or splitless injection depending on the sample concentration and the solvent used. The majority of the problems encountered in GC and GC–MS can be attributed to the injection, and it is worthwhile to be careful when selecting the setup and the running conditions. In general, split injection is simpler and more tolerant to matrix components (nonvolatile material), but splitless injection can produce really fine chromatography if conditions that facilitate solvent effects are used (remember to insert a retention gab—a length of deactivated fused silica tube similar to the column—between the injector and the column). Normally, the gas flow is optimized for the best injection/separation and should always be checked, e.g., by injection of methane. The oven program is generally used to optimize the separation time and to get narrow peaks. Samples are typically injected at low temperatures (solvent effects require temperature around 20°C below the boiling point of the solvent at column pressure), and then the temperature is increased to elute compounds having higher boiling points. Optimal separation power and retention-time stability is often found in the 2–4 degree per minute range. Please note that complex or very rapid temperature gradient (>10–20 degree per minute) can make the methods and retention times unstable, as it is impossible to reach thermal equilibrium in the column even in the best ovens as the heat is transferred by air which is impossible to reproduce stable over time on different instruments. The column eluent is normally eluted directly into the EI ion source of the mass spectrometer.

In *liquid chromatography*, there are far more options to choose from when planning the analytical procedure. First of all, there are several separation principles that can be used: ion chromatography, distribution chromatography (reversed phase chromatography), adsorption chromatography, size exclusion chromatography, etc. These basic principles can even be combined. Furthermore, liquid chromatography can be done from nanoscale (using nanoliters per miute flow injecting nanoliter samples) to process scale (using liters per minute injecting liters (kg) of samples) and with a multitude of detectors. For simplicity, only analytical distribution chromatography using reversed phase columns is discussed here as it is one of the most important techniques in metabolomics, but the other techniques are equally important in biotechnology. As discussed previously, reversed phase chromatography is based on an apolar stationary phase with the separation done by polar solvent (gradient). Numerous columns are available for reversed phase chromatography, and they come in many different designs, sizes, types of packaging material, and stationary phases. The most popular packaging material is porous spherical silica particles in the 2–10 μm range and coated with the stationary phase, but many other materials based on, e.g., polymers and monolithic structures are available. Particularly, columns based on silica particles coated with octyldecyl chains (C-18 chains) are very versatile and are widely used. However, a C-18 column is not just a C-18 column. Besides the size of the column (diameter and length), the performance is governed by differences in the silica particles (e.g., size and form, pore size, and volume), amount of phase bound to the surface, and the endcapping used to deactivate the uncoated silica surface. There can be significant differences in the selectivity between two columns that on paper may look similar—changing to another brand of column can sometimes help to solve a difficult separation problem. Having chosen a column, the next step is to select a mobile phase that matches the column and has the required selectivity to separate compounds of interest. In reversed-phase chromatography, the mobile phase is nearly always based on water as the polar component, and an organic solvent normally acetonitrile, methanol, or 2-propanol as the apolar component (the "strong eluent"). These can be used in mixtures, and modifiers are commonly added to the solvents, e.g., phosphoric buffers, trifluoric-acetic acid, formic acid, acetic acid, ammonia, and their salts. To control the selectivity, the solvent composition is changed during the run in a gradient, starting with the weakest eluting solvent normally called A (the one with the lowest elution power normally with a high content of water) and slowly changing to a stronger organic solvent called B. Complex elution patterns with more than two solvents can be used to solve complex separations. The mobile phase has to be chosen to match the detectors; thus UV transparent solvents are necessary for UV-detector, and volatile and electrospray compatible modifiers are needed for LC–MS, see Section 4.5.3. The latter excludes, in general, phosphoric buffers and also higher concentration of other volatile buffers, in particular, the use of the strong acid trifluoric-acetic acid with LC–MS. Running analysis by liquid chromatography is, in general, not that difficult when a suitable separation system has to be chosen, if adequate consideration is given to the samples and operation of the instrument:

- The samples have to be free of particles including crystals of sample components, and the sample solvent should be completely mixable with the solvent in the column at the time of injection. For good separation of early eluting components, the sample should be dissolved in the mobile phase used at the start of the run.
- The eluents and modifiers should be high-grade chemicals, free of particles as these may block tubing and columns, and also free of contaminants as these will give a high back-ground in the analysis that may blur the analysis or even obscure compounds of interest.
- In gradient analysis, adequate time should be allowed for the HPLC system and column to reach the starting conditions and equilibrate before the next sample is injected. The volume of a typical column may be 2 ml and if operated at 0.3 ml/min, it takes several minutes to flush a column. Also, remember to consider the volume in the pump and injector.
- The pluming of the HPLC-system should be done with respect to the flow rate used; thus narrow bore tubing and dead-volume should be used between the injector, and the detector should be minimized.

- The flow rate and the maximal injection volume should be matched to the column, e.g., a 2 mm internal diameter column is typically operated at flow rates around 0.3 ml/min, and this allows injection of up to 3–5 µl before the separation efficiency deteriorates (if late eluting compounds are of primary interest, the injection volume can be increased).

There are several technical issues that need to be checked and controlled to get a good and reliable HPLC method running, but it is outside the scope of this book, but guidelines can be found in most analytical textbooks. However, an operator of an HPLC system should always check for (excluding the detector) leaks, pulsation in flow and pressure, pressure limits, tube diameter, wear of seals, injector wash, and sample carryover. Most modern HPLCs are very reliable and easy to handle if the basic rules described above are combined with common sense.

4.7.2 Mass Spectrometry

As with all modern instruments, developments in electronics and computers have resulted in very high-performance mass spectrometers that are relatively easier to operate. In MS, the vacuum system is one of the critical parts, and should carefully be operated and maintained according to the instructions from the manufacture. As long as the vacuum is maintained, the mass spectrometer is quite robust, but it may give poor results if not operated correctly.

The first step is to get a good tuning, that is, to get a narrow well-focused ion beam through the instrument. This is usually done by leaking or infusing a reference compound into the ion source, thereby obtaining a beam of well-known ions. Lenses and parameters are then adjusted to optimize the beam width and the intensity either automatically or manually. In most cases, a set of criterion has to be met before the tuning is accepted. Next, a reference compound giving a series of different ions is analyzed to produce a spectrum used to calibrate the mass scale—the obtained spectrum is compared with a calculated reference spectrum, and a calibration function is calculated. In most instruments, the tuning and calibration is quite stable but drifts and changes in electronics, temperature, and contamination will require that the instrument is tuned and calibrated regularly. Also, high resolution and accurate mass determination require frequent tuning and calibration.

GC–MS is generally easy, and there are only a few parameters to consider in the mass spectrometer. The ion source conditions are nearly always the same, thus electron impact ionization at 70 eV and the source temperature should be chosen so that build-up of contaminants is minimized. It is important that the scan rate match the peak width of the chromatography and, of course, that the mass range is selected to cover the expected ions. In general, at least 5–10 spectra are required to get a good detection of a chromatographic peak, but more spectra may be needed for quantification and for efficient use of deconvolution (see Section 4.7 and Chapter 5). As the peak width in a good GC can be less than 2 s, a high scan rate is normally required.

Liquid chromatography with electrospray MS is almost becoming a routine technique like GC–MS. As described above, the instrument needs to be tuned and calibrated, which is done on suitable mixture of reference compounds. Then the instrument is just like a HPLC. However, the eluents and modifiers have to be volatile as they are to be evaporated in the ion source; and the source has to be able to accommodate the flow rate (typically below 0.5 ml/min, often optimal around 50 µl/min); in addition, the solvent composition has to allow ionization by electrospray; thus the ion strength, surface potential, and so forth have to be in a suitable range as discussed in Section 4.5.3. Most efforts in optimizations of ESI LC–MS are related to getting a suitable solvent composition that will not only give a stable spray but also facilitate efficient ionization of the analytes with minimal matrix effects. The spray stability depends on the solvents, on the gas flow rate, on the temperature, and on the geometry of the source, whereas the ionization efficiency depends on the chemistry, which has to be optimized together with the separation.

4.7.3 General Analytical Considerations

Analytical chemistry is as much a science as a craft. In the case of metabolome analysis, we generally start with a complex problem, namely very complex samples, and we may not know exactly what to look for. Therefore, it is important to plan the overall strategy carefully and remember that the chosen strategy will influence the results and can be as important as the actual analytical protocol. In general, lower concentration samples often produce superior results as most analytical methods perform better around 10–50 times the detection limit than near saturation. It is often better to start planning analyses by careful consideration of what kind of results are needed, and how they are going to be used/processed. However, in many situations it is more of a question as to what can be measured by the methods available, and which samples can be obtained, and so forth. In these situations, one should study the application range for the methodology carefully before venturing into a large analytical project.

No matter what analytical method and strategy is planned, it is important to test and secure the analytical system. It generally gives higher efficiency when a quality control system is implemented. Such a system is normally based on systematic analysis of quality control samples, analyzed and evaluated regularly. These samples can be authentic samples that can be reproduced, or they can be synthetic samples designed to demonstrate specific performance parameters. In any event, standards and blanks should always be included and regularly evaluated. A complete scheme for quality control should be a part of all method development projects in metabolomics as most data processing approaches, as discussed in Chapter 5, rely on results that can be compared more or less directly (see Chapter 5).

4.8 DATA EVALUATION

4.8.1 Structure of Data

The data produced in metabolome analyses can roughly be grouped into two categories: (i) *spectral data* from, e.g., mass spectrometers and UV photo spectrometers and (ii) *spectral data with a time dimension* from the preceding

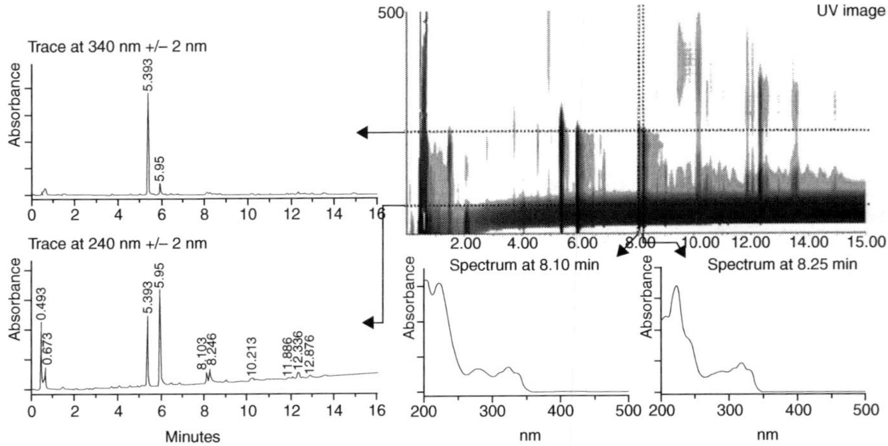

Figure 4.28 Structure of data from HPLC analysis with UV-detection. Chromatograms extracted at different wavelengths can have quite different appearance and can be efficient tools to find specific metabolites. For quantification it is crucial that the same wavelength is used a specific peak for all samples. At each time point a UV-spectrum can be extracted that may give structural information. The complete data file can be considered as an image of the sample as shown in gray-scale. (From analysis of a crude extract of the fungus *Penicillium freii* in a lab culture identical to Figure 4.29.)

compound separation technique, e.g., gas or liquid chromatography. Remember that chromatography in itself is a separation technique, thus spectrometry (or the other chromatographic detectors) is used to detect the result of the separation. The structure of results from liquid chromatography with UV-spectral detection is illustrated in Figure 4.28 and with mass spectrometric (ESI) detection in Figure 4.29.

The structure of GC–MS data is quite similar to that of LC–MS. In both cases, spectra have been collected at regular intervals, matched to the peak width of the chromatographic separation. Therefore, a spectrum has been recorded at each point in the chromatogram. On the contrary, a chromatogram is a plot of specific values taken from each spectrum and plotted as a function of time, e.g., absorption at a specific wavelength or the abundance of a specific ion. The whole data file is a matrix, where spectral information span the *y*-direction and time the *x*-direction, and the individual measurements are written in each cell. This is visualized by the images in Figures 4.28 and 4.29 where a grey-scale has been used which illustrates the values measured at each point. From these data matrices, narrow spectral bands or narrow mass ranges can be extracted, producing highly selective chromatograms as illustrated in Figures 4.28 and 4.29. These selective traces are very useful in tracking specific compounds. UV chromatograms are nearly always plotted at a specific wavelength (with a specified window around), whereas mass chromatograms are normally plotted by summarizing all ions in each spectra and plot these sums as a function of time—the so-called total ion chromatogram (TIC). In case

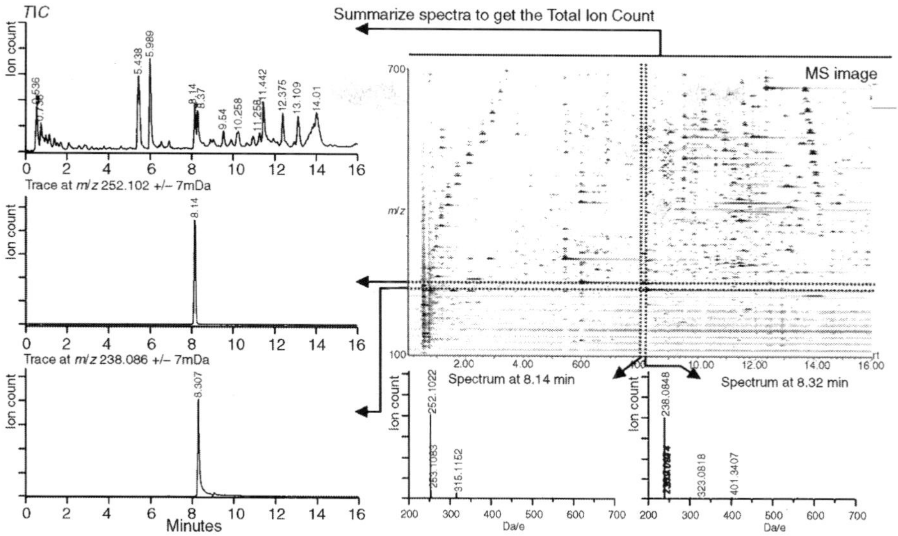

Figure 4.29 The structure of LC-MS data files. Mass spectra are collected at regular intervals, and the ion counts in each spectrum is summarized and plotted vs. time as a total ion chromatogram (TIC). A mass spectrum can be retrieved at each point a spectrum, producing mass and structure information. Very informative ion chromatograms can be extracted by plotting ion counts within a narrow mass range vs. time, here for the protonated mass of two well-known metabolites produced by *Penicillium freii*, See chapter 9. Similarly to the LC-UV data file, the full LC-MS file can be considered as an image of the sample. (From analysis of a crude extract of the fungus *Penicillium freii* in a lab culture identical to Figure 4.28.)

of LC–MS analysis, it is often more informative to plot the largest ion from each mass spectrum as a base peak chromatogram (BPC). The reason is that spectra from LC–MS analysis often contain a large number of small background ions that contribute significantly to the total sum of ions; therefore, the real contribution from smaller peaks might be hidden from the chromatogram and might blur peak detection.

Although chromatograms from gas and liquid chromatography are quite similar in structure, UV and mass spectra differ completely. UV spectra are continuous curves with maxima and minima whereas mass spectra consist of discrete values (masses), the latter is discussed in more details in Section 4.7.2 below. UV spectra are normally sampled at regular wavelength interval (e.g., 2 nm interval) with a spectral resolution set by a slit in the detector. Hence, the spectra will be aligned and will form a regular data matrix that also can be viewed as several hundred chromatograms recorded in parallel as illustrated in Figure 4.28. Mass spectra are stored in two ways as discussed in Section 4.5.7, either as continuum spectra where all data points are stored as recorded (the most raw data format) or as centroid data where the spectra are reduced to discrete mass—intensity pairs of the ions recorded in each spectrum—the latter is commonly used as it generates significantly smaller

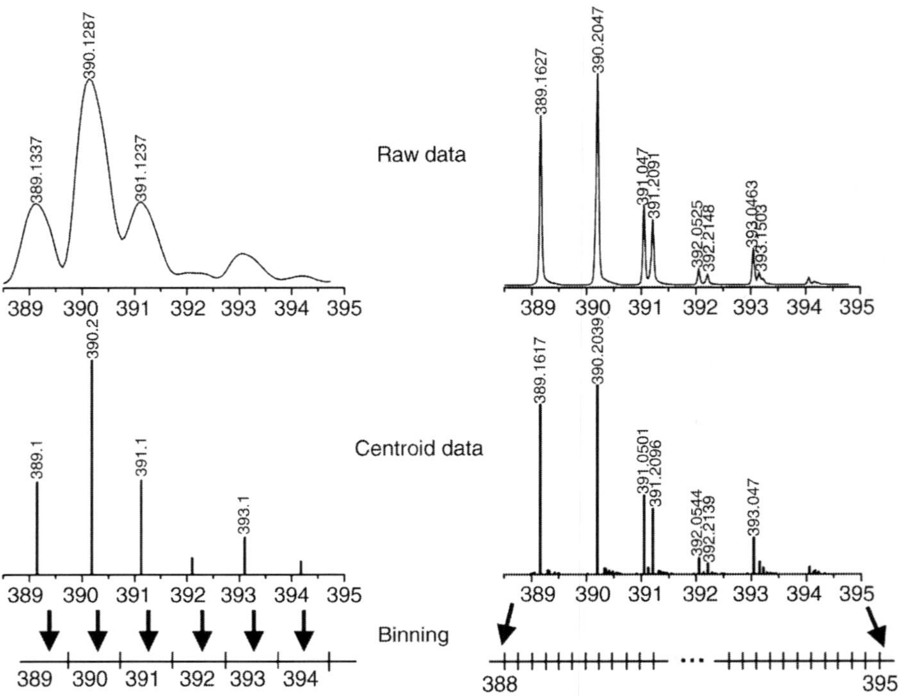

Figure 4.30 To use chemometric processing of mass spectra the variables, thus the masses need to be aligned in a grid like structure as variables. While, it is easy to design a grid for nominal mass spectra, as shown to the left, using each nominal mass as a variable, it is much more complex for high-resolution data. High-resolution data have the ions placed on a continuous scale, hence designing a grid structure for variables requires a decision of width and position of the bins matched to the resolution (or the use).

files. The masses in centroid spectra are recorded on a continuous scale and can therefore not be aligned directly but have to be binned as illustrated in Figure 4.30 to get a regular data matrix. In cases of nominal data from, e.g., quadrupole mass spectrometers binning is quite easy whereas it is not so easy for high-resolution data without loss of information. If the goal is to find specific compound producing ions of known masses, extraction of narrow ion traces around these protonated masses as illustrated in Figure 4.29 is very efficient, but more automated data processing as discussed in Section 4.7.3 and Chapter 5 normally require a regular data matrix with aligned spectra.

4.8.2 The Chromatographic Separation

It is always important to actually look at the data before more extensive data processing is applied. First of all, the standard and reference samples have to be evaluated to ensure that the key factors are as expected, e.g., peak shape, intensity, and retention time. Small variations have to be expected but they need to be small and controllable over time. Then, the real samples have to be evaluated by assessing peak shape, possible overloading, and other phenomena that deteriorate the separation efficiency. Finally, the background has to be studied to eliminate peaks from known or possible contaminants and other known defects. The latter process can be quite difficult as metabolite extracts usually result in very complex samples with many unknown peaks particularly in metabolite profiling and fingerprint analysis.

All peaks in a chromatogram may represent one or more compounds, and the latter is often the case in metabolite profiling analysis by liquid chromatography. Sometimes the number of compounds in a peak and the peak purity can be judged from evaluation of the spectra collected across the peak.

When the peaks in the chromatogram have been pre-evaluated, one may proceed to find the peaks of interest, that is, the peaks that contain relevant metabolic information or target compound information, and extract this information for further data processing.

However, it is possible to analyze the complete chromatographic data matrices directly by viewing them as images of the sample using advanced chemometric data processing as discussed in Chapter 5, but to do so, it is of utmost importance that the analytical variation is minimized and reproducibility is ensured.

4.8.3 Mass Spectral Data

In MS, the mass-to-charge ratio is determined for ions produced from sample components. Biomolecules, as those encountered in metabolome analysis, are composed of a relatively fewer elements, the most important of which are listed in Table 4.1. All

TABLE 4.1 Common Bioelements and their Isotopes Relevant for Mass Spectrometry.

Element	Isotope	Abundance (%)	Mass based on the ^{12}C standard
H, hydrogen	^{1}H	99.985	1.007825
C, carbon	^{12}C	98.93	12.000000
	^{13}C	1.07	13.003355
N, nitrogen	^{14}N	99.632	14.003074
	^{15}N	0.368	15.000109
O, oxygen	^{16}O	99.757	15.994915
	^{17}O	0.038	16.999132
	^{18}O	0.205	17.999160
P, phosphorus	^{31}P	100	30.973762
S, sulfur	^{32}S	94.93	31.972071
	^{33}S	0.76	32.971459
	^{34}S	4.29	33.967867
Cl, chlorine	^{35}Cl	75.78	34.968853
	^{37}Cl	24.22	36.965903

analytical mass spectrometers used for metabolome analysis can separate ions to at least nominal mass; some far better than that, will separate biomolecules into their isotopic composition. Therefore, the monoisotopic mass of compounds calculated from the most abundant element is always used in MS, never the average mass as used for chemical calculations (and printed on chemicals).

Looking at the elements in Table 4.1, it can be seen that the core element carbon has a valence of four and therefore forms four bonds; similarly nitrogen will form three bonds, and oxygen and sulphur two bonds. Hydrogen and chlorine can be considered as terminating elements. As nitrogen is the only element with an odd valence (three), a compound with an odd number of nitrogen (1,3,5,...) will have an odd molecular mass. From this rule, it is possible to deduce from the molecular mass if a compound contains an odd number of nitrogen (at least for low molecular mass compounds). In electrospray, these compounds will have an even ion mass as they are either protonated (+1) or sodiated (+23) in positive electrospray or deprotonated in negative electrospray, but be aware that ionizing by the ammonia ion (+14) or clusters with nitrogen-containing compounds, e.g., acetonitril (+41) from the solvent will change the mass from even to odd (or the other way round).

About 1.1% of all carbon is the ^{13}C isotope; therefore, a distinct isotopic pattern will be seen from all organic molecules in the mass spectra. The intensity ratio between the ion composed from purely ^{12}C carbon and the ones containing one ^{13}C atom (thus with a mass one higher) can be used to predict the elementary composition. However, isotopes from other elements, e.g., oxygen, nitrogen, and sulphur have to be taken into account to get a precise estimate, see McLafferty (1993) for further details. Also note that chlorine produces a distinct isotopic pattern with the m and $m + 2$ ions in a ratio of approximately 3:1.

EI mass spectra as collected from GC–MS are, in general, rich in compound-specific fragment ions that are very useful in identifying the structure. Several libraries of EI-mass spectra are available (NIST, WILEY, MSRI, see their websites) and these are very helpful, but do require some manual evaluation and common sense, see McLafferty (1993).

As discussed in Section 4.5.3, *ESI mass spectra* will show relatively fewer ions from the gentle ionization in the electrospray process. In general, small molecules will be protonated or sodiated in positive ESI, i.e., as $M + H^+$ or $M + Na^+$ ions and deprotonated in negative mode $[M—H]^-$, where M means a monoisotopic molecule. Table 4.2 summarizes some of the most common ions to look for in an ESI mass spectrum.

Electrospray MS can be used to analyze complex samples without a separation step taking advantage of the limited fragmentation. The resulting spectrum can be seen as a mass profile of the sample. However, dealing with these mass profiles requires some consideration as matrix effects (see Section 4.5.3) can seriously disturb the picture, and also results in clusters between different sample molecules. Despite these problems, direct infusion of crude samples has been demonstrated to be an efficient tool in metabolite profiling and taxonomy. This is illustrated in a case story in the second part of this book.

TABLE 4.2 Major Ions and Clusters Seen in Liquid Chromatography Electrospray Ionization Mass Spectrometry.

	Positive ESI		Negative ESI	
	Structure	Change nominal Mass change (Da/e)	Structure	Change nominal Mass change (Da/e)
Adducts	$[M+H]^+$	+1	$[M\text{-}H]^-$	−1
	$[M+NH_4]^+$	+14	$[M+Cl]^-$	+35
	$[M+H_2O+H]^+$	+19	$[M+CHOO]^-$	+45
	$[M+Na]^+$	+23	$[M+CH_3COO]^-$	+59
	$[M+CH_3CN+H]^+$	+42	$[M+HSO_4]^-$	+97
	$[M+CH_3CN+Na]^+$	+64	$[M+H_2PO_4]^-$	+97
	$[M\text{-}H+2Na]^+$	+45		
	$[M\text{-}(n-1)H+n\mathrm{Na}]^+$	$+23n-1$		
Fragments	$[M\text{-}H_2O+H]^+$	−17	$[M\text{-}H_2O\text{-}H]^-$	−19
	$[M\text{-}H_2O+Na]^+$	+5	$[M\text{-}H_3PO_4\text{-}H]^-$	−98
	$[M\text{-}CO_2+H]^+$	−27		
	$[M\text{-}CO_2+Na]^+$	−5		
Multimers	$[2M+H]^+$	$2*m+1$	$[2M\text{-}H]^-$	$2*m-1$
	$[2M+H_2O+H]^+$	$2*m+19$		
	$[2M+NH_4]^+$	$2*m+14$		
	$[2M+Na]^+$	$2*m+23$		

M is an ion with the mass m. In general, clusters with solvent molecules should be expected. For larger molecules (>about 1000 Da) doubly charged ions have to be taken into account, seen at half their molecular mass, thus at $m/2$. Also, exchange reactions can happen, e.g., a proton being replaced by a sodium atom.

4.8.4 Exporting Data for Processing

Before analytical data can be used for more advanced metabolome analysis, the raw data has to be either converted to a general readable format and organized or preprocessed into specific results.

Direct processing by modern chemometrics of the raw data has the advantage of using all information in the data files, and one does not depend on what the analyst chooses to include or not to include. In other words, these techniques have the advantage of being completely unbiased in terms of data processing. To process the raw data files directly, these data files have to be transformed from their native instrument format to a format that is readable by the data processing software. This is often a major obstacle for the development algorithms that use raw files for advanced data processing, as neither the instrument manufactures rarely includes software that can efficiently export data files to an open format (e.g., NetCDF) nor are they willing to reveal the binary structure of the files. However, more generalized processing features are constantly added to the instrument software packages and also some third party software manufactures are launching chemometrics software among other metabolomics that can work directly for a multitude of instrument data types.

The more classical approach to extract data from chromatographic analysis is the detection of peaks and calculation of peak area. The result is a compound or peak table with retention times used for further analysis. To do so, it is necessary to decide what chromatograms to use as illustrated in Figures 4.28 and 4.29. Quite different results will be obtained from peak integration in the 220-nm chromatogram and in the 400-nm chromatogram of Figure 4.28, and absolutely no similarity in peak detection will be obtained by integration of the two ion traces in Figure 4.29. However, these different chromatographic traces can be used for both compound identification and the identification of retention times and give much more reliable integration. Most importantly, choosing the right traces can minimize the effects of overlapping chromatographic peaks. An extreme example is the two ion traces shown in Figure 4.29. These peaks cannot be distinguished in either the TIC or the BPC, whereas they are completely separated by the ion traces. All data for a specific metabolite have to be calculated from the same type of signal, i.e., from the same UV wavelength or mass trace to allow calculations, whereas data from different metabolites can be obtained from different traces. The result is, in general, a simple list of related metabolite (peak retention time)—peak area informations, ready for further processing. The disadvantage is that the user has to select what to include and not to include thereby creating a bias. On the contrary, the digestion and evaluation of the data remove a considerable amount of noise from the data and thus improve the information content.

Finally, as mentioned before, very large data sets are easily generated in metabolome analysis, and it is, therefore, crucial to plan ahead. A major investment is, in general, to put into the analysis, but poor data analysis may also waste good analytical results as well as waste the entire experiment.

4.9 BEYOND THE CORE METHODS

The focus in this chapter has been on introducing the basic and the widely used analytical methods used for metabolome analysis, but the chapter is by no means a complete or comprehensive description of the analytical techniques available today. The complexity of the metabolome is a thrilling challenge that requires all the ingenuity that can be mastered by the analytical chemists. As discussed in Chapter 2 and in the introduction to this chapter, the metabolome is very complex and cannot be measured by a single analytical technique. Therefore, it is necessary to consider multiple analytical methodologies for comprehensive metabolome studies, and in most cases to use several analytical approaches. Metabolomics, in many ways drives developments in analytical chemistry but is, at the same time, also a driving force behind developments in analytical chemistry. Chromatography and MS will by no doubt continue to play key roles in metabolome analysis for a long time to come, but other techniques and new analytical instrumentations and approaches will expand what can be achieved in metabolome analysis. A few examples of newer analytical techniques used to analyze the metabolome are briefly introduced in the following sections. Very illustrative examples of the state-of-the-art analytical approaches used in metabolome analysis can be found in the very first issue of the journal "Metabolomics," see the literature list below; for further examples the reader is referred to the analytical and metabolomics literature.

4.9.1 Developments in Chromatography

Although chromatography has been around for more than a century and column chromatography for about half a century, new columns, new chemistry, new materials as well as new instrumentations are continuously introduced for both gas and liquid chromatography. These developments together with advanced data processing (Chapter 5) have significantly improved the performance of modern chromatography. To get the latest updates on what is available in columns and instrumentation the reader is adviced to consult the catalogs from the different manufactures. Of the more recent developments in chromatography, two techniques relevant to metabolome analysis deserved to be mentioned here are as follows:

4.9.1.1 Multidimensional Chromatography. In multidimensional chromatography, the idea is to use two columns (GC or LC) with different selectivity in series. This can be done either off-line or in-line. The eluent from the first column (while peaks of interest elute) is transferred (injected) to a second column (GC or LC) with a different selectivity. This idea is not new, but has been automated more recently; therefore, it is much easily applied in metabolome analysis.

The most common multidimensional chromatography is to use an HLPC column for the first separation followed by a further separation by injection into a GC column (LC–GC) or into another HPLC column (LC–LC). A typical application of LC–LC or LC–GC is to concentrate compounds of interest while getting rid of interfering matrix components, which is widely used in analyses of complex samples. This is done by injecting the sample on the first column under conditions where all compounds of interest are retained on the column; all other compounds are then eluted to waste. When this is done, the solvent system is changed and compounds of interest are eluted to the second chromatographic system for the analytical separation. This is similar to the off-line sample preparation by SPE techniques as discussed in Chapter 3, but is done automatically by valve switching in a rather complex HPLC setup. The disadvantage is, besides the complex pluming, restriction on the solvents that can be used, as the solvents used to elute the compound from the first column will go through the second column also. To separate very complex mixtures, it is also possible to perform a full HPLC separation on the first column and then select peaks (usually what is eluting in a small time segment) and injecting these on a second column by automatic valve switching. These columns may be different and the analyses can be done using different solvent systems. Similarly, peaks eluting from an HPLC column may be fractionated and then injected in GC using normal split/splitless injection, but more efficiently injected directly by large-volume on-column injection. The disadvantage of peak selection techniques is that the peaks may elute in less than a minute from the first column, and therefore there is only one a minute to perform the separation on the second column if all the peaks from the first separation are to be analyzed on the second column. Alternatively, a multiple run setup can be used where the samples are injected several times and different parts of the first separation are transferred to the second column or the column flow can be stopped in the first column, until the second column is ready for the next peak. Either way, a considerable time is required for analysis of a sample.

More recently two-dimensional gas chromatography (often referred to as GC × GC) has been introduced where peaks eluting from one GC column are trapped and then injected on a second GC column, see Górecki et al. (2004). In GC × GC, a small time-slice of the compounds eluting from the first "normal" gas chromatographic column are collected in a cryo-trap and then injected into a new second GC column with different selectivity by rapid heating of the trap. The two columns are independent of each other and typical with different phases. GC × GC can be performed in various ways: (i) As a heart-cut technique where one peak (or a few well-separated peaks) is trapped and then reinjected. Here, both columns are optimized for separation efficiency, but a second heart-cut cannot be injected on the second column before all compounds from the first separation have eluted. Heart-cut intervals therefore have to match the run-time for the second column. (ii) Everything that elutes from the first column is sampled in regular time-slices and reinjected on the second column. Typically, lower separation efficiency is used in the first columns to allow larger time-slices (in the 3–20 s range) to be transferred to the second column, thereby allowing a longer run-time on the second column. The second column is usually done as high-speed gas chromatography with a total run-time in the range of a few seconds. By the use of proper timing and columns with different selectivity, one can obtain amazing separation efficiency. GC × GC analyses can, of course, be combined with MS delivering true 3-dimensional data where the two chromatographic separations give the first two dimensions and a mass scale adds the third dimension. However, this requires very rapid scanning, see the very illustrative example by Welthagen et al. (2005).

4.9.1.2 Ultra High Performance Liquid Chromatography (UPLC). UPLC is the result of technical developments more than of new analytical principles. As discussed previously, longer narrow bore columns packed with the smallest possible particles will give the highest separation efficiency. The smallest particle currently used in normal analytical HPLC columns is around 3 μm, and these are packed in columns with a diameter in the range from 1–4 mm to about 30 cm in length. This will give a back-pressure up to around 40 MPa, which is the upper limit of most HPLC pumps. To increase separation efficiency in HPLC, long narrow bore columns packed with very small particles (1–2 μm) have recently been introduced. These columns will have a very high back-pressure that usually require reduced column flow, thus operated using micro or nanoflow techniques. Quite recently, HPLC systems capable of working at very high pressures (>300–400 MPa) have become available, along with ultra high-pressure columns, packed with 1–2 μm particle. The results are as predicted—an amazing separation efficiency that approaches what is seen on a good GC column. However, these very high-pressure chromatographs are technically more sensitive systems that require careful operation and maintenance compared with what is required for classical HPLC. UPLC is fully compatible with MS and follows the principle and theory as "classical" liquid chromatography.

4.9.2 Capillary Electrophoresis

CE is a separation technique that is comparable to chromatography, but it is based on entirely different separation principles. In the simplest form, the CE separation system is established by placing the ends of a fused silica capillary (30–200 µm inner diameter and 30–100 cm long) in a vial containing buffer solutions. A high voltage, in the range of 10–50 kV, is applied across the capillary by placing an electrode in each buffer vial. A CE system coupled with a mass spectrometer as illustrated in Figure 4.31a, however with the outlet, is connected to an MS interface rather than a buffer vial.

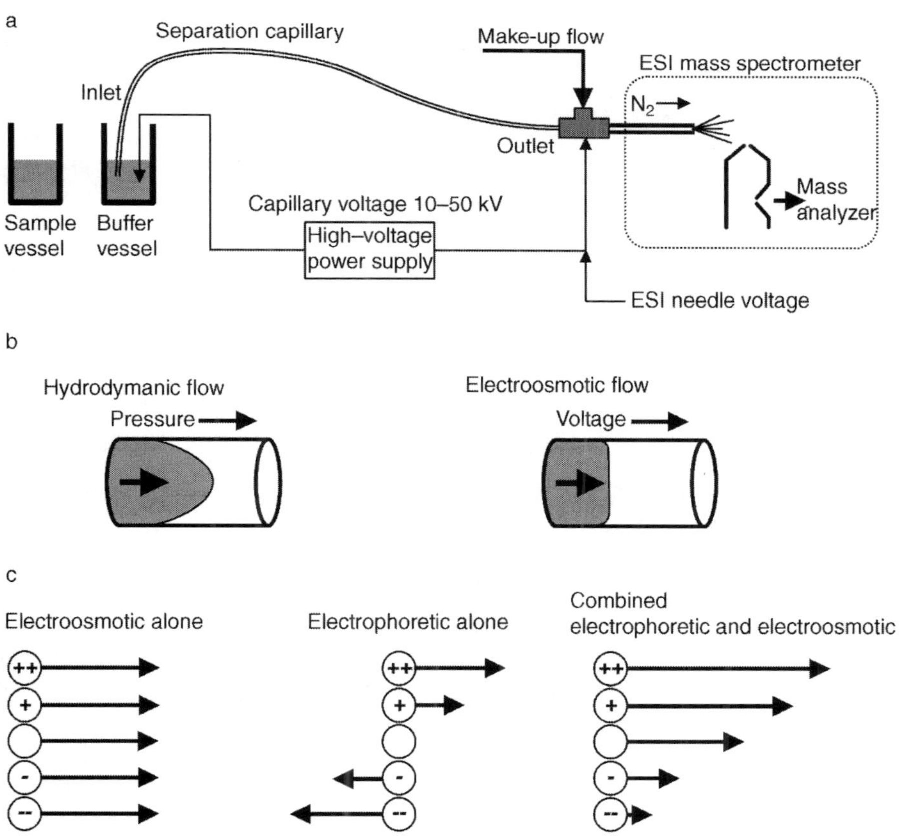

Figure 4.31 (a) Overview of a capillary electrophoresis mass spectrometry setup, see text for details. (b) The flow profiles from a normal hydrodynamic laminar flow and from electroosmotic flow the latter showing a very sharp profile giving much flow related dispersion. (c) Migration of ions in CZE—the effect due to the larger electroosmotic flow—to the electric potential and the combined effect.

The voltage leads to migration of buffer ions through the capillary and to a charging of the capillary wall. By polarization of solvent molecules, the charged wall will lead to a solvent flow through the capillary, called an electroosmotic flow. The flow profile of the electroosmotic flow is illustrated in Figure 4.31b. Compared with the laminar flow profile seen in an HPLC system, the electroosmotic flow profile gives much less dispersion than seen in HPLC, a prerequisite for high separation efficiency. The sample is introduced into the capillary by placing the inlet end into a sample vial and injecting by applying either a pressure difference or a voltage across the capillary. The separation of the analytes is, in the simple form, achieved by the small difference in their electrophoretic mobility combined with their migration properties due to the electroosmotic flow. When the voltage is switched on, the ions start migrating through the capillary because of both the electroosmotic flow and the potential. Figure 4.31c shows that if the electroosmotic flow was the only mechanism, all analytes will migrate at the same speed as the electroosmotic flow; if we have the electrophoresis alone, the anions will migrate to the cathode and vice versa; the greater the mobility the faster the migration. As the electroosmotic flow is often larger than the electrophoretic velocity, both cations and anions will migrate in the same direction, e.g., toward the anode, but the cations will migrate faster than the electroosmotic flow (thus reach the outlet first), and anions will migrate slower, see Figure 4.31c. The neutral molecules will follow the electroosmotic flow and mark the boundary between anions and cations; however, neutral analytes are not separated. This technique is generally called capillary zone electrophoresis (CZE). As separation of neutral analytes cannot be done by CZE, addition of a detergent to the buffer system (e.g., sodium dodecyl sulfate, SDS) allows the formation of micelles with the neutral analytes. The micelles can then be separated as described above. This is often referred to as micellar electrokinetic capillary chromatography. By using chiral detergents, it is even possible to achieve chiral separation. Besides the use of detergents, CE can be performed in many other variations using different buffer systems, additives, wall-coated capillaries similar to those used in GC, gel-filled capillaries, and so forth. The results obtained from CE look quite similar to those from chromatography and are called electropherograms—a well-optimized CE system can deliver amazing separation efficiency, reaching more than 10^5 theoretical plates. Many primary metabolites of importance for metabolomics are well suited for analysis by CE as they are easily ionizable in a buffer and therefore can be separated by CZE. Illustrative example can be found in Ishii et al. (2005). Another advantage is that CE only requires small amounts of sample (in nanoliter range) delivering a fascinating absolute sensitivity whereas the concentration sensitivity is in the same range as HPLC. CE is mostly used with UV and detection by laser-induced fluorescence, but can equally well be coupled with a mass spectrometer as illustrated in Figure 4.31a. However, the CE–MS coupling is not technically straightforward as both CE and the electrospray source require high voltages, and the solvent flow through the capillary (the electroosmotic flow) is too low to form a stable electrospray. Therefore, in most CE-electrospray interfaces, a makeup flow is added at the capillary exit to form a liquid junction between the CE and the mass spectrometer. Furthermore, as discussed in Sections 4.5.3, the use of ESI limits the use of buffers and ions in solvents and the use of detergents may seriously affect the ionization of analytes due to matrix effects. Fortunately, the use of makeup flow can be used to limit these effects. Unfortunately, CE methods are not so easy to develop, and it requires considerable experiences to develop and optimize CE and CE–MS methods.

4.9.3 Tandem MS and Advanced Scanning Techniques

As ESI does not produce many fragment ions with structural information, a range of MS techniques have been developed where fragmentation is induced by collision with an inert gas. This can be done in ion-trap instruments as described in Section 4.5.5 or in so-called tandem mass spectrometers. All the mass analyzers described in Section 4.5 can be combined to a tandem mass spectrometer, where two mass analyzers are combined with a collision cell in between. The collision cell is, in most instruments, a small quadrupole (or hexapole) filled with an inert gas (nitrogen or argon) and used in RF mode as discussed in Section 4.5.4. In the collision cell (often referred to by a small q, whereas separating quadrupoles are referred to by Q), ions are accelerated to kinetic energies in the range from 10 to 50 eV leading to fragmentation on impact with gas molecules. The most popular combinations are the triple quadrupole mass spectrometer (QqQ) with two normal quadrupoles mass analyzers around the collision cell and the quadrupole TOF (QqTOF or QTOF) mass spectrometer. Many other combinations are in use: ion-trap-time-of-flight (trap-TOF), two TOF analyzers (TOF–TOF), quadrupole-ion-trap (QqTrap), and an ion-trap combined with a Fourier-transform ion cyclotron resonance mass analyzer (the latter also called FT–ICR–MS or just FT–MS which is a ultrahigh resolution/accuracy mass analyzer). All these MS–MS combinations can, of course, be used with chromatography and CE as any other MS technique described in Section 4.5. Depending on configuration, MS–MS instruments can be used for more advanced analysis either for structure elucidation or for obtaining very high specificity and sensitivity in target analysis.

In analytical chemistry, MS-MS instruments are used in three different analytical modes where the mass analyzers MS1 and MS2 are used independently, as illustrated in Figure 4.32 for daughter scans, multiple reaction (neutral loss) monitoring, and parent scans.

Daughter scans (Figure 4.32a) are typically used to identify ions and interpretation of mass spectra. Here the first mass analyzer MS1 is used to select a single ion, which is further fragmented in the collision cell. The second mass analyzer is then used to record a mass spectrum of the fragments obtained. A daughter spectrum will show how a specific ion will fragment and this pattern can be used to elucidate the structure if unknown, or to find the relations between ions in a normal spectrum; thus, select which ions are fragmented from the selected specific ion. All MS–MS combinations can be used for daughter scans, including the ion-trap analyzer alone, see Section 4.5.5; however, high mass accuracy may not be obtained on instruments that require internal mass calibration, e.g., TOF–MS as these are not transmitted through MS1.

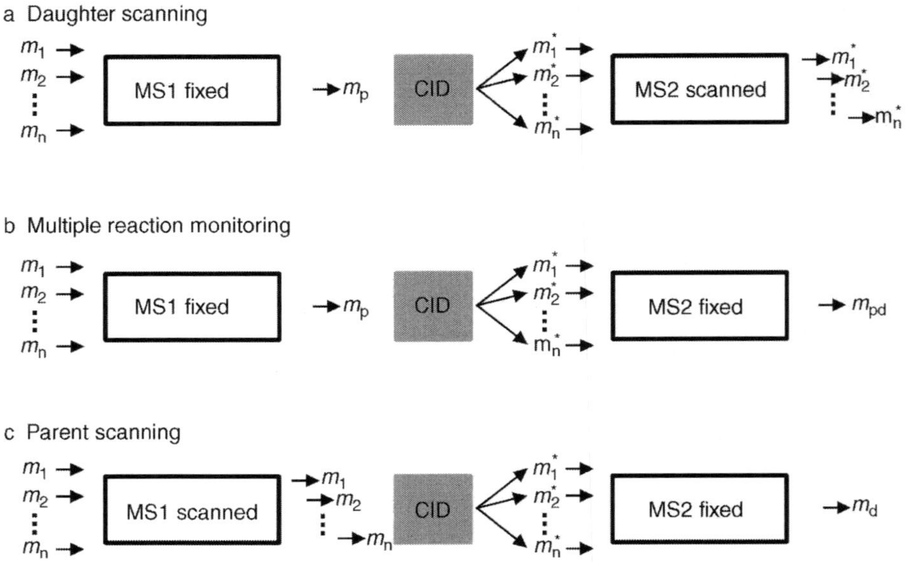

Figure 4.32 Scan techniques used for tandem mass spectrometry. (a) Daughter scanning typically used of structure elucidation and interpretation, (b) MRM scanning used for very selective analysis of target compounds, and (c) parent scanning used to find groups of related compounds with the same fragmentation.

Multiple reaction (neutral loss) monitoring (MRM-analysis, Figure 4.32b) is one of the most efficient techniques for very high selectivity target analysis; however, it can also be used for other purposes. Here, only these masses are allowed to pass MS1 as in daughter scan but only one of the fragments is allowed to pass MS2, thus both mass analyzers are fixed to only transmit-specific ions with a given difference. MRM corresponds to extracting single ion traces from a daughter scan analysis. The very high selectivity arises from the fact that we require that a specific ion m_d loose a specific neutral fragment to become m_{dp} that only a few compounds will do. Moreover, if this is combined with a required retention time, the specificity will be very high. MRM can be used to find all ions that loose a specific neutral fragment; it could be the loss of CO_2, by doing a linked scanning where MS1 and MS2 are scanned at the same rate, but with a specific mass difference (e.g., 44 Da). This technique is called neutral loss scanning. MRM and neutral loss scanning are most efficiently done on MS–MS configuration where both analyzers are scanned, typically a QqQ instrument. Other instruments (e.g., ion-traps, Q–TOF–MS or FT–MS) are of limited use for MRM and neutral loss analysis, as the second analyzers always collect full spectra, hence requiring full scan-time for each selected parent ion. MRM or neutral loss traces can be produced by extraction of single ion trace from these full daughter (MS2) spectra but at the cost of very slow scanning and using a lot of disk space.

Parent scanning is where MS1 scans normally, but only a selected ion fragment is allowed to pass the second MS2. This can be very useful in finding compounds that produce a characteristic fragment, like the McLafferty rearrangement ion at *m/z* 74 seen in EI spectra of methylated fatty acids. If we do a parent scanning GC–MS analysis of a methylated sample, then by the fragment ion of *m/z* 74 we can be able to find the fatty acids candidates (*m/z* 74 is a common rearrangement ion produced by many long-chained fatty acids). Parent scanning requires, as MRM/neutral loss scanning, that the MS2 analyzer is scanned, e.g., a triple quadrupole instrument (QqQ).

4.9.4 NMR Spectrometry

NMR spectroscopy is one of the most efficient techniques of measuring very specific molecular properties that can be used to elucidate the structure of the molecules. NMR measures the spin and magnetic moment properties of the nuclei in a molecule, and these properties depend on the environment of the nuclei experience. These properties can be measured in complex mixtures, using suitable conditions; therefore NMR have attached much attention for metabolome analysis.

Nuclei are rotating around an axis and thus have the property of spin; hence they will have angular momentum. The nuclei of most interest in biology are the hydrogen isotope ^{1}H (99.98% abundance), the carbon isotope ^{13}C (1.11% abundance), and the phosphor isotope ^{31}P (100% abundance). All these nuclei will have a spin quantum number of 1/2, thus can be in two spin states $-1/2$ and $+1/2$. Moreover, a spinning charge will create a magnetic field similar to that created when electrons flow through a wire, and as spin quantum numbers, will have two quantum magnetic states. This magnetic field is orientated along the spinning axis of the nucleus. If a nucleus is placed in a strong magnetic field, it will align itself with the external field in one of the two directions depending on the magnetic moment of the nucleus. The potential energy in a quantum state of $+1/2$ is lower than that in a quantum state of $-1/2$, thus nuclei in $+1/2$ normally predominate. However, the number of nuclei in each of the two states depends on the temperature. Transition between these two states can be brought about by absorption of energy that can be supplied by electromagnetic radiation, hence by a radio-frequency signal where the energy (frequency) is proportional to the magnetic field strength. Furthermore, it can be shown that the amount of energy absorb is proportional to the number of nuclei. A nucleus may be shielded by the environment of electrons, as these electrons also possess a magnetic moment, hence change the magnetic field sensed by the nucleus, and it may be affected by the magnetic moment of other nuclei in the neighborhood. The result is that the energy required to excite a specific nucleus depends on the local environment. An NMR spectrum is normally created by radiating the sample with a short pulse of high-energy radio frequencies (typically in the range 100–1000 MHz depending on the field strength) that excite all nuclei. Rather than measuring the absorption at each frequency, the energy emitted when the nuclei return to the low-energy state is measured as a *free induction decay* (FID) signal. By a Fourier transformation of the FID signal, the decay can be converted to a pattern of frequencies emitted representing the different energy emissions from the different nuclei when they return to the low-energy state. Usually, the scale is calibrated to the frequency of reference compounds and frequencies are converted to parts per million (ppm) of the radiation frequency to ease the comparing results between instruments. Therefore, an NMR spectrum is normally plotted as the ppm-value (often called chemical shift) vs. the intensity. The different nuclei ^{1}H, ^{13}C, or ^{31}P cannot be measured in the same spectrum, as they require significantly different frequencies, which usually require different instrument setup.

NMR is mostly used in structure elucidation of compounds where these compounds are dissolved in solvents that do not interfere with the NMR signals. In case of proton spectra, solvents without protons are preferred, e.g., deuterium-water (D_2O) or chloroform. The sample is placed in an NMR tube and then placed in the magnet. To ensure homogenous signals, the sample tubes rotate rapidly and the temperature is carefully controlled. However, it is also possible to record NMR spectra of complex crude samples thereby gaining knowledge of compound classes, and in some cases also about single compounds.

In the simple form, an NMR spectrum shows at which chemical shift the nuclei studied will absorb energy. The more shielded a nucleus is, the higher is the chemical shift; thus a proton will be found at low ppm if it is in simple hydrocarbon, and at much higher ppm if it is sitting on a benzene ring. In modern high-resolution NMR, it is possible to distinguish between very small differences. A signal from, e.g., a proton may be split into multiple signals by coupling with adjacent protons on neighboring carbon nuclei. This adds to the complexity but is also a tool to elucidate the environment of that particular proton.

When studying complex samples, it is possible to use the numerous different NMR techniques that have been developed during the last decade. These techniques allow selective decoupling of the signal from specific nuclei by radiating these nuclei with radio frequency energy that quenches their signal; thereby, a relation in complex spectra can be found. NMR allows pinpointing specific compounds, e.g., amino acids, some carbohydrates, and phosphor compounds (e.g., ATP) from their chemical shift values. These techniques are quite useful in metabolome analysis as NMR can give a sample profile in a relatively shorter time that allows the quantification of many important metabolites; see the illustrative example by Lenz et al. (2005). The disadvantage of NMR is that the sensitivity is much lower compared with MS, but as NMR is nondestructive, it is possible to collect sample scan over long time, thereby increasing the sensitivity. NMR can also be coupled with HPLC, but to record NMR spectra of the eluent, a stop-flow technique is often applied, stopping the pump to allow more time for NMR measurement.

4.10 FURTHER READING

Numerous textbooks are published each year giving anything from the basic introduction to advanced discussion of all analytical topics discussed in this chapter. The reader is advised to review libraries and bookshops for the latest new publications in analytical chemistry. The references selected below are all long-lasting key reference books in the various areas.

REFERENCES

Drozd J. 1981. *Chemical Derivatization in Gas Chromatography* (Journal of Chromatography library), Elsevier Science Ltd., ISBN: 0444419179, Burlington, MA, USA.

Giddings JC. 2002. *Dynamics of Chromatography: Principles and Theory*, CRC, ISBN: 0824712250, Danvers, MA, USA.

Górecki T, Harynuk J, Panic O. 2004. The evolution of comprehensive two-dimensional gas chromatography (GC × GC). J Sep Sci 27:359–379.

Grob, K. Jr. 1987. *On-Column Injection in Capillary Gas Chromatography: Basic Technique, Retention Gaps, Solvent Effects* (Chromatographic methods) (1st edition), Hüthig Verlag, ISBN: 3778515519, Weinheim, Germany.

Grob, K. Jr. 2001. *Split and Splitless Injection for Quantitative Gas Chromatography: Concepts, Processes, Practical Guidelines, Sources of Error* (4th edition), Wiley-VCH, ISBN: 3527298797, Weinheim, Germany.

Ishii N, Soga T, Tomita M. 2005. Metabolome analysis and metabolic simulation. Metabolomics 1:29–37.

Jönsson JA. 1987. *Chromatographic Theory and Basic Principles* (Chromatographic Science) CRC, ISBN: 0824776739, Danvers, MA, USA.

Lenz EM, Weeks JM, Lindon JC, Osborn D, Nicholson JK. 2005. Qualitative high field ^{1}H-NMR spectroscopy for characterization of endogenous metabolites in earthworms with biochemical biomarker potential. Metabolomics 1:123–136.

McLafferty FW. 1993. *Interpretation of Mass Spectra* (4th edition), University Science Books, ISBN: 0935702253, Berkeley, CA, USA.

Neue UD. 1997. *HPLC Columns: Theory, Technology, and Practice*, Wiley-VCH, ISBN: 0471190373, Weinheim, Germany.

Toyo'oka T. 1999. *Modern Derivatization Methods for Separation Science*, John Wiley & Sons, ISBN: 0471983640, New Jersey, NJ, USA.

Welthagen W, Shellie RA, Spranger J, Ristow M, Zimmermann R, Fiehn O. 2005. Comprehensive two-dimensional gas chromatography-time-of-flight mass spectrometry (GC × GC-TOF) for high-resolution metabolomics: Biomarker discovery on spleen tissue extracts of obese NZO compared to lean C57BL/6 mice. Metabolomics 1:65–7.

5

DATA ANALYSIS

BY MICHAEL A. E. HANSEN

This chapter will introduce the principles of some of the most commonly applied techniques used when analyzing metabolomics data. All of the methods described here can be used to analyze data obtained from analytical instrumentation described in Chapter 4. Irrespective of the analytical technique used, the analysis of the data is essentially performed in three stages. Initially, the raw data need to be preprocessed to convert them into a suitable form as described in Sections 5.1–5.6. Secondly, it may be useful to subject these modified data to data reduction so that only the most relevant input variables are used in the subsequent data analysis (Section 5.8). Finally, the objective of the last stage of the data analysis is to find patterns within the data, which give useful biological information that can be used to generate hypotheses that can be further tested and refined (Sections 5.9 and 5.10). The chapter is ended with a short introduction to different tools available for automation, library search, and data evaluation (Section 5.11).

5.1 ORGANIZING THE DATA

Once the data have been generated, the output has to be organized in a reasonable and intuitive structure. Fortunately, most of the software managing the instruments organizes data into a folder-structure where the raw data from each analysis of the individual samples are stored as subfolders within one single folder collecting all results for that run—a structure that can be adapted. Next, all relevant information or metadata we have about the samples and the experimental conditions has to be assembled into a table (Brown et al., 2005). This links each

Metabolome Analysis: An Introduction, by Silas G. Villas-Bôas, Ute Roessner, Michael A. E. Hansen, Jorn Smedsgaard and Jens Nielsen

of the raw data files to information available prior to the statistical analysis and may include information like: identifier (a unique label), strain/species/mutant, medium/carbon source, growth conditions, data location, date of experiment, experimenter, etc. All of which are metadata that may (or may not) play a role on the outcome of the analysis, and could be used either as direct input to the statistical analysis or as information to help us understand outliers. In the metabolomics society, standard definitions are being discussed (Jenkins et al, 2004, Jenkins et al, 2005), defining a minimum criterion of the types of information that has to follow data and several projects for the description of metabolomics experiments, and their results have been initiated, e.g., the ArMet project (http://www.armet.org).

Having prepared the information available, the next step is to get the data out of the data-files, which might be difficult for some types of raw-data. Fortunately, tools for extracting data from most instrumental software vendors exist as part of the programs. Often the converted data are converted into a nonproprietary format as, e.g., NetCDF (http://www.unidata.ucar.edu/software/netcdf) that can be imported by most commonly available statistical software programs as, e.g., Matlab (http://www.mathworks.com) or R (http://www.r-project.org).

5.2 SCALES OF MEASUREMENT

Before we look at the various ways of analyzing, presenting, and discussing metabolite data, we need to clarify on which scale the data exist as analytical data come in many sizes and scales. Hence, an efficient data analysis requires knowledge about these properties. It is often these properties that determine the procedures selected for the further statistical analysis.

As illustrated in Figure 5.1, there are at least two ways to classify different types of data. The distinction between the types of data can have an additional level when taking the differences of data and scales into account (see Anderberg, 1973 and Gordon, 1999). The main points are summarized below.

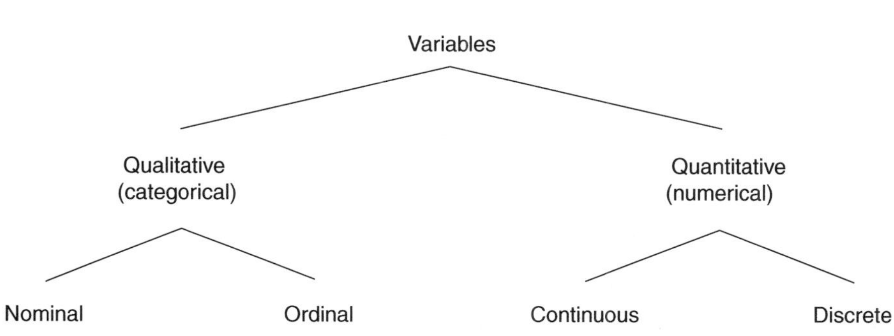

Figure 5.1 Scales of measurement. The figure illustrates the different types of data in generalized terms.

5.2.1 Qualitative Data

At the overall level we distinguish between qualitative data and quantitative data. The term qualitative comes from the word "quality," indicating a property, characteristic feature, or attribute. These are variables on which individuals differ in kind, and cannot be interpreted in terms of "how much of a difference." Analysis of qualitative data is not as simple as one would think. Although it does not require complicated statistical techniques normally used in quantitative analysis, it can be quite challenging to handle large amounts of data in a thoroughly systematic and relevant manner.

Qualitative data can be segregated into two additional categories:

5.2.1.1 Nominal Scale. Data are classified into distinct groups in which no ordering is implied. The groups can be identified by numbers, but mathematical operations cannot be performed on these numbers as they represent classes.

5.2.1.2 Ordinal Scale. Data are classified into distinct groups and ranked, i.e., the order is important. The data can be numbers. However, differences between the numbers indicating ordinal rank are not meaningful.

5.2.2 Quantitative Data

The term quantitative comes from the word "quantity," indicating amount, measure, number, size, etc. Quantitative data are always a list of numerical values where the numbers are representing an actually measured numerical quantity.

The distinction between discrete and continuous variables is quite important from a methodological point of view. Methods for solving problems involving continuous variables almost always are based on concepts from calculus, whereas methods for solving problems involving discrete variables are often solved by simple arithmetic or algebra. Both discrete and continuous variables are used in metabolomics, although continuous variables are quite a bit more common.

Quantitative variables can be segregated into two additional categories:

5.2.2.1 Continuous. The possible values of a continuous variable form an unbroken set of decimal values, with at most a finite number of distinct gaps. Continuous variables usually result from measurements made relative to a standard scale of size.

5.2.2.2 Discrete. The values of discrete variables form a set of distinct, isolated quantities. Observations that result from counting objects or items give discrete data, since only whole number values can arise.

5.3 DATA STRUCTURES

The structure of the data is independent of the data type we have chosen. In the far most cases our dataset consists of several observations, where each observation is a vector

$$\mathbf{x} = [x_1 \quad \ldots \quad x_m \quad \ldots \quad x_M]$$

containing M variables (sometimes also referred to as features or variates) extracted from each data file. This observation might be a whole spectrum or it may contain information derived from the sample, such as the presence or absence of certain ions, that is the qualitative description, and in the quantitative case the abundance of the ions. It can also be other factors such as colony growth diameter, number of colonies, etc. In other words, measurements that are not derived from, say, the spectra, but still are elements that we would like to include in our analysis, because we think they have an influence on our analysis. Using this notation, each variable spans out in one dimension in an M-dimensional space and the observation $\mathbf{x}$ is a point in this (hyper-dimensional) space. The words vector, point, and observation are used interchangeably.

In the case where we have several observations, we refer to the nth observation as

$$\mathbf{x}_n = [x_{n1} \quad \ldots \quad x_{nm} \quad \ldots \quad x_{nM}].$$

Finally, if we have N observations, all of the observations can be written into one matrix

$$\mathbf{X} = \begin{bmatrix} \mathbf{x}_1 \\ \vdots \\ \mathbf{x}_n \\ \vdots \\ \mathbf{x}_N \end{bmatrix} = \begin{bmatrix} x_{11} & \cdots & x_{1m} & \cdots & x_{1M} \\ \vdots & \ddots & \vdots & \ddots & \vdots \\ x_{n1} & \cdots & x_{nm} & \cdots & x_{nM} \\ \vdots & \ddots & \vdots & \ddots & \vdots \\ x_{N1} & \cdots & x_{Nm} & \cdots & x_{NM} \end{bmatrix}$$

in which each row is an observation and each column is corresponding to each of the variables. In this matrix each of the N rows are observations in an M-dimensional space spanned out by each of the variables.

Whereas the $\mathbf{X}$ matrix is said to contain the *explanatory variables*, some of the columns available from the table containing the so-called "external" information as described in Section 5.1 (containing all of the prior information) can be regarded as part of the *response matrix* $\mathbf{Y}$

$$\mathbf{Y} = \begin{bmatrix} \mathbf{y}_1 \\ \vdots \\ \mathbf{y}_n \\ \vdots \\ \mathbf{y}_N \end{bmatrix} = \begin{bmatrix} y_{11} & \cdots & y_{1p} & \cdots & y_{1P} \\ \vdots & \ddots & \vdots & \ddots & \vdots \\ y_{n1} & \cdots & y_{np} & \cdots & y_{nP} \\ \vdots & \ddots & \vdots & \ddots & \vdots \\ y_{N1} & \cdots & y_{Np} & \cdots & y_{NP} \end{bmatrix}$$

In this matrix, each row corresponds to the same sample as for the rows in $\mathbf{X}$, except that now the columns contain responses or information that we would like to evaluate

X against. In Sections 5.7 and 5.9, we use **Y** for classification. It is clear that all the information gathered in the table, when organizing the data, might not be relevant, and hence, we have P responses that may explain group information according to mutant, growth temperature, etc. Parts (columns) of **Y** will be used later in this chapter.

When analyzing data obtained from some of the analytical methods described in Chapter 4, the nature of the output has the same shape as the **X** matrix when the data are generated. As described in Section 4.7, data from a (binned) mass spectrum can be regarded as a vector $\mathbf{x} = [x_1 \quad \ldots \quad x_m \quad \ldots \quad x_M]$ in which each of the bins corresponds to a specific mass, and the value of x_m is the abundance/count of the ions detected within the specific mass range.

The following notation will be used throughout the chapter: vectors are denoted by lower-case bold face letters, as in **x**, and the individual components are identified using indices; thus x_i is the ith component of the vector **x**. Upper case bold letters are used to identify matrices, such as in **X**.

5.4 PREPROCESSING OF DATA

Although some of the preprocessing principles have already been mentioned previously, such as the binning principle described in Chapter 4, there are other important topics that have to be addressed before the data are prepared for further analysis.

In the following, these principles will be illustrated using data obtained from direct-infusion ESI-MS data and HPLC UV–VIS–DAD, but these methods are also applicable to most other types of spectroscopic data.

5.4.1 Calibration of Data

Working with raw data, it is important to know that some signals are normally collected as raw detector signals. In these cases, it is important to know whether the signal has to be calibrated before further processing, or the nature of the detector is yielding fully comparable signals across samples. As for the profile mass spectra, these data are stored together with a crude calibration. For the TOF instrument, the crude calibration is based on determination of the efficient flight length (the so-called L_{teff} value).

The crude calibration will normally ensure correct unit masses, but an additional external calibration is always performed prior to analyses. Generally, this is done by analyzing a reference mixture. For example, a polyethylene glycol (PEG) solution, from which about 30 ions are used to estimate a calibration polynomial (1st to 5th order) by using a calculated PEG spectrum. The calibration parameters are stored along with the raw data and applied to the mass spectra as these are read by the software. If not yet corrected the data is corrected by applying a Pth order polynomial

$$[m/z]_{\text{calibrated}} = \sum_{p=0}^{P} a_p [m/z]_{\text{raw}}^p$$

For centroid mass spectra, this calibration is often applied before data is stored. Therefore, these data do not need to be calibrated before any further processing.

In some cases, as for data from HPLC–UV–VIS DAD, calibration is not necessary due to the nature of the detector.

5.4.2 Combining Profile Scans

For some of the direct spectrometric measurement methods (e.g., direct-infusion ESI–MS), all spectra collected during the infusion of the sample contain more or less the same information. In these cases, an improvement of the signal to noise ratio can be obtained by combining the redundant spectra into a single one representing the true MS profile for the sample.

Within a time window Δt each mass spectrum contains a sequence of regularly distributed data points along the mass axis together with a corresponding intensity (Figure 5.2a). As these data points are sampled at equal intervals, they can be combined point-by-point, retaining the spectral information, and reducing the noise. The combination can be done in several ways, either by, e.g., calculating the average intensity (Figure 5.2b), calculating a trimmed mean value, or using other statistical methods. Only averaging is available in most commercial software.

If the spectra are not obtained through a direct spectrometric measurement method, but have been separated initially by either LC or GC, then a combination of the scans is unnecessary and this step can be discarded from the preprocessing.

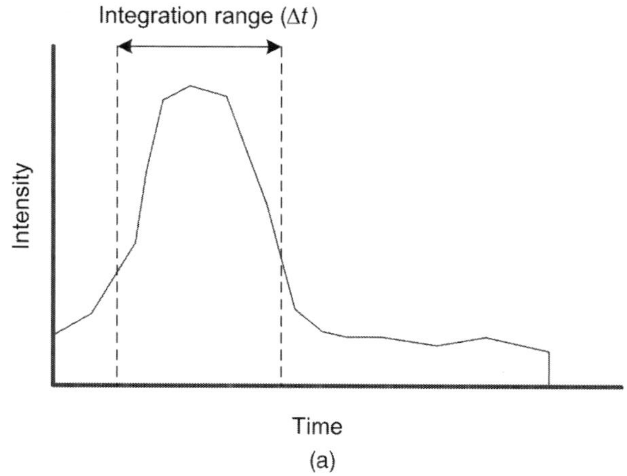

Figure 5.2 (a) Elution profile for the direct infusion ESI-MS. In order to improve the signal to noise ratio only scans within the time injection interval, Δt, is used to calculate a spectrum representing the sample. (b) shows the collected spectra within Δt plotted for the peak lying in the interval m/z 282–282.5. The mean profile is illustrated as the thick black line in the plot and could be regarded as the best suggestion to the peak. (See color plates.)

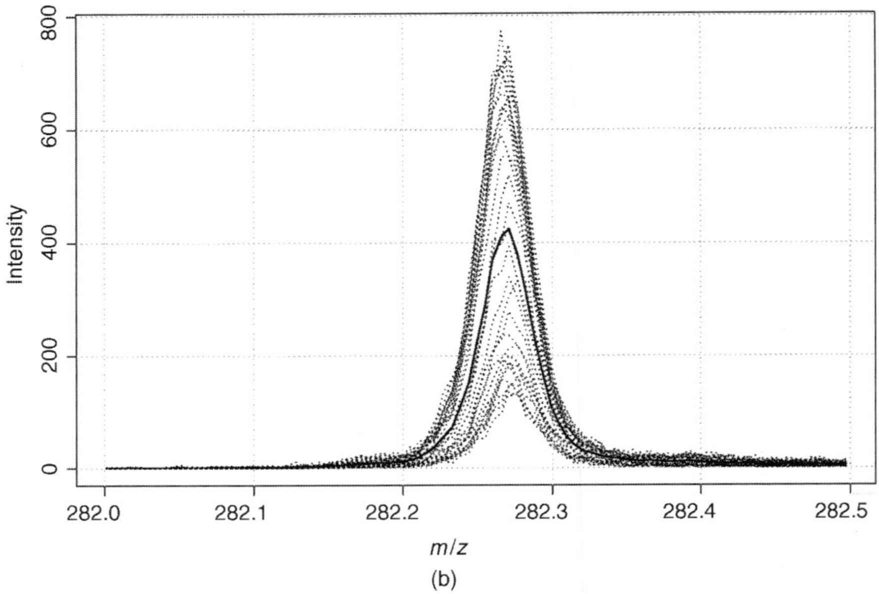

Figure 5.2 (*Continued*)

5.4.3 Filtering

Another important step is the improvement of the signal to noise ratio for the spectra. Most of the existing noise-removal techniques are based on moving window filters with fixed filter values, and implementations are available in most of the commercial software packages. The moving average filter is a simple Low Pass FIR (finite impulse response) filter commonly used for smoothing an array of data (Antoniou, 1993 and Mitra, 1998). As mentioned, this filter works as a low-pass filter removing the high-frequency spikes from the spectrum. Figure 5.3 illustrates the principle of the moving average filter.

The moving average filter can be imagined as a window of a certain size (in this case seven) moving along the spectrum, one element at a time. The middle element of the window (in this case element number 3) is replaced with the average of all elements in the window (see Figure 5.3). However, it is important to remember the

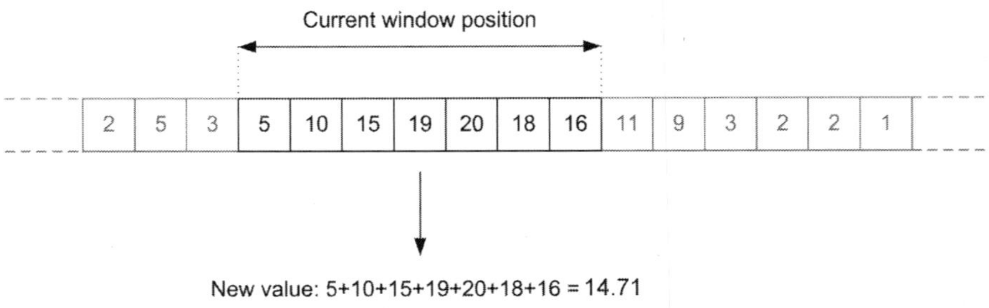

Figure 5.3 The moving average principle illustrated by a 7 point window size.

value of new elements and not make the replacement until the window has passed. This must be done since all averages shall be based on the original data in the array. When the ends of the spectrum are filtered and parts of the window are outside the spectrum, the averaging must be done on fewer elements than when the entire window is inside the array. This implementation leaves the ends of the array unfiltered. For a 7-point filter, this means that when n elements are filtered, elements 1, 2, 3, and $n-2$, $n-1$, n remain unchanged when filtering is complete. For many applications, this is no problem. Alternatively, the profiles can be padded with the values found at the end, or padded with zeros.

The larger the window is, the more peaks will be eliminated—including peaks that would not be regarded as noise. Furthermore, smoothing by fixed filters with symmetric properties does not preserve the height and width (i.e., the area) of a peak and the (centroid) position if the peak is skewed. Some of the algorithms can be made adaptive based on measured peak properties, such as, e.g., intensity or width. Figure 5.4a illustrates the problem.

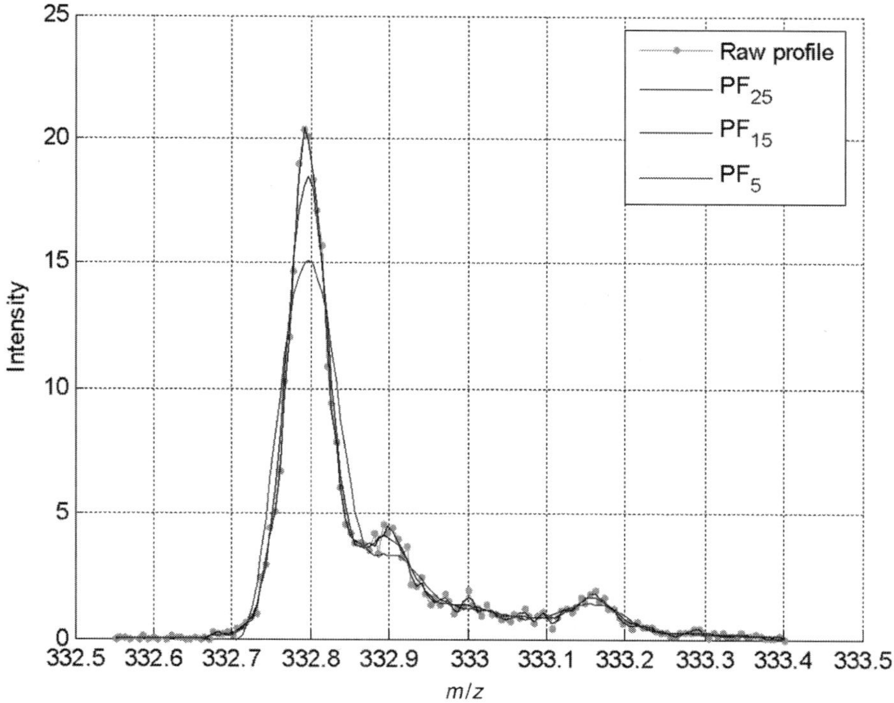

Figure 5.4 (a) Results of a moving average filter for different widths 25, 15, and 5 points. We see that the intensity of the peak is reduced even when a small size window is applied, and skewed when applying larger kernels no matter what size window is used. (b) Results of a polynomial filter of the same MS profile for different widths 25, 15, and 5 points and the polynomial of the order 3. With this filter the filtered profile maintains its shape almost all window sizes except from 25. This indicates that in this example the optimal size of window lies between 15 and 25. (See color plates.)

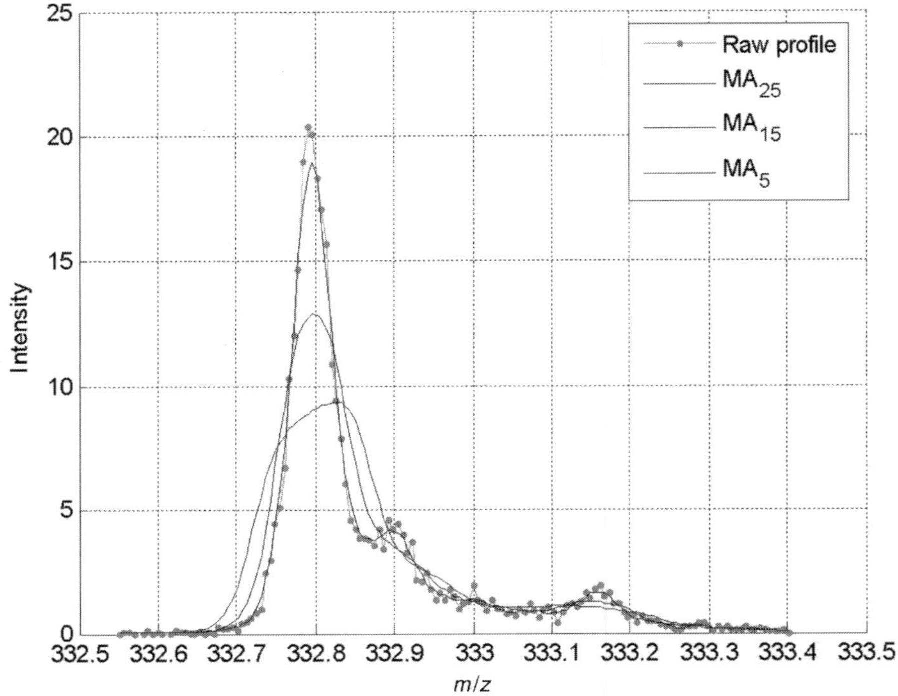

(b)

Figure 5.4 *(Continued)*

To accommodate for this problem, the spectrum can be approximated locally by a higher order polynomial (of order d) within a moving window (see Figure 5.4b). This filtering method is closely related to the so-called Savitsky–Golay filter available in most of the instrumental software packages. In the following a short description of how the polynomial filter calculates the filtered values is given.

Given a profile (e.g., a mass spectrum profile) with the data point intensities, $\hat{\mathbf{i}} = \hat{\imath}(m)$ (as in Figure 5.3), we can estimate the filtered spectrum $\hat{\mathbf{i}}' = \hat{\imath}'(m)$ by finding the solution to

$$\begin{aligned}\hat{\imath}'(m_k) &= \alpha\left(m_k\right) + \sum_{j=1}^{d} \beta_j(m_k) m_k^j \\ &= \alpha(m_k) + \beta_1(m_k)\, m_k^1 + \beta_2(m_k)\, m_k^2 + \ldots + \hat{\beta}_d(m_k)\, m_k^d\end{aligned}$$

minimizing

$$\min_{\alpha(m_k), \beta_j(m_k), j=1,\ldots,d} \sum_{m_n \in \mathbb{N}(m_k)} K_\lambda\left(m_k, m_n\right) \left[\hat{\imath}(m_n) - \alpha(m_k) - \sum_{j=1}^{d} \beta_j(m_k) m_n^j \right]^2$$

where $\mathbb{N}(m_k)$ is the neighboring region to mass m_k, λ the size of the window along the, e.g., m/z axis, and $K_\lambda(m_k, m_n)$ is a function that weights each of the data points within the window. Leaving out $K_\lambda(m_k, m_n)$ (or just setting $K_\lambda(m_k, m_n) = 1$ for all m_k and m_n), the moving average filter regards each data point in the data window to be equally important when calculating the average (filtered) value. So the reason for introducing the weighting function $K_\lambda(m_k, m_n)$ is motivated by the fact that the filter should place more emphasis on the closest data to m_k.

In other words, a new filtered value $\hat{\imath}'(m_k)$ is estimated by three steps (see Figure 5.3):

(i) Placing a window of size λ with $\hat{\imath}(m_k)$ in the center.

(ii) Estimating the parameters to the polynomial of order d, based on the intensities within the window. The intensities within the window are weighted in such a way that points close to the center m_k are assigned higher weight than those more remote from m_k.

(iii) Finally, the polynomial is evaluated at the center location m_k giving us the filtered value, $\hat{\imath}'(m_k)$.

Several good weighting functions can be used. In this example, the Epanechinikov function is chosen as the weighting scheme (Hastie et al., 2001). The function is given by

$$K_\lambda(m_k, m) = D\left(\frac{|m - m_k|}{\lambda}\right) \quad \text{where} \quad D(m) = \begin{cases} \frac{3}{4}(1 - m^2) & \text{if } |m| \le 1 \\ 0 & \text{otherwise} \end{cases}$$

In this equation the width λ should be determined by the resolution of the spectrum in such a way that two close but separate mass peaks will not be mixed together. The equation is a (bell shaped) weight function, and is applied on to all $\hat{\imath}(m_k)$ observations within a surrounding area of m_k. The resolution has to be given or estimated. Other weighting schemes that could be applied include the Gaussuan function.

In the filtering procedure described above, the estimation of the polynomial parameters can be solved using standard *weighted linear least squares*.

$$\hat{\imath}'(m_k) = b(m_k)^t \left(\mathbf{X}^t \mathbf{W}(m_k) \mathbf{X}\right)^{-1} \mathbf{X}^t \mathbf{W}(m_k) \hat{\mathbf{\imath}}$$

where $b(m_k)^t = (1, m, \ldots, m^d)$, t is the transpose of the design matrix $\mathbf{X}$ with ith row $b(m_i)$, and $\mathbf{W}$ is the weighting matrix with the ith diagonal element $K(m_k, m_i)$. Although this expression looks complex, what it does—for one value of $\hat{\imath}(m_k)$—is estimating the filter parameters within a region around $\hat{\imath}(m_k)$, and then calculating the filtered value.

The local linear regression automatically modifies the filter to correct the bias exactly to Nth order, a phenomenon dubbed as *automatic kernel carpentry*.

5.4.4 Centroid Calculation

Centroid mass spectra are described by a series of masses $m_t = \{m_t, \ldots, m^t_{K_t}\}$ with the corresponding intensities $i_t = \{i_t, \ldots, i^t_{K_t}\}$.

Going from a continuum data to a centroid data is done by finding the center of each ion peak at a specific height, typical in the range of 50–80% of the peak height. This process involves peak detection, validation, and finding of the centroid in the mass domain and the corresponding intensity as either the peak area or height. Most often the peak centroid is found at 50% of the maximum peak height, also determining the peak width (full width half maximum, FWHM) (see Figures 4.27 and 5.5).

5.4.5 Internal Mass Scale Correction

To obtain high accuracy one or more internal mass references are needed (e.g., lock-mass) to correct small variations in the mass scale. A compound can be added to the sample to serve as an internal mass reference, or sample components of known accurate mass $m_{\text{lock}} = \{m_{\text{lock},n}\}$, $n = 1, \ldots, N$ can be used. If an ion mass from m_{lock} is located in a spectrum within a tolerance window Δ_m, it will be used to move the mass scale by linearly correcting all masses so that the peak is at its correct mass value.

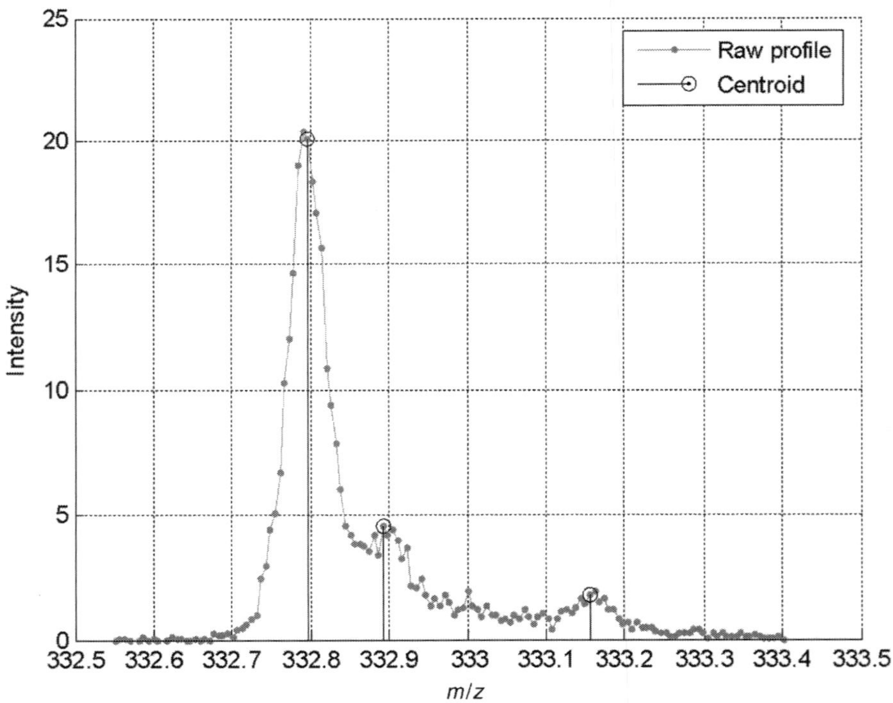

Figure 5.5 Centroid estimation of the profile filtered with a polynomial of the order 3 and window size 15.

5.4.6 Binning

We now have a list of centroid mass spectra described by a series of masses $m_t = \{m_t, \ldots, m_{K_t}^t\}$ and intensities $i_t = \{i_t, \ldots, i_{K_t}^t\}$. When comparing several observations, we will find that the centroid masses (in high resolution) will both vary in the number of detected peaks and their locations. In order to obtain a variable structure as described in Section 5.2, the centroid data is projected onto a grid with fixed bin sizes (see Figure 4.30). This is done in the following steps

(i) For each of the centroid masses, detect the mass interval that they fall within (corresponding to a specific bin).
(ii) For each of the bins, add the intensities of the corresponding centroids. Alternatively, if more than one centroid falls in a bin, one can choose to take the largest.

Finally, we have a vector of bins $\mathbf{x} = [x_1 \ldots x_m \ldots x_M]$ as described in Section 5.2 as that of the spectrum at a given resolution (reflected by the bin width).

5.4.7 Baseline Correction

Data from analytical instruments generally consist of the "real information" superimposed on a "noisy" background. In case of chromatographic data, the part recorded when only carrier gas or solvent elute from the column is called the baseline (from the IUPAC compendium of technical terminology). The baseline, or background, can be either flat, linear with a positive or negative slope, curved, or a combination of all three. It is mainly characterized by the fact that it does not vary as quickly as the peaks do.

Baseline correction is performed in order to eliminate the effect of these variations from the signal during the analysis. The chromatograms may also contain baseline variations due to shift in eluent composition or due to column bleed temperature during the analysis. In some cases, it is necessary to correct three types of baseline variations: random variations in each individual variable (e.g., between the diodes in the detector array) as these can seriously affect the correlation calculation for noise-only areas, or small peaks, especially incase of compounds determined by only a few of the variables (e.g., only shows absorption at a few wavelengths, or a few masses in their mass spectra). Baseline variations during analysis will also prevent the normalization (height scaling) to enhance data.

Consider an example where data are collected from an HPLC separation with a UV detector as illustrated in Chapter 4. Here, UV-spectra are collected at a fixed time interval as the chromatographic separation progresses. These data can be given as $y_i = y(t_i)$, for $i = 1, \ldots, M$, where y_i is the signal measured at a specific wavelength to the retention time t_i for which $i = 1, \ldots, M$ is the number of measurements in the profile (see Figure 4.28). The measured absorbance y_i can be expressed as the sum of the signal and the baseline, $x_i = x(t_i)$ and $g_i = g(t_i)$, respectively. This gives us the following equation for the measured signal

$$y(t) = x(t) + g(t) + \varepsilon(t),$$

in which $\varepsilon(t)$ is a random noise contribution assumed to be normally distributed.

In all baseline correction algorithms, it is the goal to estimate the background $g(t)$, which then is subtracted from the original chromatogram.

Often the background is approximated as a polynomial of the order of P.

$$g(t) = \beta_0 + \beta_1 t + \beta_2 t^2 + \cdots + \beta_P t^P$$

If we have a "flat" baseline then $g(t)$ is a constant (P=0), $g(t) = \beta_0$, whereas a slanted background (Figure 5.6a) can be expressed as a line (P=1), $g(\mathrm{t}) = \beta_0 + \beta_1 t$, and finally a curved baseline (Figure 5.6b) could be expressed as a second order polynomial (P=2) by $g(t) = \beta_0 + \beta_1 \mathrm{t} + \beta_2 t^2$. It is the task to estimate the parameters $\beta = \{\beta_0, \beta_1, \beta_2, \ldots, \beta_P$ in $g(t)$ such a way that it optimizes a criterion chosen to give the best fit to the background. In most algorithms, the background is estimated by a least-squares polynomial fitting performed on a user-selected subset of points belonging to the background. Providing that the points are selected correctly, the fitting yields satisfactory results. This can be attributed to the ability of the polynomial model to represent a wide class of backgrounds.

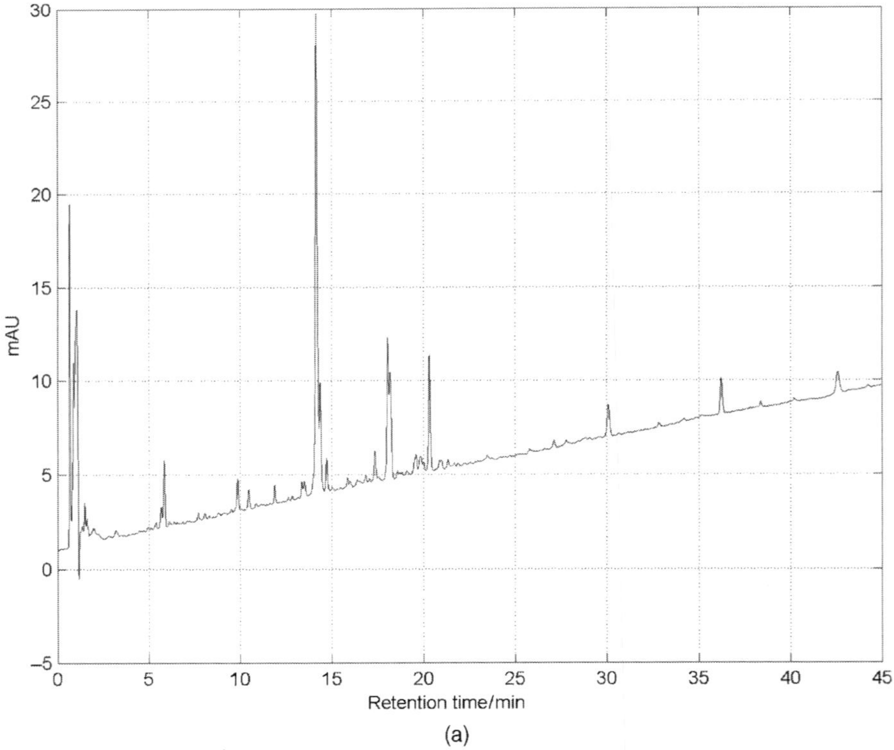

Figure 5.6 Illustration of the drift in baseline. The Figure (a) illustrates the behavior of a close to linear baseline, whereas the Figure (b) shows an example of a more complex (non-linear) baseline.

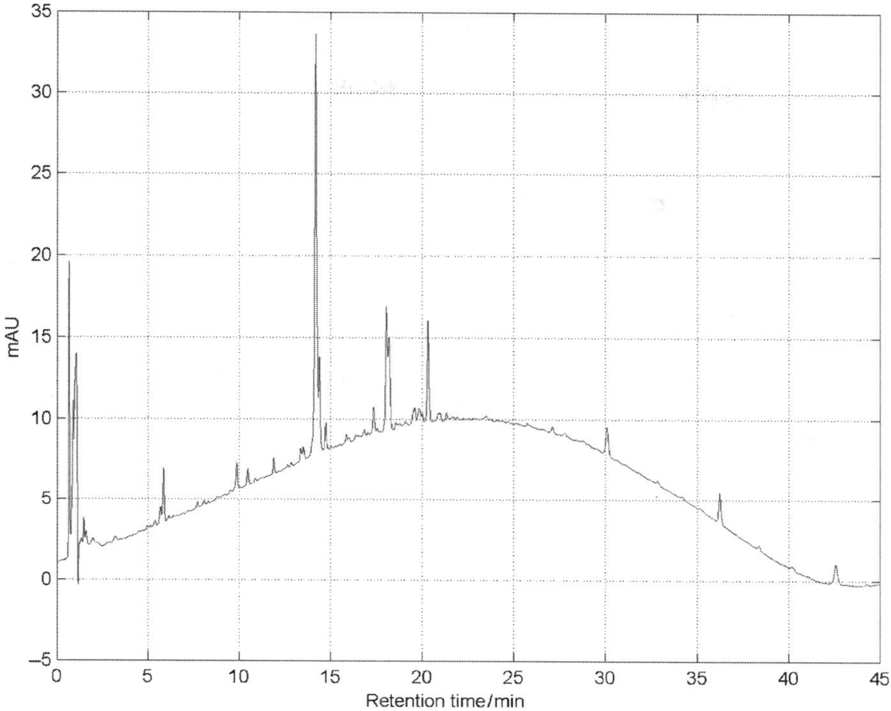

(b)

Figure 5.6 *(Continued)*

Two more or less different approaches based on piecewise linear correction are presented in the following, and also a description to how the background can be estimated using a polynomial model.

5.4.7.1 Piecewise Linear Background Estimation. This is a rather simple method where one wavelength is corrected at a time, by first finding the minimum point in a window of a specified width on the time axis for all possible window displacements. Data points found as local minima within position of this window will be considered to as a baseline point, and an estimate of the baseline for the current trace is calculated by linear interpolation between those baseline points that fulfill a set of criteria, e.g., number of window placements where they occur. The resulting piecewise linear function are then subtracted from the measured profile (e.g., at the current wavelength), yielding a baseline corrected profile.

The values between two local minima found to the retention times ta and tb is calculated by interpolation. First, we calculate the parameters for the line joining the points (ta, y(ta)) and (tb, y(tb))

$$\hat{a} = \frac{y(t_a) - y(t_b)}{t_a - t_b} \quad \text{and} \quad \hat{b} = y(t_a) - \hat{a} \cdot t_a$$

Within the interval the background is estimated by

$$g(t) = \hat{a} \cdot t + \hat{b} \quad \text{for} \quad t \in [t_a; t_b]$$

Figure 5.7 shows the result after baseline correction of the chromatographic profile shown in Figure 5.6a by piecewise linear background estimation algorithm. Figure 5.7a shows the entire profile (blue), the local minima found (marks: "*"), and the estimated profile (red). Figure 5.6b shows the resulting profile after having subtracted the background.

An advantage of the piecewise linear background subtraction method is that it is simple and fast to compute, however, it tends to be sensitive to high frequent changes in baseline. This problem is illustrated in Figure 5.7a,b, clearly seen at the beginning of the chromatogram where it contains abrupt changes, giving rise to an unfortunate artifact in the background estimate. But for slowly varying backgrounds, the piecewise linear background estimate can be very efficient.

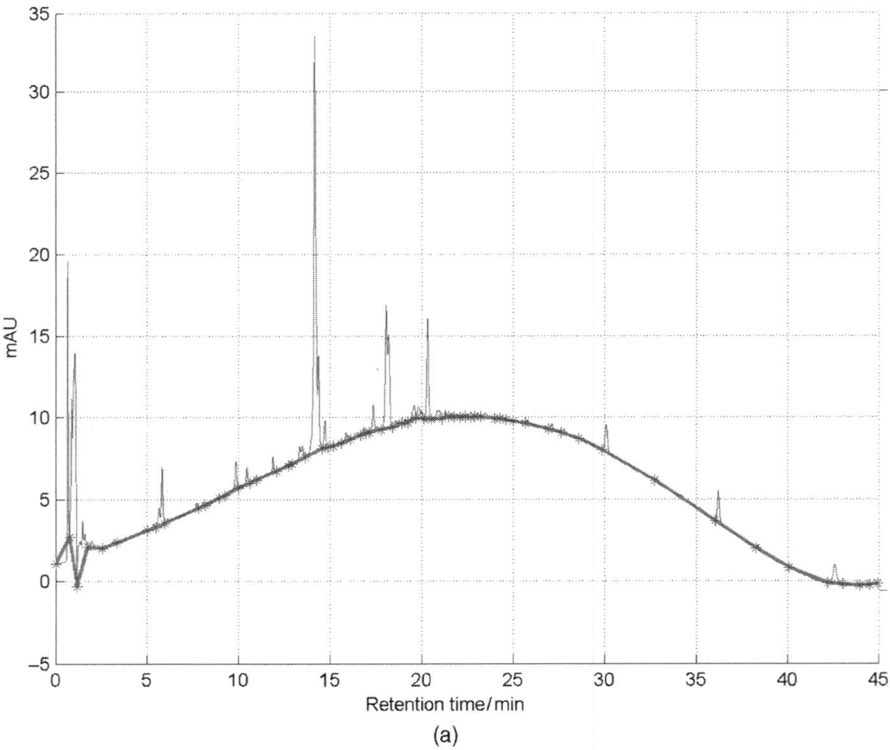

Figure 5.7 Illustration of the piecewise linear baseline correction. Figure (a) shows the chromatogram (blue line) and estimated local minima (marked with '*'). Between the segments defined by these local minima the background is estimated as lines (red line). Figure (b) shows the result after having subtracted the background from the chromatogram. (See color plates.)

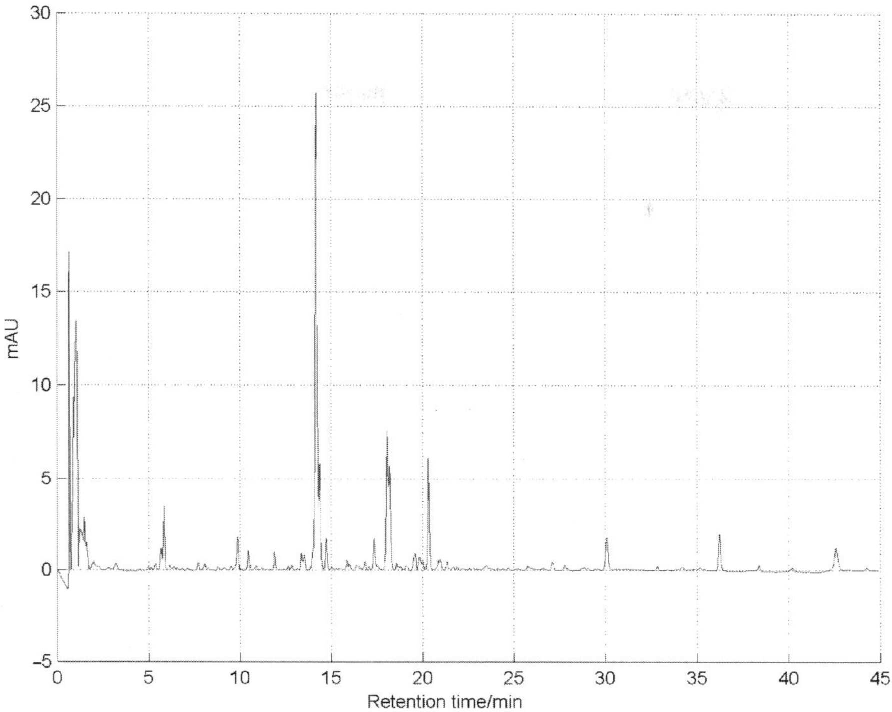

(b)

Figure 5.7 *(Continued)*

5.4.7.2 Polynomial Background Estimation. An alternative to the relatively simple piecewise linear background estimation is using a higher order (i.e., polynomial) background estimate. A polynomial equation of the order P is chosen to estimate the background based on the local minima selected by the moving window. The solution to the polynomial can be found by the ordinary least squares solution

$$\begin{bmatrix} g_1 \\ g_2 \\ \vdots \\ g_N \end{bmatrix} = \begin{bmatrix} 1 & t_1 & t_1^2 & \cdots & t_1^P \\ 1 & t_2 & t_2^2 & \cdots & t_2^P \\ \vdots & \vdots & \vdots & \ddots & \vdots \\ 1 & t_N & t_N^2 & \cdots & t_N^P \end{bmatrix} \begin{bmatrix} \beta_0 \\ \beta_1 \\ \beta_2 \\ \vdots \\ \beta_P \end{bmatrix}$$

In matrix notation the equation for a polynomial fit is given by

$$\mathbf{g} = \mathbf{T}\boldsymbol{\beta}$$

This can be solved by premultiplying by the matrix transpose $\mathbf{T}^t$ (meaning the transpose of T)

$$\mathbf{T}^t\mathbf{g} = \mathbf{T}^t\mathbf{T}\boldsymbol{\beta}$$

This equation can be solved numerically, or $\mathbf{T}^t\mathbf{T}$ can be inverted directly if it is well formed to yield the solution vector

$$\hat{\boldsymbol{\beta}} = (\mathbf{T}^t\mathbf{T})^{-1}\mathbf{T}^t\mathbf{g}$$

Setting $P = 1$ in the above equations reproduces the linear solution.

As can be seen in Figure 5.8 the polynomial background estimation creates a "smooth" fit, where extreme deviations does not have the same impact on the estimation as was the case for the piecewise linear baseline correction. This is easily seen in the "noisy" beginning of the chromatogram shown in Figures 5.7a and 5.8.

Other methods may be considered for the background estimation. The more recent wavelet transformation has become a useful tool (Depczynski, 1997; Cai,

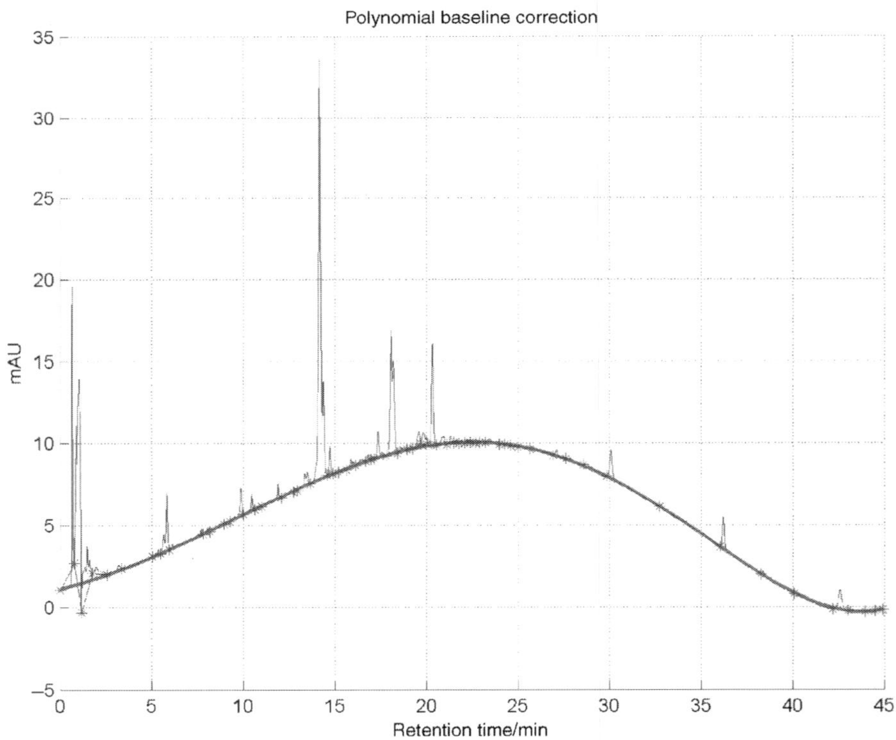

Figure 5.8 Illustration of the polynomial baseline estimation. The figure shows the chromatogram (blue line) and estimated local minima (marked with '*'). The background is estimated from these points as a 5th order ($P = 5$) polynomial. (See color plates.)

2001; Tan, 2002; Liu et al., 2003) for background removal. The method is based on applying a "wavelet transform" to the different traces, from which the wavelet coefficients are computed, and then separated from the background supposed to be in the low-frequency part (approximation coefficients) and from the peaks (and noise) supposed to be in the high-frequency part (detail coefficients). The main shortcoming of such an approach is that it implicitly supposes that the background is well separated (in the transformed domain) from the rest of the signal.

5.4.8 Chromatographic Profile Matching

An important part of chromatographic data analysis is often to compare chromatographic profiles from multiple samples. This is preferably done by some sort of pattern recognition routines, for example, fingerprinting of flavor components in coffee, of oil components in forensic investigations, or taxonomy of microorganisms. The disadvantage of peak detection and integration and of the introduction of a subjective peak selection can be avoided by using all collected data points in the multivariate statistical analysis.

In chromatography, retention time variations are a serious impediment to the successful application of automated pattern recognition methods or chemometrics. This hampers possibility for objective classification of chromatographic data, because errors in peak alignment are additional sources of signal variations that easily dominate the true variations in the data, e.g., due to chemical differences. Retention time variations are due to subtle, random, and often unavoidable changes and variations over time in instrument parameters (Figure 5.9).

Pressure, temperature, solvent composition, column aging, and flow fluctuations may be the cause for an analyte to elute at different retention times in replicate runs. Even with implementing advanced instrumentation with electronic pressure control, subtle run-to-run retention time shifting can be small but is always present, and must be taken into account to successfully apply chemometric methods. Matrix effects and stationary phase decomposition may also be the cause variation in retention time. The main reason is that most pattern recognition techniques and chemometric is based on point-to-point comparison for successful analysis.

To overcome the problem with shifts in retention time it is necessary to align the chromatograms to obtain full concordance between the eluted components. Some alignment algorithms operate by aligning specific features in the data. In general, the methods can be categorized into two major groups: those that align chromatograms based on peak information, and those who use the full chromatographic information to do the alignment.

Many of the available alignment algorithms do not require knowledge or identification of peaks. These algorithms contain some level of dynamic programming where iterated shifts are evaluated by calculating a distance between a sample and target chromatogram using some specific metric. That matching metric, or correlation, returns the optimal retention time correction for the sample. These algorithms fall in various categories: dynamic time warping (DTW), genetic algorithms, partial linear fit, and minimization of residuals.

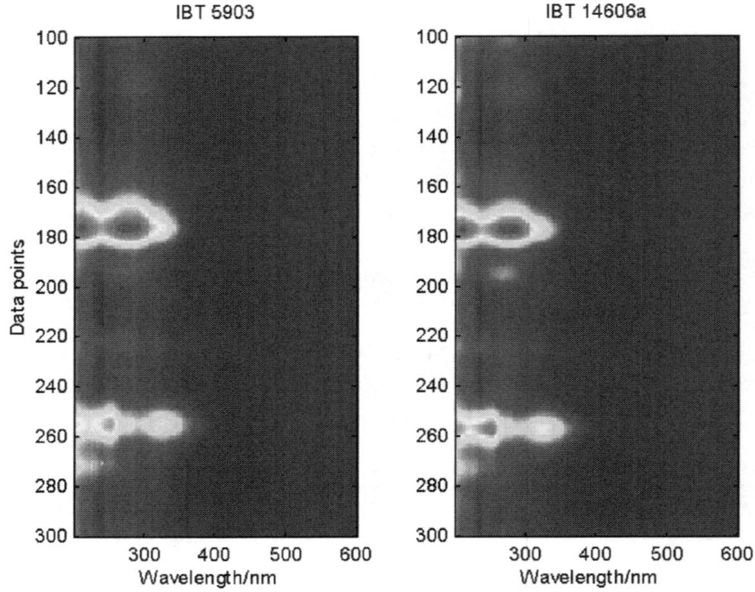

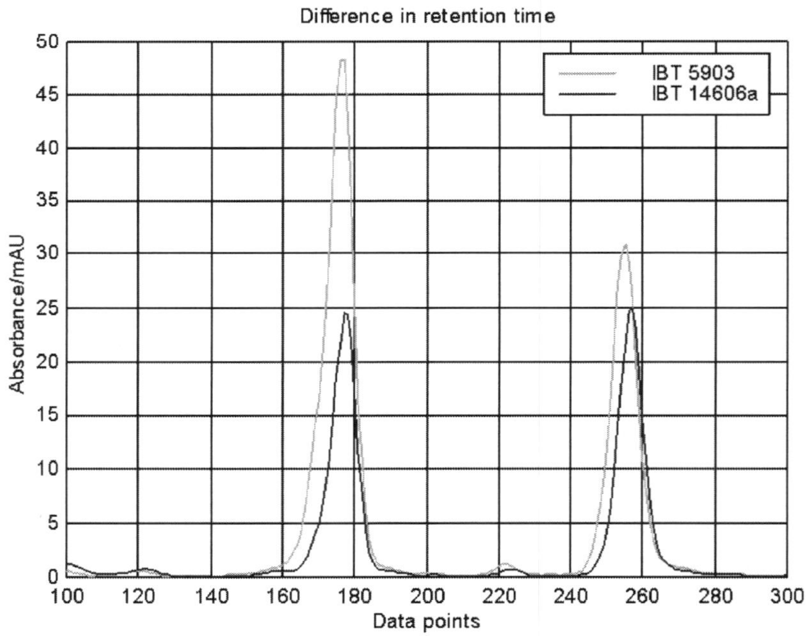

Figure 5.9 Illustration of the problem with shifts in retention time between two HPLC runs. Figure (a) shows a section of the UV absorbance of two complex fungal metabolite extracts containing two peaks. The color illustrate the amount of absorbed light going from low absorbance (blue) to higher absorbance (red). In Figure (b) the two traces along 230 nm are plotted. From the figures we see that there is a significant difference between the peak maxima for the two profiles. It is the aim of the aligning algorithm to correct for these shifts in retention time. (See color plates.)

Two different warping algorithms have received much attention in recent years for the alignment of time trajectories, chromatographic profiles, and spectra (Reiner et al., 1979; Wang and Isenhour, 1987; Pravdova et al., 2002). The first method, the DTW, was initially formulated for aligning frequency spectra of words pronounced by different speakers for recognition purposes (Itakura, 1975; Sakoe and Chiba, 1978). The more recent approach for aligning signals, the correlation optimized warping (COW), was proposed in 1998 as a means to correct chromatograms for retention time shifts prior to multivariate modeling (Nielsen et al., 1998).

5.4.8.1 Dynamic Time Warping. DTW synchronizes similar features in sets of signals using dynamic programming. DTW nonlinearly warp two signals in such a way that similar events are aligned and a minimum distance between them is obtained. Consider two profiles signals R (length L_R) and T (length L_T). A plot is constructed with the T signal in the x-axis and R in the y-axis. The algorithm constructs a path such that corresponding events in signals R and T are linked. When this path is known, it can be used to align the signals.

To find the path, a grid with size $L_T \times L_R$ is constructed and a sequence F of K points through the grid is denoted as

$$F = \{c(1), c(2), \ldots, c(k), \ldots, c(K)\}$$

where

$$c(k) = \left[i(k), j(k)\right]$$

and i and j denote the time index of T and R, respectively.

Each point $c(k)$ in the grid is described by a pair of indices and indicates a position in the grid. The sequence F can be viewed as a path on the grid. One searches for a sequence F^* that optimally matches the two signals so that a cumulative distance between them is minimized and an optimal path through the grid is found.

There are two versions of the DTW algorithm that can be used to construct the path, namely a symmetric and an asymmetric one. In the symmetric algorithm both signals, R and T, are considered as equally important and the time indexes i and j are mapped onto a common time index k (the two above equations). The optimal path passes through all the points of both signals and their roles can be reversed (i.e., T can be placed on the vertical axis and R on the horizontal axis). When the position of the signals is interchanged, the same optimal path and minimum distance are reached. In the asymmetric algorithm, the two signals are not considered as equally important; one of the signals is taken as a reference. If their roles are interchanged, a different path and minimum distance will be obtained. The time index of the signal placed on the vertical axis, R, is mapped onto the time index of the trajectory placed on the horizontal axis, T.

The time index k is then the time index i of the signal T and the optimal path contains exactly L_T points.

5.4.8.2 Correlation Optimized Warping. To correct for misalignments or shifts in discrete data signals, the COW procedure was introduced by Nielsen et al. (1998). It is a piecewise or segmented data preprocessing method (operating on one sample record at a time) aimed to align a sample data vector against a reference vector by allowing limited changes in each segment lengths in the sample vector. The ratio between the number of points in the reference vector, N, and the selected segment length I determines the number of segments, or rather the number of segment borders. An equal number of segments (borders) are specified on the sample vector. The maximum increase or decrease of sample segment length is controlled by the so-called slack parameter t. When the number of time-points in a corresponding sample and reference segment differs, the former is linearly interpolated in order to create a segment of equal length.

In COW, the different segment lengths on the sample vector are selected (or when the borders are shifted thus "warped") so as to optimize the overall correlation between sample and reference in each segment. The problem is solved by breaking down the global problem in a segment-wise correlation optimization by means of a dynamic programming algorithm (DP) (Nielsen et al., 1998; Hillier and Liebernan, 2001). The solution space of this optimization is defined by two parameters: the number of segment borders $I + 1$ and the length of the slack area t. Both parameters have to be given to the algorithm.

COW may be regarded as a special case of DTW where additional constraints are added to reduce the search space for the optimal warping and to employ correlation coefficient as optimization criterion (Tomasi et al., 2004) (see Figure 5.10).

Both the DTW and COW are useful tools for aligning different types of signals. The DTW can be used for correction of peak linear and nonlinear shifts in NIR spectra and for retention time shifts in chromatograms. Unfortunately, in some cases the distance measurement used by the DTW is not the best for similarity measurement in aligning. The correlation coefficient offers a better similarity measure, but some limitations still exists, for instance in baseline correction.

5.5 DECONVOLUTION OF SPECTROSCOPIC DATA

Deconvolution means the separation of corresponding fragments to one mass spectrum and thus for a single compound. It is a powerful mathematical tool for

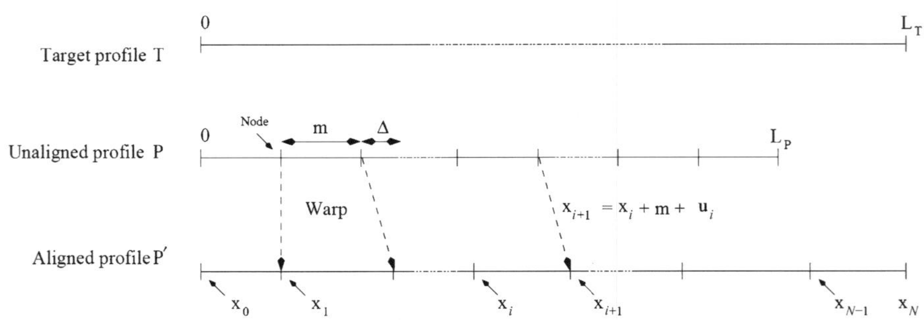

Figure 5.10 Illustration of the principle behind the correlation optimized warping.

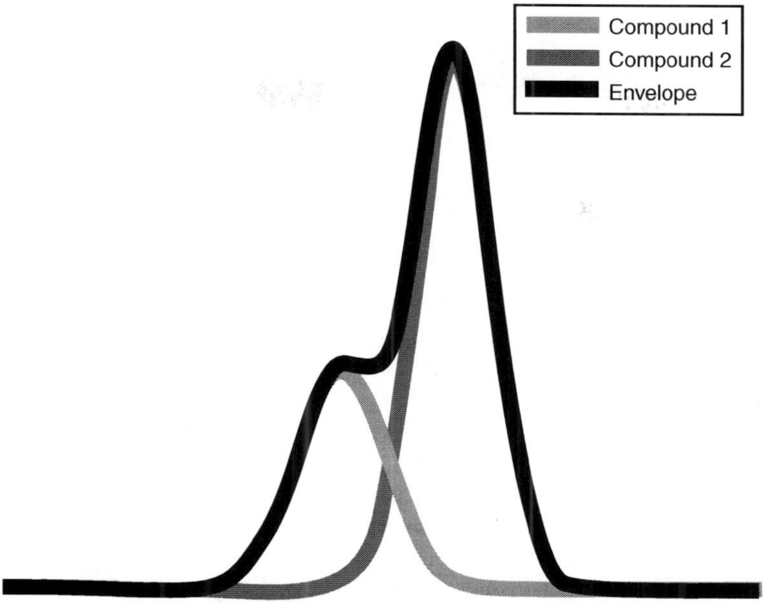

Figure 5.11 Schematic illustration of the deconvolution problem. If two compounds elute at the approximately same time they will overlap and give rise to an "artificial" spectrum being a sum of the two. (See color plates.)

enhancing the selectivity offered by chemical methods. An important application is the separation of a complex chromatographic signal in its individual contributions, when partial coelution is obtained due to an insufficient separation power of the chromatographic system (see Figure 5.11).

As a result, compounds hidden within a peak cluster can be quantified with relatively small errors.

Deconvolution can be achieved either in an automated fashion by the software packages provided with most GC–MS instruments (Pegasus, Leco, St. Jospehs, USA) or by applying separate software, such as AMDIS (http://chemdata.nist.gov/mass-spc/amdis; National Institute of Standards and Technology, Gaithersburg, USA).

5.6 DATA STANDARDIZATION (NORMALIZATION)

In some cases it is interesting to look at the relative amounts of different compounds, thus the relative differences between samples, and not necessarily the absolute amounts. In these cases, it is necessary to remove the effect of the total amount from the analysis. This type of correction is commonly known as normalization, standardization, and sometimes multiplicative correction of the data. Data standardization is the process of making all data of the same type, or class conform to an established convention or procedure to ensure consistency and comparability across different types of variables.

The ordinary preprocessing of the data before, e.g., a principal component analysis (PCA) (Section 5.7.1), the normal procedure is to subtract the mean value from the variables (center) and divide by the standard deviation (scale); another way of standardizing data.

For a comprehensive discussion of different techniques and references, please refer to Podani (1994) and Stein and Scott (1994)[1].

Data scaling is usually the first step of data transformation (dimensionality reduction), chemical similarity searching, feature extraction, hypothesis generation, and other types of machine learning.

After the initial preprocessing methods the data are cleaned and obtained in a form suitable for analysis. The steps that can be taken from here are all based upon the fact that we have data in the X matrix shape described in Section 5.3.

5.7 DATA TRANSFORMATIONS

In problems with many dimensions (with M >> N in Section 5.3), it can be necessary to reduce the effective dimension to employ some of the more efficient methods that work best for lower dimensions. Often, the variables (the columns in X) used to represent the observations (the rows in X) are not always independent, and may be correlated. Based on the redundant information spread out in the features, these can well be approximated by "projections" into a lower dimensionality space. Many of the techniques used for data reduction and visualization of multivariate data are based on a so-called decomposition of X followed by a projection of the data onto the axes defined by the extracted factors.

One of the most popular techniques used for dimensionality reduction is the PCA, which will be described in detail in the following section. Other dimensionality reduction methods can also be employed, including factor analysis, projection persuit, wavelet transforms and methods like feature histograms, and independent components analysis. These methods all have in common the property that they allow efficient characterization of a low-dimensional subspace with the overall space of raw measurements.

5.7.1 Principal Component Analysis

PCA is a technique that can be used to simplify a dataset by reducing the dimensionality as described above. More formally, it is a linear transformation (rotation of data) that chooses a new coordinate system for the dataset such that the greatest variance by any projection of the data is found on the first axis – called the first principal component (PC) – the second largest variance on the second axis, and so on. PCA can be used to reduce the dimension of data while retaining those characteristics of the dataset that contribute mostly to the variance by eliminating the higher principal components, by a more or less heuristic decision. These characteristics retained may

[1]Specifically about mass spectrometry.

be the "most important," but this is not necessarily the case and depends on the application. In the following, the mathematics behind the PCA is described in detail.

As described, the objective of the PCA is to find linear combinations (orthonormal projections—meaning that they have orthogonal unit vectors) of the original variables in our **X** matrix

$$\mathbf{X} = \begin{bmatrix} \mathbf{x}_1 \\ \vdots \\ \mathbf{x}_n \\ \vdots \\ \mathbf{x}_N \end{bmatrix} = \begin{bmatrix} x_{11} & \cdots & x_{1m} & \cdots & x_{1M} \\ \vdots & \ddots & \vdots & \ddots & \vdots \\ x_{n1} & \cdots & x_{nm} & \cdots & x_{nM} \\ \vdots & \ddots & \vdots & \ddots & \vdots \\ x_{N1} & \cdots & x_{Nm} & \cdots & x_{NM} \end{bmatrix}$$

maximizing the variance. Here it is assumed that each of the columns of **X** are standardized to have zero mean and unit variance. If the linear combination is denoted by the vector $\mathbf{a} = [a_1, a_2, \ldots, a_M]^t$ then it is the goal to choose a to maximize the variance of the elements of $\mathbf{z} = \mathbf{Xa}$. The variance of **z** may be written as

$$\text{var}(\mathbf{z}) = \frac{1}{N-1}\mathbf{a}^t\mathbf{X}^t\mathbf{Xa}$$

Because **X** is standardized, the term $1/(N-1)\mathbf{X}'\mathbf{X}$ is just the sample correlation matrix **R**, yielding $\text{var}(\mathbf{z}) = \mathbf{a}'\mathbf{Ra}$. We then obtain the covariance matrix, and **R** will be substituted with Σ in the above equation.

To understand what a covariance matrix is, one first needs to understand what covariance is. The covariance of two variables or columns in **X**, say, **a** and **b**, can be defined as the tendency to vary together. Statistics tells us that one can describe the variation in the data with standard deviation—a value that tells us something about variability around the mean. In the same way, the covariance ($\text{Cov}[\boldsymbol{x}_{\cdot i}, \boldsymbol{x}_{\cdot j}]$) can describe the variability—as the product of the averages of the deviation of data points from the mean (of that dataset). The resulting $\text{Cov}[\boldsymbol{x}_{\cdot i}, \boldsymbol{x}_{\cdot j}]$ value will be larger than 0 if $\boldsymbol{x}_{\cdot i}$ and $\boldsymbol{x}_{\cdot j}$ tend to increase together, below 0 if they tend to decrease together, and 0 if they are independent. The covariance matrix, Σ, of **X** is merely a collection of the covariance's between all variables in the form of a $M \times M$ matrix:

$$\Sigma = \text{Cov}(\mathbf{X}) = \begin{bmatrix} \text{Cov}(x_{g1}, x_{g1}) & \cdots & \text{Cov}(x_{g1}, x_{gM}) \\ \vdots & \ddots & \vdots \\ \text{Cov}(x_{gM}, x_{g1}) & \cdots & \text{Cov}(x_{gM}, x_{gM}) \end{bmatrix}$$

$$\Sigma = \begin{bmatrix} \text{Var}(x_{g1}) & \cdots & \text{Cov}(x_{g1}, x_{gM}) \\ \vdots & \ddots & \vdots \\ \text{Cov}(x_{gM}, x_{g1}) & \cdots & \text{Var}(x_{gM}) \end{bmatrix}$$

where x_{gm} means the mth column in **X**. The diagonal of the covariance matrix corresponds to the variance of the x_{gm}. Said in other words Σ explain how data is spread out in the M-dimensional space, and it is possible to obtain the *correlation matrix*, **R** with the elements r_{ij}, by dividing each of the elements in with the product of the variances

$$r_{ij} = \frac{\mathrm{Cov}(x_{gi}, x_{gj})}{\mathrm{Var}(x_{gi})\mathrm{Var}(x_{gj})}$$

Because we can choose the components of **a** to be arbitrarily large and thereby obtain infinite variance ($\mathrm{var}(\mathbf{z}) = \infty$), a constraint is applied saying that the length of the vector **a** has to be one ($\mathbf{a}'\mathbf{a} = 1$). The solution to this optimization problem is known to be called the eigenvalue–eigenvector problem stated as

$$(\mathbf{R} - \lambda\mathbf{I})\mathbf{a} = 0$$

where the vector **a** is called an eigenvector and the scalar λ is called an eigenvalue. Provided, that the matrix **R** has full rank (thus there is no perfect multi-colinearity among the observed variables, **X**), then the solution will consist of M positive eigenvalues and associated eigenvectors.

Figure 5.12 illustrates the principle of the PCA in a simple two-dimensional case. Here the $\hat{x}_1$ and $\hat{x}_2$ coordinate-system span out two dimensions in which observations

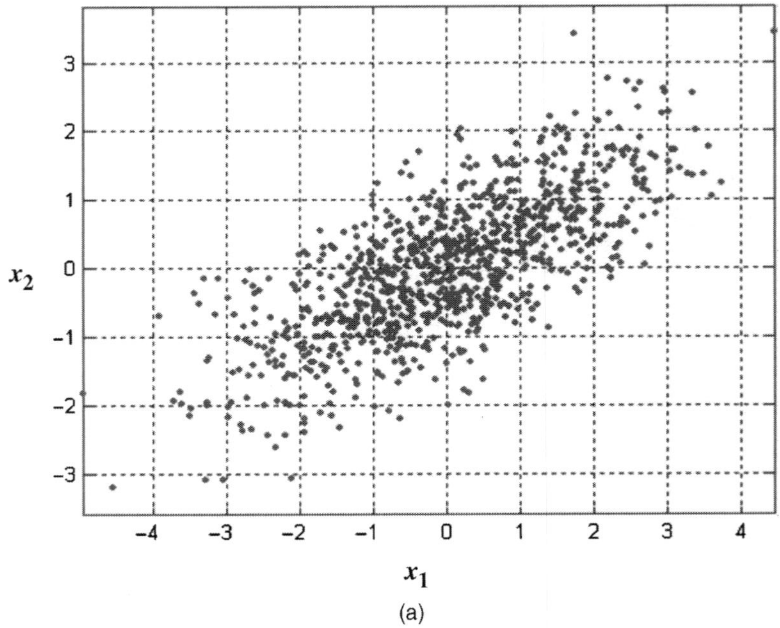

Figure 5.12 Principal component analysis (PCA) example. The figure illustrates the transformation of data according to the directions with large variation.

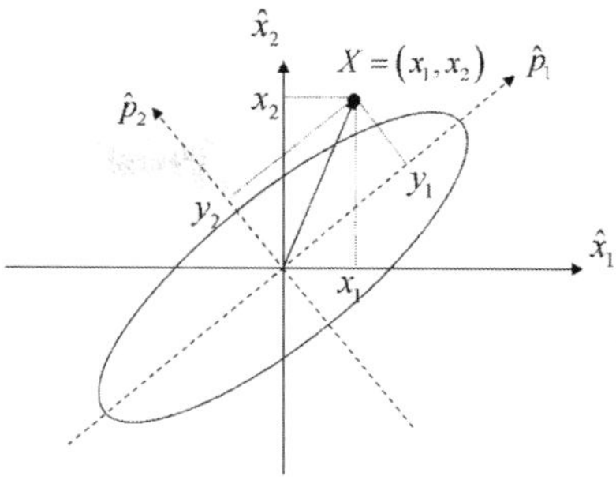

(b)

Figure 5.12 (*Continued*)

are measured (Figure 5.12a). The data has been centered to have zero mean. The covariance between the measurements are summarized by the ellipsis drawn in Figure 5.12b. The new coordinate system (the eigenvectors) found by the PCA are plotted in Figure 5.12b as $\hat{p}_1$ and $\hat{p}_2$.

The PCA has some interesting properties. First, it is important to note that the eigenvalues $\lambda_1, \lambda_2, \ldots, \lambda_M$ are exactly the same as the variances for the M principal components. The consequence is that the ith principal component (PC_i) contains

$$p_i = \frac{\lambda_i}{\sum_{m=1}^{M} \lambda_m} \cdot 100 = \frac{\lambda_i}{\lambda_1 + \lambda_2 + \cdots + \lambda_M} \cdot 100$$

percent of the total variance in the data. This can be used to reduce the dimensionality of the data, since one might choose to retain only the principal components describing, e.g., 98% of the total variation in the data.

For the PCA the eigenvectors are called the loadings, and the projections are called the scores.

The PCA rotates and projects data onto a new coordinate system spanned out by the eigenvectors, and the eigenvectors are found according to directions in data along which the variance is described decreasingly.

Often when analyzing metabolite data, additional related qualitative information exists which can be used to couple species, mutant, or other nominal characters to each profile. In these cases an alternative transformation approach can be used to find projections of the data using that extra information—not by explaining the variance but the class variation.

5.7.2 Fisher Discriminant Analysis

Discriminant analysis is in general used to classify information to achieve the clearest possible separation or discrimination between groups, or tightest relations within groups (Figure 5.13).

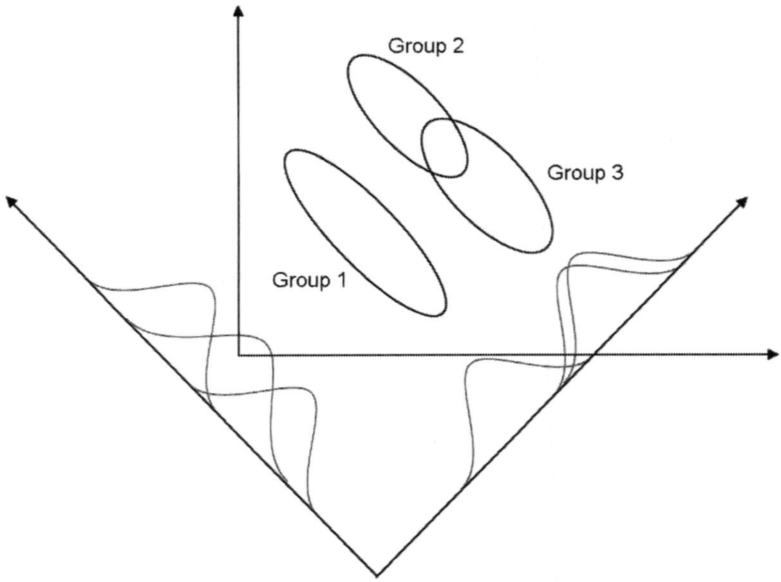

Figure 5.13 Stylized scatter plot for three-group discriminant analysis problem. (See color plates.)

As was the case for the PCA, the mathematical problem is the eigenvector-reduction of a real, symmetric matrix. The eigenvalues represent the discriminating power of the associated eigenvectors.

Assuming that we have observations divided into G groups. Each of these groups could in the optimal case be separated in a space of at most G-1 dimensions, one dimension to separate each group. In the simple case where we have two groups, we would need one dimension; in the case of three we would need two, etc. This will be the number of discriminating axes or factors that can be obtained in a common practical situation, when $N > M > G$ (where N is the number of rows (observations), and M the number of columns (variables) of the input data matrix, $\mathbf{X}$). There is one eigenvalue for each discriminant function. Letting Σ_W denote the within-group covariance and Σ_B denote the between-group covariance matrix, the problem for the discriminant function is to find projections in the data that maximizes the ratio between the between-group variance and the within-group variance or the so-called Rayleigh coefficient (or Fisher's criteria)

$$J(\mathbf{a}) = \frac{\mathbf{a}^t \Sigma_B \mathbf{a}}{\mathbf{a}^t \Sigma_W \mathbf{a}}$$

Solving this equation for **a** yields the solution

$$(\Sigma_W^{-1} \Sigma_B - \lambda \mathbf{I})\mathbf{a} = 0$$

which can be identified as the all-too-familiar structure of an eigenvalue–eigenvector problem.

As for the PCA, a set of eigenvectors (discriminant functions) and eigenvalues is obtained. The ratio of the eigenvalues obtained indicates the relative discriminating power of the discriminant functions. For example, if the ratio of two eigenvalues is 1.6, then the first discriminant function explains 60% more between-group variance in the dependent categories than does the second discriminant function.

5.8 SIMILARITIES AND DISTANCES BETWEEN DATA

If data can be represented as points in an appropriate space, dissimilar entries are regarded as distant from each other, and similar entries close to each other. In such a space, a distance function $d_{ij} = d(\mathbf{x}_i, \mathbf{x}_j)$ captures such differences taking two observations $\mathbf{x}_i$ and $\mathbf{x}_j$ as input.

5.8.1 Continuous Functions

This section presents different quantitative dissimilarity measures, ranging from the more common to the more special, and providing their mathematical form.

5.8.1.1 Weighted $\mathbf{L}_p$-Norm. For continuous data, it is most common to calculate the dissimilarity between two patterns using the L_p-norm ($\|\cdot\|_p$)

$$d(\mathbf{x}_i,\mathbf{x}_j) = \|\mathbf{w}(\mathbf{x}_i - \mathbf{x}_j)\|_p = \left[\sum_{\forall k} w_k \, |x_{ik} - x_{jk}|^p\right]^{1/p}$$

For $\mathbf{w} = \mathbf{1}$, the most widely used are the 1-norm, 2-norm, and ∞-norm ($\|(\mathbf{x}_i - \mathbf{x}_j)\|_\infty = \max |x_{in} - x_{jn}|$, for $n = 1,\ldots,N$) referred to as the City-block or Manhattan distance, the Euclidian, and the Chebychev distances. Figure 5.14 illustrates the behavior of L_p for $p = \{1, 2, 3, \infty\}$. These do, however, depend strongly on the scales on which the features are measured. One way to minimize this strong dependence is by standardization, where data is rescaled to have zero mean and *unit variance*. Standardization is often used prior to many multivariate analysis methods, such as, e.g., PCA, and is done in particular when the individual features (variables) exists on different scales.

5.8.1.2 Mahalanobis. A generalization of the Euclidean distance, defined in terms of the covariance matrix Σ

$$d(\mathbf{x}_i,\mathbf{x}_j) = \frac{1}{|\det \Sigma|^p}(\mathbf{x}_i - \mathbf{x}_j)^t \Sigma^{-1}(\mathbf{x}_i - \mathbf{x}_j)$$

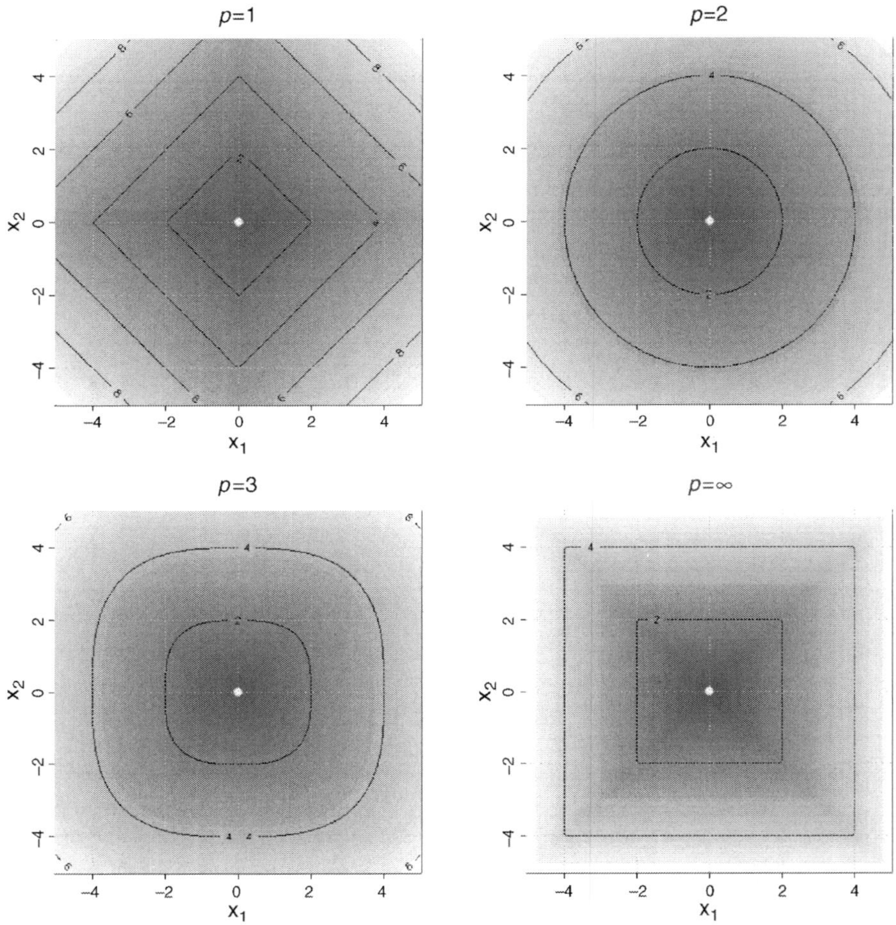

Figure 5.14 The behavior of the L_p norm for different values of p in a two-dimensional space. The intensities (contours) illustrate the L_p distances relative to the center point (0,0).

Σ^{-1} is the matrix inverse of Σ, and the superscript "t" denotes transposed. If Σ is the identity matrix **I**, the Mahalanobis distance reduces to the squared Euclidean distance (L_2-norm).

5.8.1.3 Generalized Euclidean. In a further generalization of the Mahalanobis distance where the matrix **W** is positive definite but not necessarily the inverse of a covariance matrix, the multiplicative factor is omitted

$$d(\mathbf{x}_i, \mathbf{x}_j) = (\mathbf{x}_i - \mathbf{x}_j)^t \mathbf{W} (\mathbf{x}_i - \mathbf{x}_j)$$

5.8.1.4 Correlation. The correlation similarity measure is the covariance, divided by the variances, and takes values between -1 and 1.

$$d(\mathbf{x}_i,\mathbf{x}_j)=\mathrm{corr}(\mathbf{x}_i,\mathbf{x}_j)=\frac{\sum_{\forall k}(x_{ik}-\bar{x}_i)(x_{jk}-\bar{x}_j)}{\sqrt{\sum_{\forall k}(x_{ik}-\bar{x}_i)^2}\sqrt{\sum_{\forall k}(x_{jk}-\bar{x}_j)^2}}$$

With this measure, the relative direction of the two observation vectors is important. The correlation similarity is closely related to the cosine of the angle between the two observations measured from their center of mean.

5.8.1.5 The Angle. Is defined as

$$d(\mathbf{x}_i,\mathbf{x}_j)=\mathrm{corr}(\mathbf{x}_i,\mathbf{x}_j)=\frac{\sum_{\forall k}x_{ik}x_{jk}}{\sqrt{\sum_{\forall k}x_{ik}^2}\sqrt{\sum_{\forall k}x_{jk}^2}}$$

which is the cosine of the angle between the two observation vectors measured from orego and takes values in the interval of -1 to 1.

The distance function concept can be extended to embrace more specialized applications.

5.8.1.6 Relative Entropy. This (information-theoretical) quantity is defined for probability distributions, as

$$d(\mathbf{x}_i \parallel \mathbf{x}_j)=\sum_{\forall k}x_{ik}\log\frac{x_{ik}}{x_{jk}}.$$

The relative entropy is only meaningful if the entries of $\mathbf{x}_i$ and $\mathbf{x}_j$ are non-negative and $\sum_{\forall k}x_{ik}=\sum_{\forall k}x_{jk}=1$. This metric is often used for database retrieval purposes, where the first argument should be a query vector, and the second argument the vector from the database.

5.8.1.7 χ^2-Distance. It is defined only for probability distributions as

$$d(\mathbf{x}_i,\mathbf{x}_j)=\sum_{\forall k}\frac{x_{ik}^2-x_{jk}^2}{x_{jk}^2}.$$

It lends itself to a natural interpretation only if the entries of $\mathbf{x}_i$ and $\mathbf{x}_j$ are non-negative and $\sum_{\forall k}x_{ik}=\sum_{\forall k}x_{jk}=1$.

		Observation j		
		1	0	
Observation i	1	a	b	a+b
	0	c	d	c+d
		a+c	b+d	K

Figure 5.15 Contingency table of the outcome when comparing K binary variables between two observations x_{ik} and x_{jk}. a denotes the number of variables that are "1" for both objects, b denote the number of variables that are "1" for x_{ik} and "0" for x_{jk}, c denote the number that are "0" for x_{ik} and "1" for x_{jk}, and finally d denotes the number that are "0" for both observations. Finally, $K = a + b + c + d$.

5.8.2 Binary Functions

Whereas most of the above-described distance measures are applied on to the quantitative data, a special case is that of having qualitative (binary) outcome: if the binary variable x_i belongs to only two states, e.g., $x_i \in \{0, 1\}$ and if a set of entries are described by such K binary variables, e.g., presence or absence of specific metabolites in a fungal extract.

If we have a pair of observations $\mathbf{x}_i = \{x_{ik}\}$ and $\mathbf{x}_j = \{x_{jk}\}$, relations between the presence and absence of each single metabolite in both species can be established as illustrated in Figure 5.15.

There are many measures of the (dis)similarity between binary variables. In the following we describe some of the most common.

5.8.2.1 Simple Matching Coefficient. Constructing a similarity measure from the above "components" is intuitive, e.g., all matches $(c + d)$ relative to all possibilities, i.e., matches plus mismatches $(c + d) + (a + b)$, yields

$$d(\mathbf{x}_i, \mathbf{x}_j) = \frac{c+d}{a+b+c+d}$$

called the simple matching coefficient (Sneath and Sokal, 1973). Here, equal weight is given to matches and mismatches.

5.8.2.2 Jaccard. When absence of a feature in both objects is deemed to convey no information, then d should not occur in a similarity measure. Omitting d from the simple matching coefficient, one obtains the Jaccard (alias Tanimoto) similarity measure.

TABLE 5.1 Table of Binary (Dis)similarity Measures.

Name	Function
Simple matching coefficient	$d(\mathbf{x}_i,\mathbf{x}_j)=\dfrac{c+d}{a+b+c+d}$
Jaccard	$d(\mathbf{x}_i,\mathbf{x}_j)=\dfrac{c}{a+b+c}$
Hamming, Manhattan, taxi-cab, City-block	$d(\mathbf{x}_i,\mathbf{x}_j)=a+b$
Dice	$d(\mathbf{x}_i,\mathbf{x}_j)=\dfrac{c}{0.5[(a+c)+(b+c)]}$
Yule	$d(\mathbf{x}_i,\mathbf{x}_j)=\dfrac{cd-ab}{cd+ab}$
Euclidian	$d(\mathbf{x}_i,\mathbf{x}_j)=\sqrt{a+b}$
Variance	$d(\mathbf{x}_i,\mathbf{x}_j)=\dfrac{a+b}{4(a+b+c+d)}$
Pattern difference	$d(\mathbf{x}_i,\mathbf{x}_j)=\dfrac{ab}{(a+b+c+d)^2}$

$$d(\mathbf{x}_i,\mathbf{x}_j)=\frac{c}{a+b+c}$$

Table 5.1 lists some of the distance measures that are recommended in situations when the coding by "1" or "0" is arbitrary (i.e., if the binary variable is in fact nominal) or if double zeros are considered to be as significant carriers of information as double "0."

Methods for the analysis of binary response variables and related topics can be found in Sneath and Sokal (1973), McCullagh and Nelder (1997), and Cox and Snell (1989).

Example: Please consider the simple case containing four observations. Each observation consists of 10 binary measurements. In this example, it is not important what each of the binary measurements indicate, and you are welcome to use your imagination.

$$\mathbf{X}=\begin{bmatrix}1&1&1&1&1&0&0&0&0&1\\1&1&1&1&1&0&0&1&0&1\\1&1&1&1&0&1&1&1&1&1\\1&0&0&1&0&1&1&1&1&1\end{bmatrix}$$

The task is to calculate the binary Euclidian distance among all four observations (see Table 5.1).

The binary Euclidian distance gives us the following distance matrix:

$$\mathbf{D} = \begin{bmatrix} 0 & 1.0 & 2.2 & 2.6 \\ 1.0 & 0 & 2.0 & 2.4 \\ 2.2 & 2.0 & 0 & 1.4 \\ 2.6 & 2.4 & 1.4 & 0 \end{bmatrix}$$

This distance matrix depicts the interrelationship between all points in **X** (or the reduced space) and can be used as input to, e.g., clustering algorithms.

5.9 CLUSTERING TECHNIQUES

Clustering can be considered the most important unsupervised learning problem used to find structures in a collection of unlabeled observations. A loose definition of clustering could be "the process of organizing objects into groups whose members are similar in some way." A cluster is therefore a collection of objects which are "similar" to them and are "dissimilar" to the objects belonging to other clusters.

In general two different types of clustering methods exist: the hierarchical and nonhierarchical methods. Hierarchical clustering algorithms typically organize data in tree structures with main clusters containing subclusters that contain even smaller clusters and so on. Nonhierarchical clustering, on the contrary, partitions data on one level only.

The different algorithms often have different parameters that the user needs to choose. For instance, an algorithm might want to know how similar two objects must be to be part of the same cluster, or the user might have to decide how many clusters the algorithm should produce. Furthermore, the user must decide what kind of similarity or distance measurement to use.

Common to all clustering algorithms is the distance measure between data points. If the components in the data vectors are all on the same physical (comparable) scale, then the simple Euclidean distance metric is sufficient to successfully group similar observations. However, even in well-behaved cases the Euclidean distance can sometimes be misleading.

5.9.1 Hierarchical Clustering

Hierarchical clustering can be divided into agglomerative (bottom-up) and divisive clustering (top-down) (Anderberg, 1973; Hartigan, 1975; Kaufman and Rousseeuw, 1990). Divisive clustering starts with one big cluster containing all data, and proceeds by dividing this cluster into successively smaller clusters. Agglomerative clustering starts with the individual objects, joining more and more together, creating bigger and bigger clusters.

Hierarchical clustering has more or less become the standard clustering method for most biological data. The agglomerative variant works as follows:

(i) The similarity between each pair of objects is calculated.
(ii) The two most similar objects are merged together to create a cluster.
(iii) The similarity between this cluster and all other objects is calculated.
(iv) Steps 2 and 3 are repeated, fusing together objects and objects, objects and clusters, or clusters and clusters, until all are contained in one cluster.

The result is a so-called dendrogram—a tree diagram where the clustering on different levels is visualized.

Hierarchical agglomerative methods are often characterized by the shape of the clusters they tend to find. Given a distance matrix $d(\mathbf{x}_i, \mathbf{x}_j)$ (see Section 5.8) between objects, there are various ways to define the distance between two clusters C_k and C_l. Different hierarchical clustering algorithms implement different distance measures. Among others, there are:

5.9.1.1 Single Linkage. Single linkage defines the distance between the objects C_k and C_l as

$$\min_{\mathbf{x}_i \in C_k, \mathbf{x}_j \in C_l} d(\mathbf{x}_i, \mathbf{x}_j),$$

i.e., the shortest distance between any pair of objects belonging to C_k andC_l, respectively.

5.9.1.2 Complete Linkage. Complete linkage uses the largest distance between any pair of objects belonging to C_k and C_l, respectively, i.e.,

$$\max_{\mathbf{x}_i \in C_k, \mathbf{x}_j \in C_l} d(\mathbf{x}_i, \mathbf{x}_j).$$

Furthermore, Sneath and Sokal (1973) proposed several other linkage methods which can be briefly summarized.

5.9.1.3 Unweighted Pair-Group Average (UPGMA). The distance between two clusters is calculated as the average distance between all pairs of objects in the two different clusters. This method is also very efficient when the objects form natural distinct "clumps," however, it performs equally well with elongated, "chain" type clusters.

5.9.1.4 Weighted Pair-Group Average (WPGMA). This method is identical to the UPGMA method, except that in the computations, the size of the respective clusters (i.e., the number of objects contained in them) is used as a weight. Thus, this method (rather than the previous method) should be used when cluster sizes are suspected to be very uneven.

5.9.1.5 Unweighted Pair-Group Centroid (UPGMC). The centroid of a cluster is the average point in the multidimensional space defined by the dimensions. In a sense, it is the center of gravity for the respective cluster. In this method, the distance between two clusters is determined as the difference between centroids.

5.9.1.6 Weighted Pair-Group Centroid (Median). This method (WPGMC) is identical to the previous one, except that weighting is introduced into the computations to take into consideration differences in cluster sizes (i.e., the number of objects contained in them). Thus, when there are (or one suspects there to be) considerable differences in cluster sizes, this method is preferable to the previous one.

5.9.1.7 Ward's Method. This method (proposed in 1963 by Ward) is distinct from all other methods because it uses an analysis of variance approach to evaluate the distances between clusters. In short, this method attempts to minimize the sum of squares (SS) of any two (hypothetical) clusters that can be formed at each step. In general, this method is regarded as very efficient; however, it tends to create small clusters.

A supplementary overview of different hierarchical clustering methods, and descriptions of reaching a consensus between several clustering's can be found in Hubert (1974), Baker and Hubert (1975), Gordon (1987), and Gordon (1999). Alternative methods for hierarchical clustering can be found in Kleiner and Hartigan (1981).

Example: To illustrate how the hierarchical clustering works, we now do a hierarchical clustering of the observations based on the distance matrix calculated in the example in Section 5.8. We use single linkage to join together clusters.

The Euclidian distance gave us the following distance matrix:

$$\mathbf{D} = \begin{bmatrix} 0 & 1.0 & 2.2 & 2.6 \\ 1.0 & 0 & 2.0 & 2.4 \\ 2.2 & 2.0 & 0 & 1.4 \\ 2.6 & 2.4 & 1.4 & 0 \end{bmatrix}$$

Initially, all observations are treated as single clusters. The distance matrix is then used to do a hierarchical clustering in the following steps (see Figure 5.16):

(a) $D(r, s) = \text{Min } \{D(i,j)$: where object i is in cluster "r" and object j is cluster "s"$\}$ = **1.0** (A and B). Now A and B has been merged into one new cluster.

(b) $D(r, s) = \text{Min } \{D(i,j)$: where object i is in cluster r and object j is cluster $s\}$ = **1.4** (C and D) (*Remark*: the distance from C and D to the "red" cluster is in the range of 2.0–2.6).
Now C and D have been merged into another cluster.

(c) $D(r, s) = \text{Min } \{D(i,j)$: Where object i is in cluster r and object j is cluster $s\}$ = **2.0** (AB and CD) (*Remark*: the distance from C and D to the "red" cluster is in the range of 2.0–2.6)

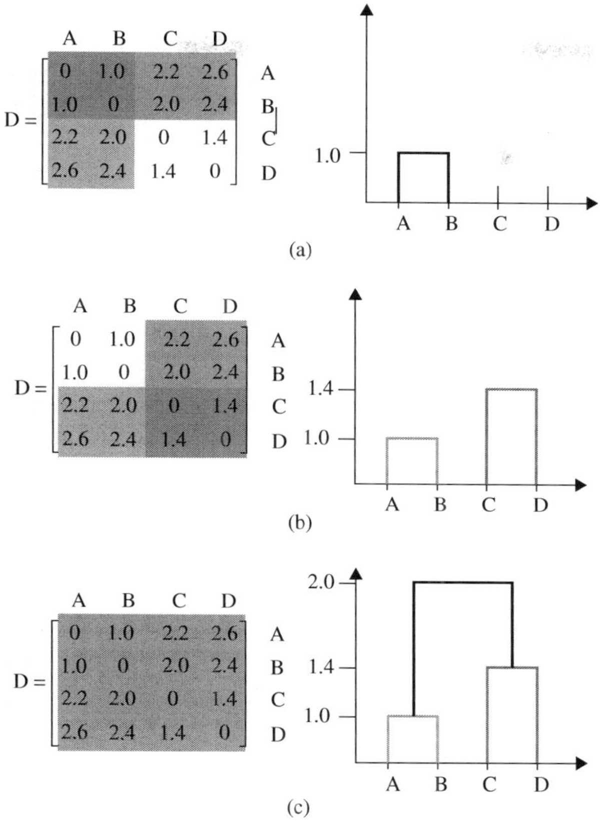

Figure 5.16 Illustration of the hierarchical clustering method using single linkage. (See color plates.)

Finally, we have merged all observations into one cluster. The result can be seen in Figure 5.16c (right figure).

5.9.2 *k*-Means Clustering

A nonhierarchical approach to clustering is to specify a desired number of clusters, say, *k*, then assign each case (object) to one of the *k* clusters so as to minimize the measure of dispersion within the clusters. A very common way to measure the ability to separate between clusters is by the sum of distances from the mean of each cluster. The problem can be set up as an integer-programming problem, but because solving integer programs with a large number of variables is time consuming, therefore, clusters are often computed using a fast, heuristic method that generally produces good (but not necessarily optimal) solutions. The *k*-means algorithm is one such method.

k-Means training starts with a single cluster, with the mean of the data used as a center. This cluster is split into two and the means of the new clusters are calculated and used as centers. These two clusters are again split and the process continues iteratively until the specified number of clusters is obtained. If the specified number of clusters is not a power of two, then the nearest power of two above the number specified is chosen, and then the least important clusters are removed and the remaining clusters are again iteratively trained to get the required number of clusters. Alternatively, the user can specify a random start algorithm that generates k cluster centers randomly, and goes ahead by fitting the data points in those clusters. This process is repeated for as many random starts as specified by the user until the best start value is found. The outputs based on this value are displayed.

5.10 CLASSIFICATION TECHNIQUES

Classification is a prediction or learning problem by which the variables are predicted assuming that one of the K unordered values, $Y \in \{c_1, c_2, \ldots, c_K\}$, arbitrarily can be labeled as $\{1,2,\ldots, K\}$ or sometimes $\{0,1,2,\ldots, K - 1\}$. The K values correspond to K predefined classes, e.g., tumor class, bacteria type, fungal specie, mutant, etc.

The task is to classify an object into one of the K classes on the basis of the observed measurements X,

$$\mathbf{X} = \begin{bmatrix} \mathbf{x}_1 \\ \vdots \\ \mathbf{x}_n \\ \vdots \\ \mathbf{x}_N \end{bmatrix} = \begin{bmatrix} x_{11} & \cdots & x_{1m} & \cdots & x_{1M} \\ \vdots & \ddots & \vdots & \ddots & \vdots \\ x_{n1} & \cdots & x_{nm} & \cdots & x_{nM} \\ \vdots & \ddots & \vdots & \ddots & \vdots \\ x_{N1} & \cdots & x_{Nm} & \cdots & x_{NM} \end{bmatrix}$$

i.e., predict the classes Y from X.

A classifier or predictor is a function, g, that for all K classes is a mapping from the space spanned out by all variables measured for each observation into the integers $\{1,2,\ldots, K\}$. In other words, a classifier partitions the space into K disjoint and exhaustive subsets, $\{A_1, A_2,\ldots, A_K\}$, in such a way that a sample of, e.g., an expression profile $\mathbf{x} = \{x_1, x_2,\ldots, x_M\} \in A_k$, will be predicted to be in class k. A formal way to write this mapping is

$$g:\mathbf{x} \rightarrow \{1,2,\ldots, K\}$$

which corresponds to say that the function g takes an observation, $\mathbf{x}$, that is supposed to belong to one of the K classes, $\mathbf{x} \in A_k$, and assigns it to one of these K labels, $\hat{y} = g(\mathbf{x}) = k$. Classifiers are built from past experience, i.e., from observations which are known to belong to certain classes. Such observations comprise the learning (training) set

$$L = \{(\mathbf{x}_1, y_1),\ldots,(\mathbf{x}_N, y_N)\}$$

containing pairs of known relations between class and characters. The classifier is then built based upon the information about these relations. In the following we give an introduction to how the classifier can be built.

5.10.1 Decision Theory

Classification can be viewed as a statistical decision theory problem. Let us assume that the observations are independently and identically distributed from an unknown multivariate distribution. The class k prior, or proportion of objects of class k in the population, is denoted as $\pi_k = p(Y = k)$. Objects in class k have feature vectors with class conditional density $p_k(\mathbf{x}) = p(\mathbf{x} \mid Y = k)$.

If (unrealistically) both π_k and $p_k(\mathbf{x})$ are known, this problem has a solution—the Bayes rule. This unrealistic situation also delimits the upper bounds of the performance of classifiers. In the more realistic setting where these quantities are not known—the Bayes risk.

In order to obtain a solution to the problem, a loss-function needs to be added. The loss function $L(i, j)$ simply elaborates the loss incurred if a class i case is erroneously classified as belonging to class j. The risk function for a classifier is the expected loss when using it to classify, that is,

$$R(g) = E\left[L(Y, g(\mathbf{x}))\right] = \sum_{\forall k} E\left[L(k, g(\mathbf{x})) \mid Y = k\right]\pi_k$$
$$= \sum_{k} \int L(k, g(\mathbf{x})) p_k(\mathbf{x}) \pi_k$$

Typically, $L(i, i) = 0$ (correct classification), and in many cases the loss is symmetric thus having $L(i, j) = 1$ for $i \neq j$, and therefore, an error of one type is equivalent to making an error of a different type. Then the risk can be simplified to the misclassification rate

$$p\left(g(\mathbf{x}) \neq Y\right) = \sum_{k} \int_{g(\mathbf{x}) \neq k} p_k(\mathbf{x}) \pi_k$$

However, in some important cases such as diagnosis, the loss function is not symmetric.

In the unlikely situation where the classes have conditional densities $p_k(\mathbf{x}) = p(\mathbf{x} \mid Y = k)$ and the class priors $\pi_k = p(Y = k)$ are known, then

$$p(k \mid \mathbf{x}) = \frac{\pi_k p_k(\mathbf{x})}{\sum_{\forall l} \pi_l p_l(\mathbf{x})}$$

denotes the posterior probability of class k given feature vector $\mathbf{x}$.

The Bayes rule predicts the class of an observation $\mathbf{x}$ by that of highest posterior probability

$$g_B(\mathbf{x}) = \arg\max_{k} \left[p(k \mid \mathbf{x})\right] = \arg\max_{k} \left[\frac{\pi_k p_k(\mathbf{x})}{\sum_{\forall l} \pi_l p_l(\mathbf{x})}\right]$$

The Bayes rule minimizes the total risk under a symmetric loss function—Bayes risk. In the case where the loss-function is general, i.e., has varying losses added to the different classes, the classification rule minimizes the total risk

$$g_B(\mathbf{x}) = \arg\max_j \left[\sum_{i=1}^{K} L(i, j) p(i \mid \mathbf{x}) \right]$$

Suitable adjustments can be made for other loss functions, and to accommodate the doubt and outlier classes.

5.10.2 *k*-Nearest Neighbor

Nearest neighbor methods are based on a measure of distance between observations, e.g., the Euclidean distance or one minus the correlation between two metabolite profiles. The k-nearest neighbor rule, k-NN (Fix and Hodges, 1951), classifies an observation $\mathbf{x}$ as follows

1. Find the k observations in the learning set that are closest to $\mathbf{x}$;
2. Predict the class of $\mathbf{x}$ by majority vote, i.e., choose the class that is most common among those k observations.

Note that for a large enough number of neighbor's k, the k-NN classifier suggests a simple estimate of the class posterior probabilities: the proportion of votes for each class. The class posterior probability estimates $p(k \mid \mathbf{x})$ may be used to measure confidence for individual predictions. In general, classifiers with $k = 1$ are quite successful.

The number of neighbor's k can be chosen by cross-validation. Each observation in the learning set is treated in turn as if it were in a test set: the distance to all of the other learning set samples (except itself) is computed, and it is classified by the nearest neighbor rule. The classification for each observation on the learning set is then compared to the truth, producing a cross-validation error rate. This is done for a number of k's, and the k for which the cross-validation error rate is smallest, is retained.

Several extensions being based on the k-NN classifier have been developed. Among these are the addition of a voting scheme dealing with issues of unequal class priors, differential misclassification costs, and feature selection (Brown and Koplowitz, 1979; Friedman, 1994). Finally, Hastie and Tibshirani (1996) described the discriminant adaptive nearest neighbor (DANN) procedure, in which the distance function is based on local discriminant information.

5.10.3 Tree-Based Classification

Classification trees are used to predict membership of cases or objects in the classes of a categorical dependent variable from their measurements on one or more predictor variables (Breiman et al., 1984).

The goal of classification trees is to predict or explain responses on categorical dependent variables in X, and as such, the available techniques have much in common with the techniques used in the more traditional methods of discriminant analysis and cluster analysis described earlier. The flexibility of classification trees makes them an attractive analysis option, but this is not to say that their use is recommended to the exclusion of more traditional methods. Indeed, when the typically more stringent theoretical and distributional assumptions of more traditional methods are met, the traditional methods may be preferable. But as an exploratory technique, or as a technique of last resort when traditional methods fail, classification trees are, in the opinion of many researchers, unsurpassed.

5.11 INTEGRATED TOOLS FOR AUTOMATION, LIBRARIES, AND DATA EVALUATION

One of the challenges of multi-targeted compound analysis is the development of automated chromatogram evaluation. Many software packages delivered with the GC- or LC–MS system (Xcalibur, ThermoElectron, Austin, US or HP Chemstation, Agilent, Palo Alto, US) are able to use either self-created or commercial mass spectra libraries for peak detection, identification, and integration. The limitation of these software packages are that, they search and integrate only targets, which the researcher has to know and enter into the search lists. This situation has been improved recently with the development of novel software packages for untargeted chromatogram evaluation based on mass spectral deconvolution.

Recently, other helpful commercial and free software packages have become available. Examples include MSFacts for GC–MS (Duran et al. 2003) or MetAlign for GC- and LC–MS (www.metalign.nl), which automatically import, reformat, align, correct the baseline, and export large chromatographic data sets to allow more rapid visualization and interrogation of metabolomics data. To date, these software packages are indispensable for unambiguous data extraction. Very recently, a novel software package named AnalyzerPro (www.spectralworks.com; Runcorn, Cheshire, UK) has been made available which meets the high requirements of an automatic GC–MS and also LC–MSn chromatogram evaluation. In addition to signal deconvolution, mass spectra library matching and quantification, the implementation of retention time indices (RI) for improved signal identification are beneficial. Retention times of eluted substances following chromatographic separation do change dramatically over time. Retention time indices include for their calculation a range of added time references (e.g., long-chain alkanes), and therefore provide a better prediction of the absolute retention time of the analytes. In addition, retention time indices are very stable both within and between systems, allowing valid system to system comparisons, provided that injection, separation, and ionization parameters are kept similar (Schauer et al. 2005).

REFERENCES

Anderberg MR. 1973. *Cluster Analysis for Applications* Academic Press, New York, NY.

Antoniou A. 1993. *Digital Filters: Analysis, Design, and Applications* McGraw-Hill, New York, NY.

Baker FB, Hubert LJ. 1975. Measuring the power of hierarchical cluster analysis. J Am Stat Assoc 70:31–38.

Breiman L, Friedman J, Olshen RA, Stone CJ. 1984. Classification and regression trees. Wadsworth.

Brown M, Dunn WB, Ellis DI, Goodacre R, Handl J, Knowles JD, O'Hagan S, Spasić I, Kell DB. 2005. A metabolome pipeline: From concept to data to knowledge. Metabolomics 1:39–51.

Brown TA and Kolpitz J. 1979. The Weighted Nearest Neighbor Rule for Class Dependent Samples Sizes, IEEE Trans. Information Theory, vol. 25, pp. 617–619, Sept.

Cox and Snell 1989. Analysis of Binary Data, 2nd ed. Chapman & Hall.

Duran AL, Yang J, Wang L and Sumner LW. 2003. Metabolomics Spectral Formatting, Alignment and Conversion Tools (MSFACTs). Bioinformatics 19(17): 2283–2293.

Fix E and Hodges JL. 1951. Discriminatory Analysis: Nonparametric Discrimination, Project 21-49-004, Report #4, USAF School of Aviation Medicine, Randolph Field, Texas.

Friedman JH. 1994. Flexible Metric Nearest Neighbor Classification. Technical Report 113, Stanford University Statistics Department. http://citeseer.ist.psu.edu/friedman94flexible.html

Gollmer K, Posten C. 1996. Supervision of bioprocesses using a dynamic time warping algorithm. Control Eng Pract 4:1287–1295.

Gordon AD. 1987. A review of hierarchical classification. J. Royal Stat. Soc A 150:119–137.

Gordon AD. 1999. *Classification* (2nd edition), Chapmann and Hall, London.

Hartigan J. 1975. *Clustering Algorithms* John Wiley & Sons, New York, NY.

Hastie T, Tibshirani R, Friedman J. 2001. *The Elements of Statistical Learning: Data Mining, Inference and Prediction.* Springer-Verlag, Berlin.

Hubert L. 1974. Approximate evaluation techniques for the single-link and complete link hierarchical clustering procedures. J Am Stat Assoc 69:698–704.

Hillier FS, Liebernan GJ. 2001. *Introduction to Operations Research* (7th edition), McGraw-Hill, New York.

Itakura F. 1975. Minimum prediction residual principle applied to speech recognition. IEEE Trans ASSP AS23:67–72.

Jenkins H, Johnson H, Kular B, Wang T, Hardy N. 2005. Towards supportive data collection tools for plant metabolomics. Plant Physiol 138:67–77.

Jenkins H, Hardy N, Beckmann M, Draper J, Smith AR, Taylor J, Fiehn O, Goodacre R, Bino RJ, Hall R, Kopka J, Lane GA, Lange BM, Liu JR, Mendes P, Nikolau BJ, Oliver SG, Paton NW, Rhee S, Roessner-Tunali U, Saito K, Smedsgaard J, Sumner LW, Wang T, Walsh S, Wurtele ES, Kell DB. 2004. A proposed framework for the description of plant metabolomics experiments and their results. Nature Biotechnol 22:1601–1606.

Kaufman L, Rousseeuw PJ. 1990. *Finding Groups in Data: An Introduction to Cluster Analysis*, New York: John Wiley & Sons, Inc.

Kleiner B, Hartigan JA. 1981. Representing points in many dimensions by trees and castles. J Am Stat Assoc 76:260–269.

McCullagh P and Nelder JA (Second edition 1989). Generalized Linear Models. Chapman and Hall: London. (mathematical statististics of generalized linear model). Reprinted 1997.

Mitra SK. 1998. *Digital Signal Processing: A Computer-Based Approach* Mcgraw-Hill, New York, NY.

Nielsen NPV, Carstensen JM, Smedsgaard J. 1998. Aligning of single and multiple wavelength chromatographic profiles for chemometric data analysis using correlation optimised warping. J Chromatogr A 805:17–35.

Podani J. 1994. *Multivariate Data Analysis in Ecology and Systematics* Volume 6 of Ecological Computations Series (ECS). SPB Academic Publishing bv, 2509 GC The Hague, The Netherlands.

Pravdova V, Walczak B, Massart DL. 2002. A comparison of two algorithms for warping of analytical signals. Anal Chim Acta 456:77–92.

Reiner E, Abbey LE, Moran TF, Papamichalis P, Shafer RW. 1979. Characterization of normal human cells by pyrolysis gas-chromatography mass spectrometry. Biomed Mass Spectrom 6:491–498.

Sakoe H, Chiba S. 1978. Dynamic-programming algorithm optimization for spoken word recognition. IEEE Trans ASSP 26:43–49.

Schauer N, Steinhauser D, Strelkov S, Schomburg D, Allison G, Moritz T, Lundgren K, Roessner-Tunali U, Forbes MG, Willmitzer L, Fernile AR, Kopka J. 2005. GC-MS libraries for the rapid identification of metabolites in complex biological samples. FEBS Letters, 579, 1332–1337.

Sneath PHA, Sokal RR. 1973. *Numerical taxonomy* W. H. Freeman & Co., San Francisco.

Stein SE, Scott DR. 1994. Optimization and testing of mass spectral search algorithms for compound identification. J Am Soc Mass Spectrosc 5:859–866.

Tan H-W, Brown S. 2002. Wavelet analysis applied to removing nonconstant, varying spectroscopic background in multivariate calibration. J Chemom 16:228–240.

Tomasi G, van den Bergand F, Andersson C. 2004. Correlation optimized warping and dynamic time warping as preprocessing methods for chromatographic data. J. Chemometrics. 18:231–241.

Wang CP, Isenhour TL. 1987. Time-warping algorithm applied to chromatographic peak matching gas-chromatography Fouriers-transform infrared mass-spectrometry. Anal Chem 59:649–654.

Ward JH. 1963. Hierarchical grouping to optimize an objective function. J Am Stat Assoc 58:236–244.

Liu B, Sera Y, Matsubara N, Otsuka K, Terabe S. 2003. Signal denoising and baseline correction by discrete wavelet transform for microchip capillary electrophoresis. Electrophoresis 24:3260–3265.

Depczynski U, Jetter K, Molt K, Niemfller A. 1997. The fast wavelet transform on compact intervals as a tool in chemometrics: I. Mathematical background. Chemom Intell Lab Syst 39:19–27.

Cai T, Zhang D, Ben-Amotz D. 2001. Enhanced chemical classification of Raman images using multiresolution wavelet transformation. Appl Spectrosc 55:1124–1130.

PART II

CASE STUDIES AND REVIEWS

6

YEAST METABOLOMICS: THE DISCOVERY OF NEW METABOLIC PATHWAYS IN *SACCHAROMYCES CEREVISIAE*

By Silas G. Villas-Bôas

The brewers' and bakers' yeast Saccharomyces cerevisiae was the first eukaryote to have its complete genome sequenced, and it was a turning point in molecular biology because this yeast represents a flexible experimental system for eukaryotic cell biology. The challenge is now to discover what each of the 6000 genes does, and how they are regulated in a living yeast cell. In this chapter we will review a series of metabolomics experiments that lead to the discovery of a new metabolic pathway in S. cerevisiae as well as the detection and identification of de novo metabolites in yeast culture, giving evidence of many more metabolic pathways yet to be described in this intensively studied microorganism.

6.1 INTRODUCTION

Yeast cells, especially *S. cerevisiae*, have been intensively studied because of their great importance in society as a cell factory for production of beer, wine, bread, ethanol, and many different pharmaceuticals. They are easy to manipulate genetically and to cultivate, and their many biological pathways resemble those of mammalian cells, making them a very useful model organism to study cell physiology and biochemistry. However, the importance of yeasts goes much further than being a model organism for mammalian cells. The production of ethanol by fermentation of fruit

Metabolome Analysis: An Introduction, by Silas G. Villas-Bôas, Ute Roessner, Michael A. E. Hansen, Jorn Smedsgaard and Jens Nielsen

juices or by hydrolytic breakdown of starch from cereal flours has been the most successful of human industries since ancient time. It is well recognized that the main and invariable agent of these biotechnological applications is the yeast *S. cerevisiae*.

S. cerevisiae, the famous protagonist of centuries of bread, wine, and beer making, probably the first living organism to be domesticated by the man, is one of the best known organisms on Earth, be it physiologically, genetically, morphologically, or technologically. In spite of the fact that the genome of *S. cerevisiae* was completely sequenced in 1996 (Goffeau et al., 1996), a vast number of its protein-encoding genes still have unknown functions, and our knowledge concerning how these approximately 6000 genes are regulated and the ways in which their products interact with each other, gets even narrower.

To enhance the functional analysis of the yeast genome, a large European research network, called EUROFAN (Oliver, 1996), created a library of yeast strains each of which carry a specific deletion of an ORF that encodes a protein. Today, as a result of a cooperative work between different projects (i.e., BMBF, EUROFAN I, and EUROFAN II, as part of the worldwide yeast gene deletion project), the Institute of Microbiology located at the Biocenter of the University of Frankfurt runs the **EURO**pean ***S**accharomyces **C**erevisiae* **AR**chive for **F**unctional analysis (EUROSCARF) (web.uni-frankfurt.de/fb15/mikro/euroscarf), which holds a strain collection setup for the deposit and delivery of biological materials generated in genome analysis networks. Thereby, one can easily get *S. cerevisiae* strains that carry specific single deletion in virtually every single ORF of the whole yeast genome, which make *S. cerevisiae* an excellent eukaryote model to study most biological phenomenon at the molecular level.

The present case study will go through a series of metabolite analysis of yeast samples that began with general metabolite profiling of *S. cerevisiae* cultivated at different environmental conditions and ending with ^{13}C-labeling experiments to confirm hypothesis raised from metabolite profiling data. Hereby, we will illustrate how metabolomics alone can be a powerful tool to generate hypothesis that can be later tested using a more targeted approach.

6.2 BRIEF DESCRIPTION OF THE METHODOLOGY USED

The detailed methodology used to obtain the data discussed here can be found in Villas-Bôas et al. (2005a,b) and in Devantier et al. (2005). In the following we will summarize the basic procedures used for the metabolite analysis.

6.2.1 Sample Preparation

Figure 6.1 summarizes the basis of the sample preparation procedure used for all experiments. If not stated otherwise, the samples for analysis of intracellular metabolites were harvested at mid-exponential phase using syringes, and quenched in nonbuffered cold methanol solution (−40°C). The biomass was separated from the quenching solution by centrifugation at low temperature (−20°C), and 1 ml chloroform was added to the recovered pellet and stored at −80°C before metabolite

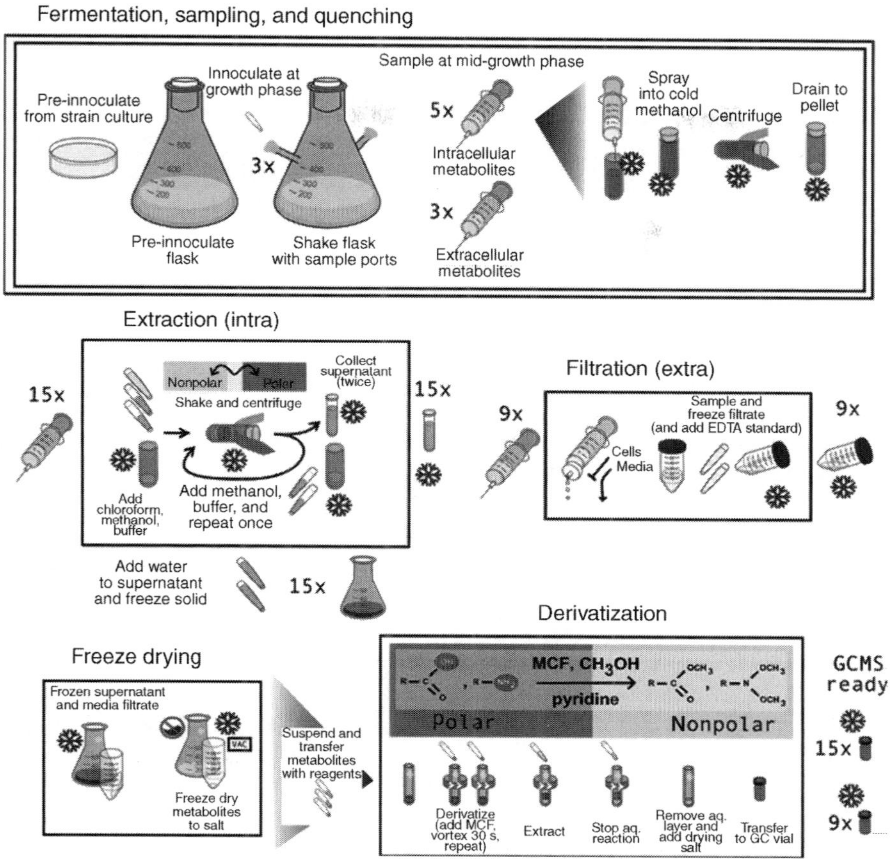

Figure 6.1 Summary of the methodology applied for the analysis of intra- and extracellular metabolites of yeasts according to Villas-Bôas et al. (2005a). Shake flasks were inoculated from the same pre-inoculum's culture at exponential growth phase. Samples were harvested at mid-exponential phase (O.D.600 nm = 5.0). Five culture suspension samples from each flask were harvested with a disposable syringe and sprayed into a cold methanol solution (−40°C) in order to quench the cellular metabolism. The cell pellets were separated from the extracellular medium by centrifugation at low temperature (−20°C). Additional three samples were harvested and filtered using Millipore membrane (0.45 μm) and the filtrate was stored at −20°C for analysis of extracellular metabolites. The intracellular metabolites were extracted from the cell pellets using a mixture of chloroform, methanol, and buffer at low temperature (−40 to −20°C). The upper polar phase from a three-phase mixture was used for the analysis of intracellular metabolites. Both samples containing intra- and extracellular metabolites were freeze-dried prior to chemical derivatization. Since the intracellular extracts contained large amount of organic solvent, distillated water was added to the samples in order to keep them frozen during the lyophilization process. The dried samples were resuspended in sodium hydroxide solution and derivatized using methylchloroformate (MCF). Fifteen samples of intracellular metabolites and nine of extracellular medium were analyzed for each condition tested. This figure was designed and kindly donated by Joel F. Moxley (Dept. of Chemical Eng./MIT/USA). (See color plates.)

extraction. For analysis of extracellular metabolites the cell culture was harvested and filtered using Millipore membrane filters (0.45 µm) and the filtrated samples were stored at −20°C prior to analysis.

The intracellular metabolites were extracted from the biomass pellet by adding additional chloroform, methanol, and buffer (PIPES + EDTA, pH 7.0), followed by rigorous shaking at low temperature (−20°C) for 45 min. The mixture was separated into three phases (nonpolar, biomass, and polar) by centrifugation at low temperature (−20°C). The polar phase was reserved for the analysis of the polar metabolites. Prior to each analysis, the extracted samples of intracellular metabolites as well as the filtered samples of spent medium were lyophilized to dryness to enhance the detection of those low-concentrated compounds.

The dried samples were re-suspended in 200 µl of sodium hydroxide solution and the alkaline suspensions were derivatized following the MCF procedure, as described in detail by Villas-Bôas et al. (2003) and summarized in Figure 6.1. MCF derivatization mainly targets metabolites containing one or more carboxylic and/or amino groups in their molecular structure, which complies about 40% of *S. cerevisiae* metabolome.

6.2.2 The Analysis

The metabolites were analyzed by GC–MS using a quadrupole mass selective detector, with electron ionization source operated at 70 eV. The GC-capillary column used to resolve the metabolite mixture was 30 m long with 250 µm i.d. and 0.15 µm film thickness. The MS was operated in scan mode for the metabolite profiling experiments and in selective ion monitoring mode for detection of ^{13}C-labelling glyoxylate. Two injection modes were applied throughout the study. Initially, the derivatized samples were injected under split mode (split ratio 1:20) and later pulsed-splitless mode was applied in order to obtain a higher sensitivity. Further details of the analytical methodology can be found in Villas-Bôas et al. (2005a,b) and Devantier et al. (2005).

6.3 EARLY DISCOVERIES

During development of the sensitive and low-discriminative analytical techniques for metabolome analysis of yeasts, several unusual or unexpected metabolites were detected at significant levels both in intra- and extracellular samples of *S. cerevisiae* wild-type strain (Villas-Bôas et al. 2005a). For instance, despite no homologous sequences for lactate biosynthetic enzymes in *S. cerevisiae* genome, lactate was observed at higher levels for both intracellular and extracellular samples. However, Martins *et al.* (2001) described the methylglyoxal catabolism in wild-type strains of *S. cerevisiae* that results in the formation of D-lactate. The authors observed an intracellular accumulation of D-lactate and demonstrated that lactate dehydrogenases (DLD1 and CYB2), involved in lactate catabolism in *S. cerevisiae*, are repressed by glucose and induced by lactate. Our study reported in Villas-Bôas *et al.* (2005a), showed that lactate is also secreted into the extracellular medium at significant levels, both under aerobic and anaerobic conditions.

Similarly, the saturated fatty acid myristate was detected at high extracellular levels in samples of *S. cerevisiae* growing anaerobically. In yeast food products or even in the vast available literature on *S. cerevisiae* physiology, no information exists about this important nutritional metabolite. In clinical trials, myristate has been shown to reduce cardiovascular disease risk (Khosla and Sundram, 1996; Loison *et al.* 2002) and lowering of the cholesterol-binding plasma low-density lipoprotein C levels, in which myristate has an important compositional role. Myristate is also present in flavor components of essential oils (Kajuwara *et al.* 1988) and spices (Kostrzewa and Karwowska, 1975). As a saturated fatty acid, myristate is involved in fatty acid acylation of proteins in higher eukaryotes (Towler and Glaser 1986). Proteins with N-terminal myristoyl-glycine residues have been also found in *S. cerevisiae*, and they are related to the biosynthesis of membrane proteins (Towler et al. 1987). Extracellular myristate can be a good indicator of oxygen depletion during *S. cerevisiae* cultivations, and its high levels may be related to the reduced biomass formation rate during anaerobic growth, which requires less acylation of proteins for membrane synthesis.

2-Oxovalerate was another unusual metabolite detected in cell extracts and spent culture medium samples of *S. cerevisiae*. Very little is known about the metabolic role of this 2-keto acid in the cell physiology. It has never been reported as part of the metabolic network of *S. cerevisiae* until its first detection during our extensive metabolite profiling of yeast cells and culture. 2-Oxovalerate is believed to be involved in the pyruvate metabolism and it can be formed from 2-propylmalate via deacetylation of acetyl-CoA [Equation (6.1)], but this reaction has not been described in *S. cerevisiae.*

$$\text{2-Propylmalate} + \text{Acetyl-CoA} \rightarrow \text{2-Oxovalerate} + \text{CoA} \qquad (6.1)$$

At last, glyoxylate was also detected during both aerobic and anaerobic growth on glucose at considerably high levels. The glyoxylate cycle is normally found to be inactive during growth on glucose as the sole carbon source due to glucose repression (Fernandez et al., 1993). The glyoxylate pathway could be unrepressed when the cell samples were collected (mid- to late exponential growth phase), which was unlikely. Therefore, this data strongly point to the presence of an alternative pathway for glyoxylate biosynthesis in *S. cerevisiae* that is not repressible by glucose and has not been described previously.

6.4 YEAST STRESS RESPONSE GIVES EVIDENCE OF ALTERNATIVE PATHWAY FOR GLYOXYLATE BIOSYNTHESIS IN *S. CEREVISIAE*

A laboratory strain and an industrial strain of *S. cerevisiae* were cultivated at high substrate concentration, also known as very high gravity fermentation (VHG), and

TABLE 6.1 Average of Intracellular Metabolite Concentrations (μmol/g Dry Cell Mass) Obtained with the MCF Method and Calculated from a Total of Eight Independently Processed Samples (Devantier et al., 2005).

	Strain1		Strain2	
	SD medium	VHG medium	SD medium	VHG medium
Glyoxylate	35.7	0.0	46.0	0.0
Glycine	10.7	45.7	11.1	20.3

SD = standard laboratory medium; VHG = very high gravity fermentation medium.

they were compared with their fermentation performance on laboratory standard medium. This study was carried out to investigate the yeast stress response to high ethanol concentrations and high osmotic stress (Devantier et al., 2005). The VHG cultivations were achieved by applying simultaneous saccharification and fermentation of 280 gl of maltodextrin as carbon source. For the standard laboratory culture medium 20 gl of glucose was used as carbon source. All cultivations were carried out under anaerobic conditions and the metabolite profiles of yeast cells (intra- and extracellular) were determined during exponential and stationary growth phases (for further details see Devantier et al., 2005).

Several significant differences were observed on the intra- and extracellular metabolite profiles of the yeast strains depending mainly on the cultivation medium and, to a lesser extent, on the genetic background. However, particularly interesting to this case study is the detection of glyoxylate only in the standard laboratory medium cultivation samples. By applying principal component analysis of the data generated in yeast stress response study, glyoxylate appeared as an outstanding variable and, interestingly, inversely related to glycine levels (Table 6.1). In other words, samples containing high levels of glyoxylate presented lower levels of glycine, and samples where glyoxylate was not detected had higher levels of glycine. Since the glyoxylate cycle is repressed during growth on glucose (Fernandez et al., 1993), one explanation could be glyoxylate formation through glycine. Although this pathway was not described in *S. cerevisiae*, it exists in several microorganisms, e.g., *Bacillus subtilis* (Job *et al.* 2002). Therefore, the yeast stress response study generates an important hypothetic answer to explain the high levels of glyoxylate observed during *S. cerevisiae* cultivation on glucose, that was worth investigating further.

6.5 BIOSYNTHESIS OF GLYOXYLATE FROM GLYCINE IN *S. CEREVISIAE*

The glyoxylate cycle (Figure 6.2) is the main and well-known pathway that leads to glyoxylate biosynthesis in *S. cerevisiae* (Chaves et al., 1997; López et al., 2004). Isocitrate lyase (Icl) is the key enzyme of the glyoxylate cycle, which bypasses the two decarboxylation steps in the TCA (tricarboxylic acids) cycle and leads to the

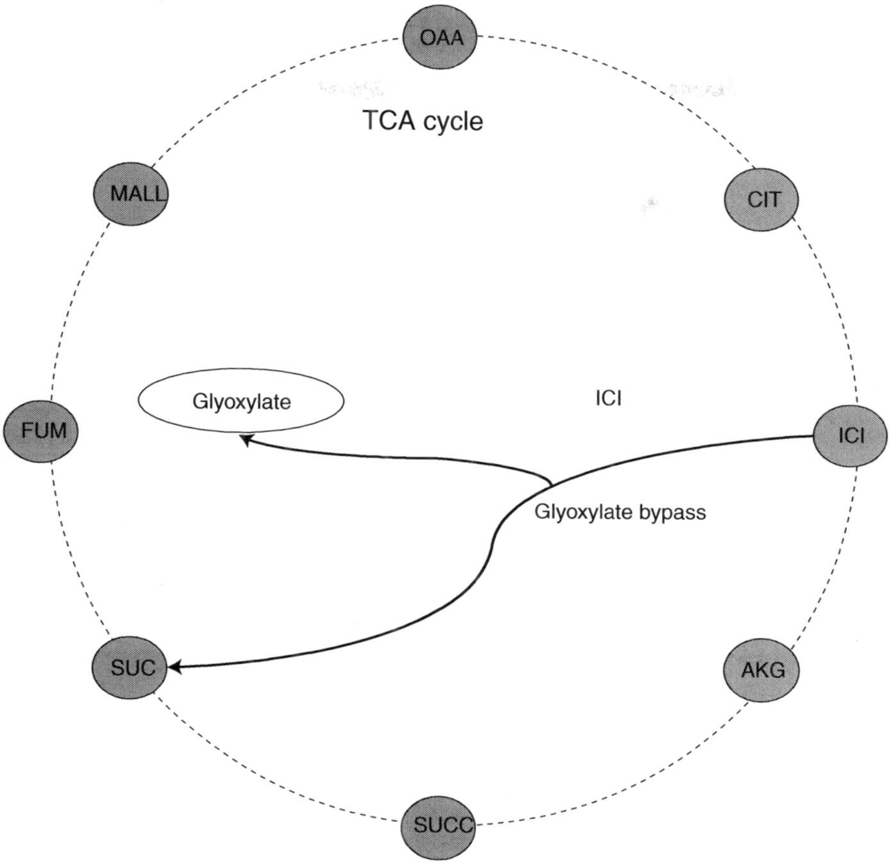

Figure 6.2 The glyoxylate cycle. Isocitrate lyase (Icl) is the key enzyme of the glyoxylate cycle, which bypasses the two decarboxylation steps in the TCA (tricarboxylic acids) cycle and leads to the synthesis of succinate (C4) and glyoxylate (C2). Abbreviations: OAA, oxaloacetate; CIT, citrate; ICI, isocitrate; AKG, 2-oxoglutarate; SUCC, succinylCoA; SUC, succinate; FUM, fumarate; MAL, malate.

synthesis of succinate (C4) and glyoxylate (C2). However, there is strong evidence in the literature about the repression of Icl by glucose (Takada and Noguchi, 1985; Fernandez et al., 1993; Maaheimo et al., 2001). Nonetheless, glyoxylate has been detected at high levels intra- and extracellularly in *S. cerevisiae* cultures growing on glucose, as described previously.

Glycine was shown to be the potential alternative precursor for glyoxylate in *S. cerevisiae* by the yeast stress response study. Biosynthesis of glyoxylate from glycine has been described in several prokaryotes such as *Bacillus subtilis* (Nishiya and Imanaka, 1998; Job et al., 2002) and *Nitrobacter agilis* (Sanders et al., 1972). However, the most well-described catabolic reaction of glycine in yeasts is its decarboxylation with subsequent conversion to serine, catalyzed by the glycine decarboxylase multienzyme complex (Gdc) as shown in the Equation (6.2) (Sinclair and Dawes, 1995). The Gdc, also known as the glycine cleavage system or glycine synthase (EC 2.1.2.10), fills a critical metabolic position connecting the metabolism of one-, two-, and three-carbon compounds and is linked to many different metabolic reactions.

$$\text{5, 10-Methylenetetrahydrofolate} + \text{Glycine} + H_2O \leftrightarrow \text{Tetrahydrofolate} + \text{L-Serine} \quad (6.2)$$

Although glycine is usually described as a poor source of nitrogen for yeasts, *S. cerevisiae* can grow on glycine as the sole nitrogen source (Sinclair and Dawes, 1995). Sinclair and Dawes (1995) have investigated yeast strains with mutations in single genes involved in glycine uptake and decarboxylation, and they found a solid indication of a second pathway for glycine assimilation in yeasts, as two of the mutants tested could not decarboxylate glycine but could still use it as the sole nitrogen source.

The putative second pathway for glycine assimilation could be a reversible reaction catalyzed by alanine:glyoxylate aminotransferase (Agt). Agt (EC 2.6.1.44) is one of three different enzymes used for glycine synthesis in *S. cerevisiae*. Glyoxylate is transaminated to glycine by Agt with a concurrent conversion of alanine to pyruvate (Figure 6.3). However, this enzyme has been reported to be repressed by glucose, and a purified enzyme preparation was demonstrated to be highly selective for using L-alanine and glyoxylate as substrate, hence there was strong evidence for irreversibility of this reaction (Takada and Noguchi, 1985).

6.5.1 Stable Isotope Labeling Experiment to Investigate Glycine Catabolism in *S. cerevisiae*

In order to investigate the formation of glyoxylate from glycine, two different *S. cerevisiae* reference strains and a mutant with a deletion in the gene that encodes

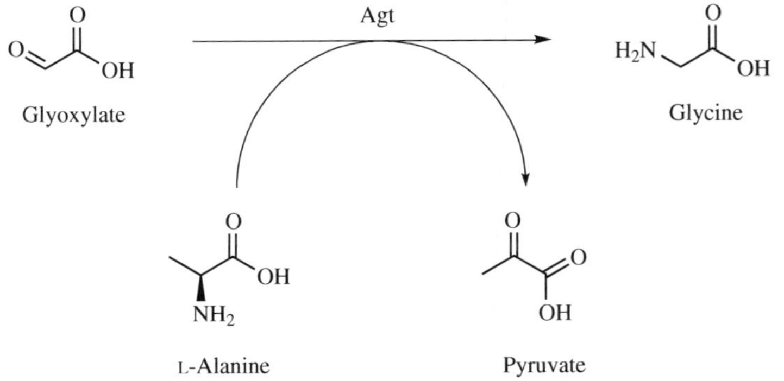

Figure 6.3 The alanine:glyoxylate aminotransferase (Agt) reaction. Agt (EC 2.6.1.44) is one of three different enzymes used for glycine synthesis in *S. cerevisiae*. Glyoxylate is transaminated to glycine by Agt with a concurrent conversion of alanine to pyruvate.

Agt were cultivated on glucose and galactose, with galactose representing a non-fermentable carbon source and, thus, imposing little carbon catabolite repression, under aerobic and anaerobic conditions. ^{13}C-(fully)-labeled glycine was used as the sole nitrogen source and its catabolism was followed by metabolite profile analysis of ^{13}C-containing compounds using GC–MS (Villas-Bôas et al., 2005b).

All the strains grew comparatively well on both media (glucose/galactose) with glycine as nitrogen source. The specific growth rates varied depending on the genetic background of the strains or on the carbon source employed. All the strains presented a higher specific growth rate when growing on galactose, suggesting that glucose repression was a cause of the lower specific growth rate of *S. cerevisiae* during growth on glucose with glycine as the sole nitrogen source. The mutant strain also grew comparatively well on minimal medium with glycine as the main nitrogen source even though its alanine:glyoxylate aminotransferase-encoding gene was deleted. Therefore, it was confirmed that it is unlikely that the catabolism of glycine involves the reversibility of the alanine:glyoxylate aminotransferase reaction.

Glyoxylate was detected and was shown to have a drastic increase in the abundance of its $m + 1$ ion in samples from all cultivations, indicating that it was a direct product/intermediate from ^{13}C-glycine metabolism. An increase in the abundance of $m + 1$ ion from 2-oxovalerate was also detected in samples from most cultivations. Decarboxylation of glycine to CO_2 and NH_4^+ by Gdc yields the activated one-carbon unit for the formation of serine via 5,10-methylene-tetrahydrofolate [Equation (6.2)]. But serine was not detected in the samples from any of the cultivations. However, serine is metabolized in *S. cerevisiae* by serine deaminase (EC 4.3.1.17) to pyruvate. Pyruvate is either transported to mitochondria or converted to alanine, valine, and leucine via 2-oxoisovalerate and isopropylmalate, or to isoleucine via 2-oxobutanoate. But a huge dilution in the labeling atoms of pyruvate and posterior intermediates is expected to occur because the main carbon source (glucose/galactose) was not labeled and, thus, the ^{13}C incorporated from glycine consisted of a fairly small fraction, possibly below the detection limit of the instrument. The pyruvate molecules did not have any labeling, but 2-oxoisovalerate, isopropylmalate, isoleucine, valine, and oxaloacetate appeared labeled in several samples. In addition, several other metabolites, including some intermediates of the TCA cycle, such as fumarate, malate, isocitrate, and citrate presented labeling in different samples from different cultivations.

Therefore, based on the ^{13}C-labelling results, it is clear that glycine can be directly oxidized to glyoxylate in *S. cerevisiae*, as demonstrated in other microorganisms (Sanders et al., 1972; Nishiya and Imanaka, 1998; Job et al., 2002).

The catabolic reaction of glycine via Gdc is believed to be repressed by glucose (Sinclair and Dawes, 1995; Piper et al., 2002), and the activity of this pathway could not be directly determined by using ^{13}C-glycine, due to the lack of serine detection in the metabolite pool. On the contrary, the growth rate of all strains on glucose medium was lower than on galactose medium, which suggests that the catabolism of glycine was more efficient in absence of glucose. Glucose could be repressing the catabolic reaction of glycine via Gdc but the cells still had the alternative pathway to metabolize glycine that was not repressible by glucose, because there was yeast growth on glucose medium with glycine as sole nitrogen source.

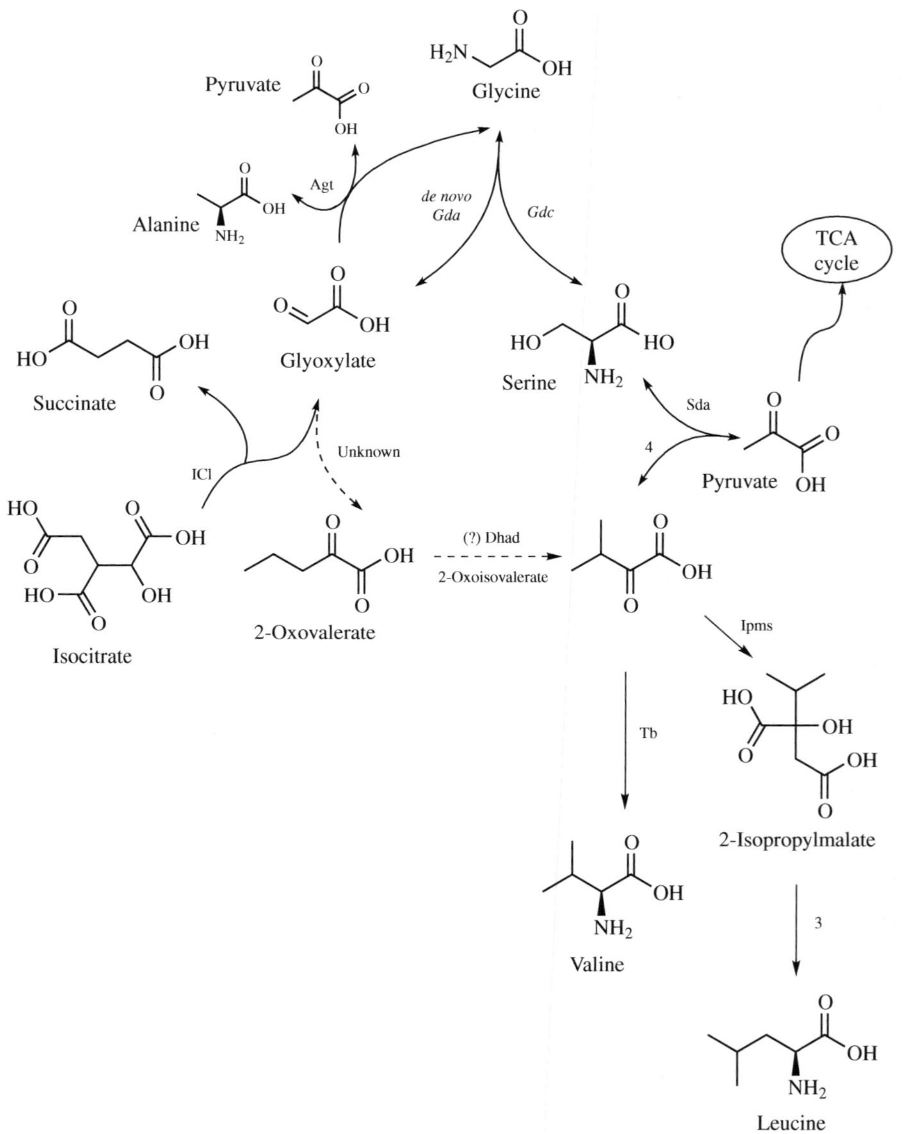

Figure 6.4 Glycine metabolism in *S. cerevisiae*. It is proven that there are at least two pathways for glycine catabolism in *S. cerevisiae*: (1) via *Gdc* and (2) via a *de novo Gda*. Based on ^{13}C-labeling experiments, it is postulated that 2-oxovalerate is synthesized from glyoxylate by an unknown reaction/enzyme with its subsequent conversion to 2-oxoisovalerate by (putatively) *Dhad*. *Gdc*:glycine decarboxylase multienzyme complex; *Sda*:serine deaminase; *Agt*: alanine:glyoxylate aminotransferase; *Gda*:glycine deaminase; *Dhad*:dihydroxy acid dehydratase; *Ipms*:isopropylmalate synthase; *Icl*:isocitrate lyase; *Tb*:transaminase B. Full arrows indicate confirmed pathways and dashed arrows indicate speculative pathways. The numbers on some arrows specify the number of reaction steps not shown in the pathway.

The direct deamination of glycine to glyoxylate did not seem to be repressed by glucose since ^{13}C-labeling was observed in glyoxylate in all cultivation conditions tested, at both aerobic and anaerobic growth conditions, and it is not a reversible Agt reaction, as the mutant with the Agt-encoding gene deleted, grew comparatively well on a medium containing glycine as the main nitrogen source and presented ^{13}C-labelling glyoxylate. Therefore, these results prove the presence of a yet nondescribed pathway for glycine catabolism and glyoxylate biosynthesis in *S. cerevisiae*. This pathway could be one that has earlier been indicated by Sinclair and Dawes (1995). But, the contribution of this pathway to the global catabolism of glycine by *S. cerevisiae* and its influence on the yeast's ability to utilize glycine as nitrogen source still need to be elucidated by further studies.

6.5.2 Data Leveraged for Speculation

It is still unclear why valine and isopropylmalate appeared labeled in several samples, while leucine did not. A possible answer could be connected to the finding that 2-oxovalerate was labeled in all samples where it was detected. Figure 6.4 shows a suggestion for the global pathways for glycine metabolism in *S. cerevisiae*, and it speculates a possible biosynthetic reaction of 2-oxovalerate and its subsequent metabolic pathways. On the basis of the labeling pattern of 2-oxovalerate, it is postulated that it is possibly synthesized from glyoxylate. Once synthesized, 2-oxovalerate could be putatively converted to 2-oxoisovalerate, the main precursor of valine by the dihydroxy-acid dehydratase (EC 4.2.1.9), which has been considered a low-specific enzyme (Limberg and Thiem, 1996).

Therefore, besides confirming the presence of a so far nondescribed metabolic pathway for glyoxylate biosynthesis and speculating on a few other unknown pathways in *S. cerevisiae*, these studies show how data from global metabolome analysis with simultaneous metabolite identification, as discussed here, can be coupled to data from isotope labeling analysis, and then be used to discover new metabolic pathways.

REFERENCES

Chaves RS, Herrero P, Ordiz I, Del Brio MA, Moreno F. 1997. Isocitrate lyase localization in *Saccharomyces cerevisiae* cells. Gene 198:165–169.

Devantier R, Scheithauer B, Villas-Bôas SG, Pedersen S, Olsson L. 2005. Metabolite profiling for analysis of yeast stress response during very high gravity ethanol fermentations. Biotechnol Bioeng 90:703–714.

Fernandez E, Fernandez M, Moreno F, Rodicio R. 1993. Transcriptional regulation of the isocitrate lyase encoding gene in *Saccharomyces cerevisiae*. FEBS Lett 333:238–242.

Goffeau A, Barrell BG, Bussey H, Davis RW Dujon B Feldmann H, Galibert F, Hoheisel JD, JACQ C, Johnston M, Louis EJ, Mewes HW, Murakami Y, Philippsen P, Tettelin H, Oliver SG. 1996. Life with 6000 genes. Science 274:546–567.

Job V, Marcone GL, Pilone MS, Pollegioni L. 2002. Glycine oxidase from *Bacillus subtilis*—characterization of a new flavoprotein. J Biol Chem 277:6985–6993.

Kajuwara T, Hatanaka A, Kawai T, Ishihara M, Tsuneya T. 1988. Study of flavour compounds of essential oil extracts from edible Japanese kelps. J Food Sci 53:960–962.

Khosla P, Sundram K. 1996. Effects of dietary fatty acid composition on plasma cholesterol. Prog Lipid Res 35:93–132.

Kostrzewa E, Karwowska K. 1975. The evaluation of aromatic and flavour properties of pimento extracts. Prace Instytutow i Laboratoriow Badawczych Przemyslu Spozywczego 25:67–74.

Limberg G, Thiem J. 1996. Synthesis of modified aldonic acids and studies of their substrate efficiency for dihydroxy acid dehydratase (DHAD). Aust J Chem 49:349–356.

Loison C, Mendy F, Serougne C, Lutton C. 2002. Dietary myristic acid modifies the HDL-cholesterol concentration and liver scavenger receptor BI expression in the hamsters. Br J Nutr 87:199–210.

López ML, Redruello B, Moreno EVF, Heinisch JJ, Rodicio R. 2004. Isocitrate lyase of the yeast *Kluyveromyces lactis* is subject to glucose repression but not to catabolite inactivation. Curr Genet 44:305–316.

Maaheimo H, Fiaux J, Çakar ZP, Bailey JE, Sauer U, Szyperski T. 2001. Central carbon metabolism of *Saccharomyces cerevisiae* explored by biosynthetic fractional ^{13}C labelling of common amino acids. Eur J Biochem 268:2464–2479.

Martins AM, Cordeiro CA, Ponces-Freire AM. 2001. *In situ* analysis of methylglyoxal metabolism in *Saccharomyces cerevisiae*. FEBS Lett 499:41–44.

Nishiya Y, Imanaka T. 1998. Purification and characterization of a novel glycine oxidase from *Bacillus subtilis*. FEBS Lett 438:263–266.

Oliver SG. 1996. A network approach to the systematic analysis of the yeast gene function. Trends Genet 12:241–242.

Piper MDM, Hong SP, Eiβing T, Sealey P, Dawes IW. 2002. Regulation of the yeast glycine cleavage genes is responsive to availability of multiples nutrients. FEMS Yeast Res 2:59–71.

Sanders HK, Becker GE, Nason A. 1972. Glycine-cytochrome *c* reductase from *Nitrobacter agilis*. J Biol Chem 247:2015–2025.

Sinclair DA, Dawes IW. 1995. Genetics of the synthesis of serine from glycine and the utilization of glycine as sole nitrogen source by *Saccharomyces cerevisiae*. Genetics 140:1213–1222.

Takada Y, Noguchi T. 1985. Characteristics of alanine:glyoxylate aminotransferase from *Saccharomyces cerevisiae*, a regulatory enzyme in the glyoxylate pathway of glycine and serine biosynthesis from tricarboxylic acid cycle intermediates. Biochem J 231:157–163.

Towler DA, Glaser L. 1986. Protein fatty acid acylation:enzymatic synthesis of an *N*-myristoylglycyl peptide. Proc Natl Acad Sci USA 83:2812–2816.

Towler DA, Adams SP, Eubanks SR, Towery DS, Jackson-Machelski E, Glaser L, Gordon JI. 1987. Purification and characterization of yeast myristoylCoA:protein *N*-myristoyltransferase. Proc Natl Acad Sci USA 84:2708–2712.

Villas-Bôas SG, Delicado DG, Åkesson M, Nielsen J. 2003. Simultaneous analysis of amino and nonamino organic acids as methyl chloroformate derivatives using gas chromatography-mass spectrometry. Anal Biochem 322:134–138.

Villas-Bôas SG, Moxley JF, Åkesson M, Stephanopoulos G, Nielsen J. 2005a. High-throughput metabolic state analysis: The missing link in integrated functional genomics of yeasts. Biochem J 388:669–677.

Villas-Bôas SG, Åkesson M, Nielsen J. 2005b. Biosynthesis of glyoxylate from glycine in *Saccharomyces cerevisiae*. FEMS Yeast Res 5:703–709.

7

MICROBIAL METABOLOMICS: RAPID SAMPLING TECHNIQUES TO INVESTIGATE INTRACELLULAR METABOLITE DYNAMICS—AN OVERVIEW

BY SILAS G. VILLAS-BÔAS

The knowledge of concentrations of intracellular metabolites is important for quantitative analysis of metabolic networks. The frequently used sampling techniques show an inherent limitation with regards to very fast response of intracellular metabolites in the millisecond range. For microbial cultivations, the time window between an induced disturbance factor and the first sample is constrained by the time necessary to obtain a homogeneous distribution of the perturbation within the bioreactor. Thus, ingenious sampling devices coupled to bioreactors have been developed to study intracellular metabolite dynamics in microbial cells, varying from manual sampling to fully automated (computer-aided) techniques. This chapter will briefly review the state-of-art of sampling devices in microbial metabolomics.

7.1 INTRODUCTION

Steady-state cultivations as well as transient analysis of intracellular metabolites belong to the well-established tools of microbial physiology and biochemistry. Recently, the information about the concentration of metabolites is also of increasing importance in metabolic engineering and functional genomics, as part of metabolomics-related studies. Intracellular metabolite concentrations play important regulatory roles

Metabolome Analysis: An Introduction, by Silas G. Villas-Bôas, Ute Roessner, Michael A. E. Hansen, Jorn Smedsgaard and Jens Nielsen

in the cellular metabolic network of microorganisms. Together with information about kinetics properties of the enzymes involved in specific pathways, knowledge of the *in vivo* concentrations of the intermediary metabolites is of fundamental importance for characterization of the microbial metabolism through kinetic modeling.

For quantitative analysis of intracellular metabolites, it is an essential prerequisite to define the physiological state of the biological system used for these measurements. Of course, this imperative requires experimental conditions and related process operations that are defined and reproducible. It is, therefore, desirable to start the dynamic experiment from a well controlled steady-state situation. Furthermore, the complexity of dynamic modeling of microbial metabolism can be reduced if regulation at the DNA level can be ignored at least within the time frame of the dynamic experiment (Weuster-Botz and de Graaf, 1996). This is possible only if dynamic experiments can be monitored on a time scale smaller than time constants for changes in intracellular enzyme concentrations (<300 ms).

Several intracellular metabolic reactions, especially catabolic reactions and reactions involved in the energy metabolism have high turnover rates as discussed in Chapter 3. Considering the reported intracellular concentrations of glycolytic intermediated and cytosolic ATP of up to millimole level (Schaefer et al., 1999), a quenching time far below 300 ms is necessary. Therefore, it is evident that classical sampling of microbial cultures by using syringes and automatic pipettes is completely inadequate to achieve inactivation times within 100 ms and to keep process operations defined and reproducible enough to study intracellular metabolite dynamics.

Sampling techniques to measure reliable intracellular metabolite concentrations of a steady-state culture can be successful only if (a) a representative sample can be taken from a controlled reactor without disturbing the steady-state metabolism of the cells; (b) a rapid inactivation of the metabolism of the sampled cells is achieved, avoiding uncontrolled reactions in the sampling device; (c) the intracellular metabolites are completely extracted and the intracellular enzymes are simultaneously denaturized; (d) the stability of the metabolites is not affected by the sampling and extraction procedure; and (e) the sampling rate is high enough to study very rapid dynamic metabolic reactions.

Research works on sampling systems focussing on measurements of metabolite dynamics on a subsecond timescale have been reported during the last 10 years, with pioneering research groups based mainly in Germany and in the Netherlands. Ingenious devices have been developed, which present *pros* and *cons* and vary from manual sampling to fully automated (computer-aided) devices. A global overview of the main sampling techniques developed to date will be presented and discussed in the following sections:

7.2 STARTING WITH A SIMPLE SAMPLING DEVICE PROPOSED BY THEOBALD ET AL. (1993)

A relatively simple sampling technique was described by Theobald et al. (1993 and 1997) which consists of a homemade sample port coupled to the bioreactor. The sample port has a dead volume of about 0.2 ml and it ends in a capillary (Figure 7.1).

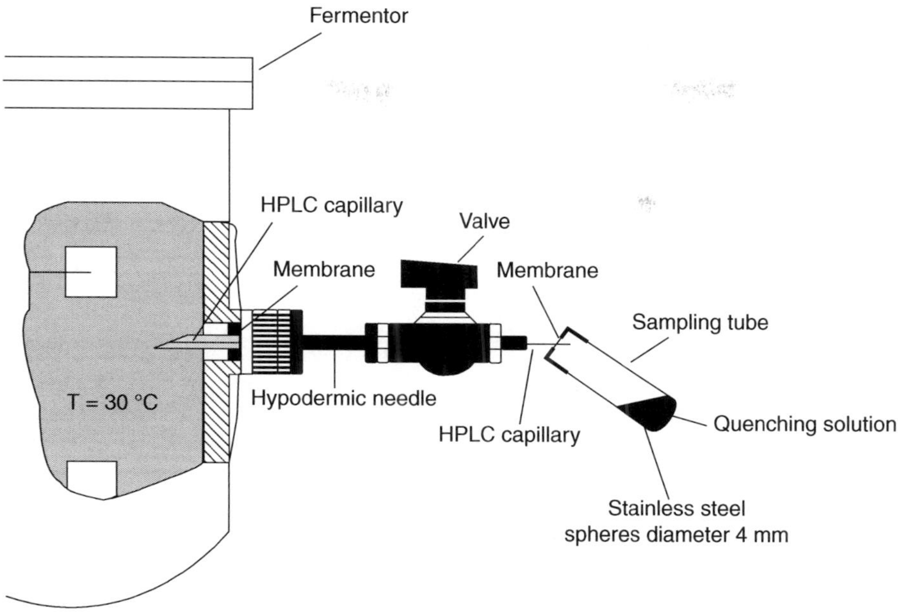

Figure 7.1 Schematic representation of the sampling device connected to the mixing zone of the bioreactor according to Theobald *et al.* (1993 and 1997). Reproduced from Analytical Biochemistry, vol. 214, *In vivo analysis of glucose-induced fast changes in yeast adenine nucleotide pool applying a rapid sampling technique*, page 32, Copyright (1993), with permission from Elsevier.

The samples are quenched manually using a sampling tube containing the quenching solution under vacuum, mounted with a holed screw cap fitted with a membrane. The vacuum is created inside the tubes by piercing the membrane with a capillary mounted on a tube connected to a vacuum pump. When the sampling-tube membrane is pierced by the port capillary, the vacuum provokes a rapid displacement of the sample from the bioreactor into the tube. The flow rate through the port was estimated to be 0.5–1.5 ml/s, resulting in a residence time of the sample in the port of less than 1 s. A short residence time is necessary to prevent a large change in the environmental conditions experienced by the cells and also to ensure a rapid transfer to the quenching solution.

However, the sampling device proposed by Theobald et al. (1993 and 1997) has an important limitation with respect to reproducibility of sampling volume. Injecting the sample by means of a needle into the evacuated and sealed test tube is susceptible to blockage of the needle and premature loss of vacuum with a subsequent deviation of the sample size.

7.3 AN IMPROVED DEVICE REPORTED BY LANGE ET AL. (2001)

Lange et al. (2001), reported an improved sampling device that offers the same advantages as the one proposed by Theobald et al. (1993 and 1997), but with a

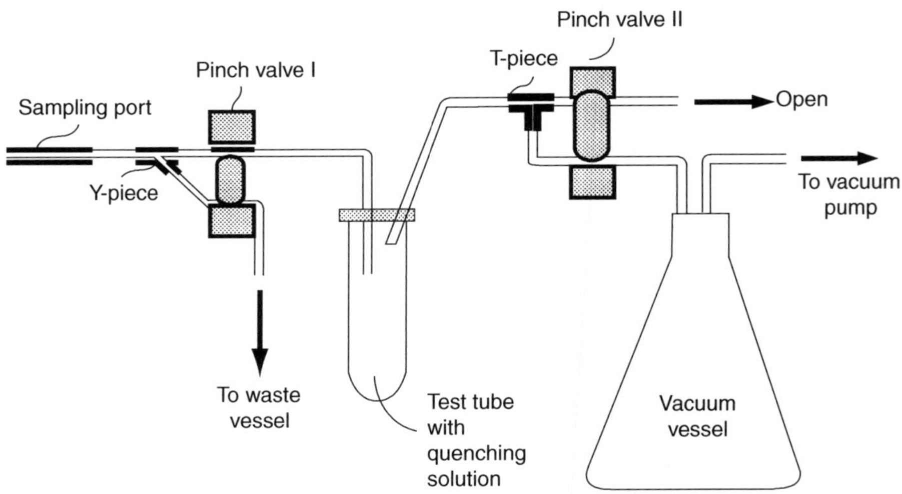

Figure 7.2 Scheme of the rapid sampling setup proposed by Lange *et al.* (2001). Reproduced from Biotechnology and Bioengineering, vol. 75, *Improved rapid sampling for in vivo kinetics of intracellular metabolites in Saccharomyces cerevisiae*, page 409, Copyright (2001), with permission from John Wiley & Sons, Inc.

better sampling reproducibility, and it also enables withdrawal of small sample sizes, which is advantageous for laboratory scale analysis. The modified system consists of a submerged capillary port with an inner diameter of 1 mm and a length of 80 mm, placed inside a stainless steel cylinder to fit a standard bioreactor port. Silicon tubing (i.d. 0.8 mm) connects the port via a Y-piece to a waste container and to the sampler tube adapter (Figure 7.2). A pinch valve directs the flow to either of them, and switching times are controlled electronically through a controlled digital counter. The tube adapter closes the top of any standard-sized test tubes airtight with a foam pad against which the tube is pushed. Two stainless steel tubes are lead through the foam closure into the test tube. During sampling, the smaller, centrally placed tube is connected to the silicon tube coming from the bioreactor. The second tube is used to evacuate the tube prior to sampling; a silicon pump tubing leads via a T-piece to a 2-l vessel, which is kept at a constant vacuum, and the other end is kept open. A second, electronically controlled pinch valve enables switching between the opening to ambient pressure and the vacuum container (Figure 7.2). With this system, the test tubes are filled with quenching solution, weighted, and if using cold or hot quenching solutions, they are set to the desired temperature prior to sampling. The tubes are weighted after sampling to determine the sample size. During sampling operation, cultivation broth is constantly flowing at a lower flow rate (e.g., 0.5 ml/s) into the waste container. After placing a tube containing the quenching solution under the tube adapter, the starting of a three-step valve operating sequence is triggered manually: 1st step, the pinch valve 2 (Figure 7.2) opens the tube leading to the vacuum container;

2nd step, 1 s later, the pinch valve 1 switches from the waste container to the tube adapter; and, 3rd step, after a further interval of around 0.7 s, both valves fall back to their starting position. The total inner volume of the sample port, the tubing, and the tube adapter is about 100 μl, of which only the 50 μl between the Y-piece and the orifice of the tube adapter contain stagnant liquid during sampling. Lange et al. (2001) obtained a sampling rate of 1.3 samples/s with about 3% variation in sample volumes.

Despite their relatively fast sampling, the devices proposed by Theobald et al. (1993 and 1997) and Lange et al. (2001) are still considered to be too slow for monitoring fast dynamic changes in microbial metabolism.

7.4 SAMPLING TUBE DEVICE BY WEUSTER-BOTZ (1997)

Weuster-Botz (1997) proposed a sampling tube device for monitoring intracellular metabolic dynamics, which was coupled to a controlled bioreactor and presented much higher sampling rates. The basic idea is to perform sampling, quenching, and extraction of intracellular metabolites continuously in a tube connected to a bioreactor. The sampling tube device was a home-built sampling probe with an inlet of 4 mm diameter for continuous sampling at the tip of the probe, an inlet of 4 mm diameter for continuous supply of quenching/extraction solution on the other side of the probe, and an outlet of 8 mm diameter connected to the sampling tube was installed into a standard connecting pipe of the stirred tank reactor (Figure 7.3). The quenching/extraction solution was able to mix with the sample continuously 3 mm from where the sample entered the tip of the sampling probe. The sampling tube was made of polyethylene, with an inside diameter of 8 mm, a length of 100 m, and was coiled with a diameter of 0.5 m.

Before starting the continuous rapid sampling, the polyethylene tube was filled with water to provide a constant pressure-driven flow of sample and quenching solution into the sampling tube (Figure 7.3a). The quenching solution receiver was connected to the sampling probe in a way that no gas is left in the connecting pipe. The continuous sampling out of the bioreactor with a microbial culture was started by opening simultaneously the diaphragm valves at the sampling probe (Figure 7.3b). A continuous flow of sample and quenching solution was achieved within a few seconds because of the pressure in the reactor and in the quenching solution receiver. After 200 s, the continuous sampling was stopped by closing the diaphragm valves. The exact flow rates of quenching solution and cultivation medium mixed with quenching solution were determined gravimetrically to calculate the dilution factor of quenching solution and to transform the position of sample in the sampling tube to the sampling time. The sampling tube was disconnected and frozen at −80°C (Figure 7.3c). To achieve single samples, the frozen and coiled wound-up sampling tube was divided into identical parts by cutting the tube. The individual parts of the tube with the frozen samples were transferred to sample flasks for thawing the sample. Selection of a suitable quenching solution that can be frozen inside the tube is important for application of this procedure. With this technique, Weuster-Botz

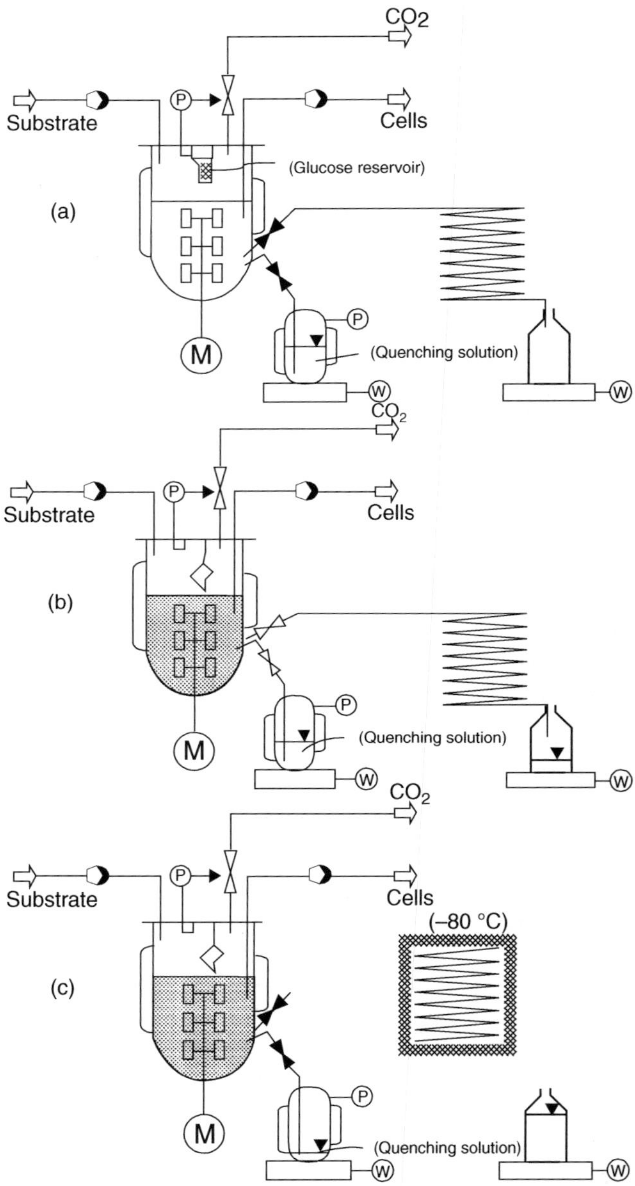

Figure 7.3 Principle of rapid sampling from a bioreactor with high sampling rate according to Weuster-Botz (1997) (a) Steady-state cultivation; (b) continuous sampling, inactivation, and extraction with perchloric acid (−40°C), in the sampling tube after glucose injection; (c) sampling tube disconnected and frozen at −80°C. Fast dynamic metabolite concentration changes are fixed at a certain position in the sampling tube (P, pressure indication, registration and control; W, weight indication and registration). Reproduced from Analytical Biochemistry, vol. 246, *Sampling tube device for monitoring intracellular metabolite dynamics*, page 226, Copyright (1997), with permission from Elsevier.

(1997) obtained a sampling rate of 13.6 ml/s using $HClO_4$ as quenching agent, with 2.8 ms time window between the sample leaving the reactor and its contact with the quenching agent. The great advantage of this technique is its high resolution in time that is achieved due to the dispersion of the samples in the tube. According to Weuster-Botz (1997), the events of 1 s in the bioreactor are distributed over a sampling tube length of about 5 m (at a tube position of 85 m). These represent about 15 individual samples (parts of the sampling tube). However, intracellular and extracellular metabolites will be invariably analyzed together since the freezing/thaw cycle disrupts the cell envelops, independently of the quenching agent in use.

7.5 FULLY AUTOMATED DEVICE BY SCHAEFER ET AL. (1999)

Schaefer et al. (1999) proposed a fully automated device for the fast quenching of microbial cultures from bioreactors that have the advantage of allowing separation of the biomass from the extracellular medium via centrifugation. This automated rapid sampling device consists of a tube with an inner diameter of 3.2 mm and a length of 130 mm connected to the outlet opening at the bottom of the bioreactor (Figure 7.4). This tube was closed by a magnetic pinch valve during cultivation. Continuous sampling out of the bioreactor was started by opening the magnetic pinch valve, and due to the pressure inside the bioreactor, the samples were sprayed continuously with a fast flow rate into individual sample flasks at the top. Sample flasks (50 ml) were fixed in transport magazines made of aluminum (Figure 7.4). The magazines were transported horizontally in a way that in every 220 ms a new sample was positioned 20 mm under the opening of the magnetic pinch valve (Figure 7.4). The transport of the magazines was facilitated by a straight-toothed gear belt moved by a step engine (see Schaefer et al., 1999 for further details). Schaefer et al. (1999) used cold methanol solution (60% v/v, −50°C) as quenching agent, and the sample flasks in the magazines were filled with the cold quenching solution before the sampling started. The magazines with the quenched samples were transferred manually into a −28°C freezer. At the end of the continuous sampling, the magnetic pinch valve of the bioreactor was closed. The volume of the added sample into each of the sample flasks was controlled gravimetrically. With this approach, it was possible to quench a sample volume of 5.0 ml and obtain an excellent standard deviation of 0.08 ml (1.6%). The sampling rate was 4.5 samples/s, and after quenching the samples were centrifuged at −20°C to separate the biomass from the extracellular medium.

7.6 THE STOPPED-FLOW TECHNIQUE BY BUZIOL ET AL. (2002)

According to Buziol et al. (2002), as far as the very fast and initial response of intracellular metabolites in the millisecond range is concerned, the techniques described by Weuster-Botz (1997) and Schaefer et al. (1999) show an inherent limitation. The time span between the disturbance and the first sample is constrained

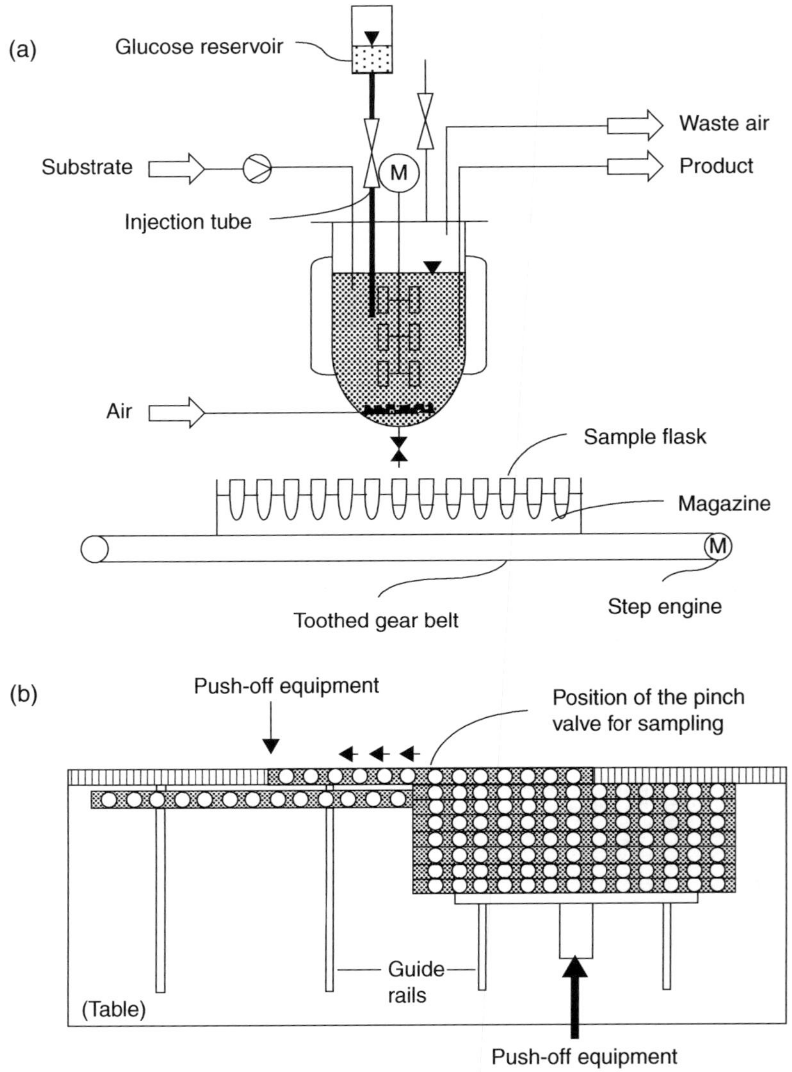

Figure 7.4 Principle of the automated sampling device coupled to a stirred bioreactor with equipment for rapid glucose injection, according to Schaefer *et al.* (1999) (a) Front view, (b) top view. Reproduced from Analytical Biochemistry, vol. 270, *Automated sampling device for monitoring intracellular metabolite dynamics*, page 90, Copyright (1999), with permission from Elsevier.

by the time required for obtaining a homogeneous distribution of the perturbation within the bioreactor. Therefore, Buziol et al. (2002) proposed a new device based on a stopped-flow technique combined with a modified rapid-freezing method. The sampling device simultaneously serving as a mixing chamber was located in a connecting piece of the bioreactor as shown schematically in Figure 7.5. A detailed

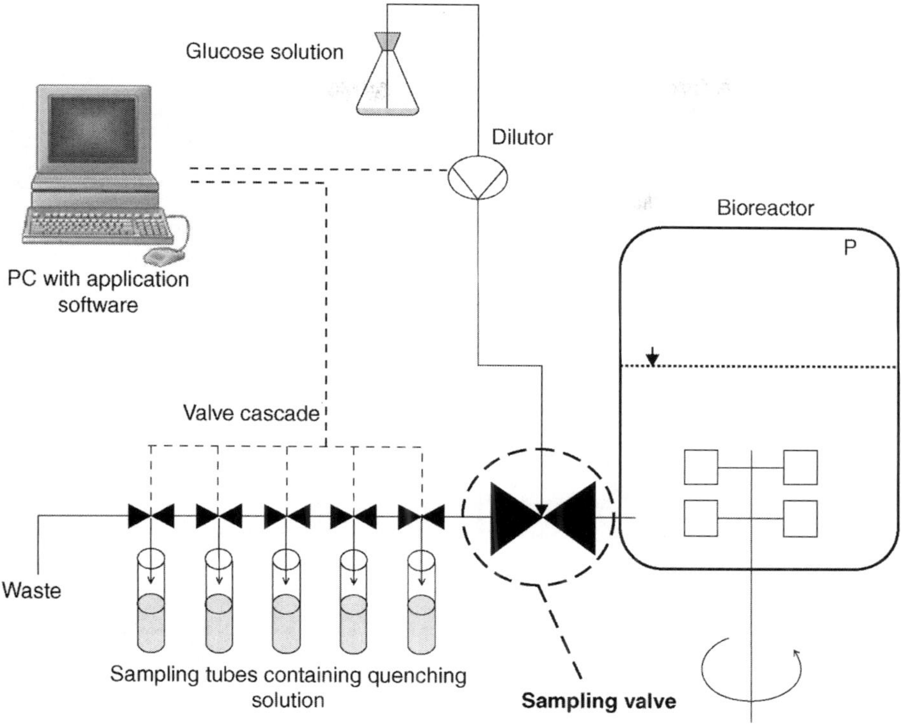

Figure 7.5 Assembly of the new bioreactor coupled rapid stopped-flow sampling technique according to Buziol *et al.* (2002). Reproduced from Biotechnology and Bioengineering, vol. 80, *New bioreactor-coupled rapid stopped-flow sampling technique for measurements of intracellular metabolite dynamics on a subsecond time scale*, page 633, Copyright (2002), with permission from John Wiley & Sons, Inc.

description of the sampling valve is found in Buziol et al. (2002). In resume, the concentrated glucose solution was pumped into the mixing chamber inside the sampling valve, and it was there mixed with the cultivation medium. The cultivation medium loaded with the concentrated glucose solution flowed through the outlet capillary toward the waste. After the capillary was flushed with the mixture of cultivation medium and glucose solution to the waste, the first sample flow was redirected through the position of valve 1 to the sampling tube containing the quenching fluid (liquid nitrogen, −196°C). The opening time of valve 1 was under control of the computer. The first valve was then closed and the mixture proceeded toward the waste to flush the capillary again to the second valve. The second valve was redirected, and the procedure (flow into the tube filled with quenching fluid) was repeated. The procedure was continued until the suspension flowed into the waste tube. According to Buziol et al. (2002), the main features of this sampling device are as follows: (i) the cultures remain at a steady-state because the organisms are stimulated by the glucose in the mixing chamber within the valve; (ii) sampling time and reaction time are decoupled; (iii) the time span between glucose stimulus and first sample can be less than 100 ms; and (iv) the method can be easily adapted to other stimuli, e.g., temperature or pH, which may lead to irreversible stress responses. The only limitations were a possible problem of oxygen limitation at aerobic growth and the impossibility of distinguishing extracellular from intracellular metabolites when using liquid nitrogen as quenching agent.

7.7 THE BIOSCOPE: A SYSTEM FOR CONTINUOUS-PULSE EXPERIMENTS

Similar to the stopped-flow technique reported by Buziol et al. (2002), but with minimized size and apparently without oxygen limitation problem, the BioScope is also based on the continuous flow principle in which only a small flow of fermentation broth is perturbed outside the fermentor instead of perturbing the whole fermentor (Visser et al., 2002). Figure 7.6 provides a schematic overview of the BioScope device according to Visser et al. (2002). The device consists of oxygen-permeable silicon tubing with an inner diameter of 0.8 mm and a wall thickness of 0.6 mm, which is connected to the fermentor. The tubing resembles a miniaturized serpentine to keep its size minimal. The BioScope consists of 20 small serpentine units between which 11 sampling ports are located. The total length of the tubing connecting the serpentine units is 6.6 m, of which 17% is straight.

The flow of fermentation broth throughout the tubing is controlled by a pump located at the beginning of the tubing. By setting up the tubing flow at a lower rate than the feed-flow of the fermentor, the steady-state is not disturbed. Different perturbations/stimuli can be applied, and the residence time between the fermentor port and the mixing point is calculated to be approximately 3 s and sampling time

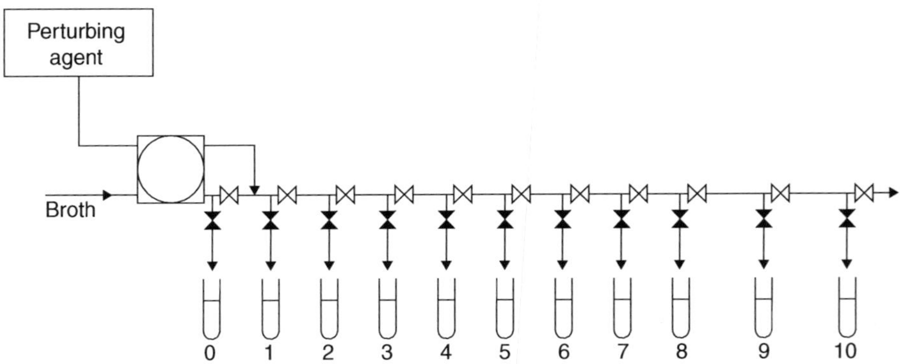

Figure 7.6 Schematic overview of the BioScope device according to Visser *et al.* (2002). Reproduced from Biotechnology and Bioengineering, vol. 79, *Rapid sampling for analysis of in vivo kinetics using the BioScope: A system for continuous-pulse experiments*, page 675, Copyright (2002), with permission from John Wiley & Sons, Inc.

lower than 100 ms. The complete set-up is located in a thermostated box, and the air temperature inside the box is controlled at the same temperature as that of the fermentor. According to Visser et al. (2002), the BioScope offers a number of advantages over the other approaches reported so far. For instance, (a) a large number of different perturbation experiments can be carried out on the same day, because the physiological state of the fermentor is not disturbed; (b) *in vivo* kinetics during fed-batch experiments and in large-scale reactors can be also investigated; (c) all metabolites of interest can be measured using samples obtained in a single experiment, because the volume of the samples is unlimited; (d) the amount of perturbing agent spent is minimal, because only a small volume of broth is perturbed; and (e) the system is completely automated.

7.8 CONCLUSIONS AND PERSPECTIVES

The development of rapid sampling techniques to investigate intracellular metabolite dynamics has achieved major advances toward automation and miniaturization of the systems. The readers must have noticed that researches in this field are anterior to the pioneering works on metabolomics and have started even before the word metabolome was created. With systems available today, samples can be harvested in less than 100 ms with excellent reproducibility and without disturbance of the physiological state of the cell in the bioreactor.

Experimental data for the dynamics of intracellular metabolite concentrations within seconds after the addition of a perturbation agent to a balanced steady-state culture are absolutely necessary to identify the parameters of dynamic models as well as metabolic flux analysis. The BioScope sampling system is likely to be a particularly valuable tool because of the possibility of achieving the highest sampling rates at short inactivation times without disturbing the steady-state of the cells, with an additional advantage to be fully automated.

However, all these developments are not easily accessible to the scientific community because they are mostly home-built devices not available commercially. Future commercialization of rapid sampling devices systems for microbial cultures, designed to attend the requisites of the metabolomics field are extremely necessary and are likely to become a technological mark toward method standardization that metabolomics is currently lacking.

REFERENCES

Buziol S, Bashir I, Baumeister A, Claaßen W, Noisommit-Rizi N, Mailinger W, Reuss M. 2002. New bioreactor-coupling rapid stopped-flow sampling technique for measurements of metabolite dynamics on a subsecond time scale. Biotechnol Bioeng 80:632–636.

Lange HC, Eman M, van Zuijlen G, Visser D, van Dam JC, Frank J, Teixeira de Mattos MJ, Heijnen JJ. 2001. Improved rapid sampling for *in vivo* kinetics of intracellular metabolites in *Saccharomyces cerevisiae*. Biotechnol Bioeng 75:406–415.

Schaefer U, Boos W, Takors R, Weuster-Botz D. 1999. Automated sampling device for monitoring intracellular metabolite dynamics. Anal Biochem 270:88–96.

Theobald U, Mailinger W, Reuss M, Rizzi M. 1993. *In vivo* analysis of glucose-induced fast changes in yeast adenine nucleotide pool applying a rapid sampling technique. Anal Biochem 214:31–37.

Theobald U, Mailinger W, Baltes M, Rizzi M, Reuss M. 1997. *In vivo* analysis of metabolic dynamics in *Saccharomyces cerevisiae*: I. Experimental observations. Biotechnol Bioeng 55:305–316.

Weuster-Botz D, de Graaf AA. 1996. Reaction engineering methods to study intracellular metabolite concentrations. Adv Biochem Eng Biotechnol 54:75–108.

Weuster-Botz D. 1997. Sampling tube device for monitoring intracellular metabolite dynamics. Anal Biochem 246:225–233.

Visser D, van Zuylen GA, van Dam JC, Oudshoorn A, Eman MR, Ras C, van Gulik WM, Frank J, van Dedem GWK, Heijnen JJ. 2002. Rapid sampling for analysis of *in vivo* kinetics using the BioScope: A system for continuous-pulse experiments. Biotechnol Bioeng 79:674–681.

8

PLANT METABOLOMICS

By Ute Roessner

This chapter gives a short summary of metabolomics applications in plant research. It has been estimated that several hundreds of, thousand different metabolic components may be produced within the plant kingdom, and they vary in their abundances by 6 orders of magnitude. Any valid metabolomics approach must be able to unbiasedly extract, separate, detect, and accurately quantify this enormous diversity of chemical compounds. These requirements dictate the challenges that are continually addressed in the field of plant metabolomics, which will be discussed in the following chapter.

8.1 INTRODUCTION

Plants play the most important part in the cycle of nature. Without plants, there could be no life on Earth. They are the primary producers that sustain all other life forms. Plants are the ultimate source of food and metabolic energy for nearly all animals who cannot manufacture their own food. Animals depend directly or indirectly on plants for their supply of food. Leaves are the main food-making part of most plants. They use the energy from sunlight and turn water and carbon dioxide into carbon sources such as sucrose, starch, proteins, or fat. Although some 3000 different plant species have been used as food by humans, 90% of the world's food comes from only 20 plant species including rice, wheat, barley, potato, tomato, soy, and pea. Green plants possess chlorophyll that allows them to capture Gibbs free energy in valuable carbon sources. Through the process of photosynthesis (Figure 8.1), plants take Gibbs free energy from the sun, carbon dioxide from the air, and water and minerals from the soil. In the process of generating storage

Metabolome Analysis: An Introduction, by Silas G. Villas-Bôas, Ute Roessner, Michael A. E. Hansen, Jorn Smedsgaard and Jens Nielsen

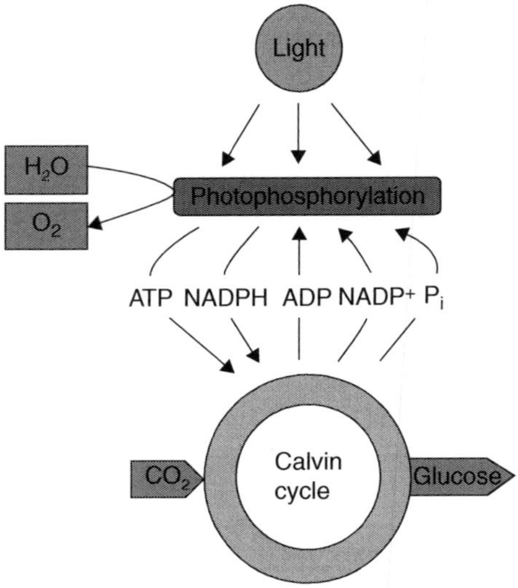

Figure 8.1 Simplified scheme of the photosynthetic process. Light energy is used for photophosphorylation using water, ADP, P_i, and $NADP^+$ producing O_2, ATP, and NADPH. These are further used in the dark reaction (Calvin cycle) for carbon fixation producing glucose. (See color plates.)

carbon sources, they release water and oxygen. Animals and other nonproducers take part in this cycle through respiration. Respiration is the process where oxygen is used by organisms to release carbon dioxide and energy from food. The cycles of photosynthesis and respiration help to maintain the earth's natural balance of oxygen, carbon dioxide, and water.

Besides foods (e.g., grains, fruits, and vegetables), plant products are vital to humans. Valuable plant products include wood and wood products, vitamins, antioxidants, fibers, drugs, oils, latex, pigments, and resins. Coal and petroleum are fossil substances of plant origin. Thus, plants provide people with not only food sources but also shelter, clothing, fuels, and the raw materials from which innumerable other products are derived. Furthermore, throughout history, plants have been of great importance to medicine. Eighty percent of all medicinal drugs originate from wild plants. In spite of all the medical advances, only 2% of the world's plant species have ever been tested for their medical potential. That means that there are many important drugs yet to be discovered, in which a metabolomics approach will be of great importance.

A plant may be microscopic in size and simple in structure, as are certain one-celled algae, or a gigantic, many-celled complex system, such as a tree. Plants are generally distinguished from animals in that they possess chlorophyll, are usually immobile, have no nervous system or sensory organs and hence do not respond to stimuli, and have rigid supporting cell walls. In addition, the anatomy of plant cells is different to those of animals. Most plant cells contain plastids and large vacuoles and, as mentioned before, are surrounded by cell walls.

The study of plant metabolism has fascinated scientists for a long time. The investigation of the ability of green tissue to fix carbon for energy storage made first great success, when Michael Tswett (1872–1920) developed the first concept and technique of chromatography for the separation of chlorophyll, xanthophyll, and carotene in 1906. About 50 years later, Melvin Calvin and Andrew Benson discovered the photosynthetic cycle, today commonly called the "Calvin cycle." But other plant-specific pathways have been under investigation for many decades, such as the starch synthetic pathway, cell wall biosynthesis, vitamin production, sucrose synthesis and recycling, amino acid biosynthesis, or fatty acid synthesis and degradation. A large number of analytical technologies have been developed for the analysis of plant metabolites in order to study plant metabolism in great detail. In addition, the development of methodologies for genetic transformation of plant genomes by mutation of transgenesis has introduced a great demand for sophisticated biochemical techniques for a detailed characterization of the effects of these genetic alterations. In addition, the interest in the determination of genetic diversity and by this chemical diversity of a large number of plant species in many different environmental situations has risen. The development of multi-parallel and/or highly sensitive analytical tools to measure cell products has made enormous progress. Most prominent amongst these new technologies has been the establishment of protocols for the determination of the expression levels of many thousands of genes in parallel (transcriptomics), the detection, identification, and quantification of the protein complement (proteomics), and the possibility of determining and identifying a large number of metabolic compounds in parallel and in a high-throughput manner (metabolomics). Metabolomics today is one of the most important tools to investigate plant metabolism, plant behavior in certain environmental conditions, or metabolic responses to genetic alterations. In the following, a short overview about the history of plant metabolomics, its particularities, and potential valuable applications will be presented.

8.2 HISTORY OF PLANT METABOLOMICS

The determination of plant metabolic compounds has already been done for many decades. As mentioned above, the work of Twsett in the beginning of the 20th century can be seen as the pioneer work in the separation of plant compounds using chromatographic techniques. By the introduction of other analytic techniques, like column chromatography or electrophoresis, the development of protocols for plant metabolite analysis has made great progress. The metabolite profiling was first mentioned in the early 1970s in the medical field where GC–MS was applied for multicomponent analysis of human urine. This concept was further followed by using not only GC–MS, but also HPLC and NMR for expansion of the types of compounds being analyzed. The interest on the concept of multi-targeted analysis of biological compounds increased dramatically and resulted in a special edition focusing on metabolite profiling of the *Journal of Chromatography* in 1986. The first report on metabolite profiling in plants was presented by Sauter et al. from BASF in 1991, where they used a GC–MS-based method as a diagnostic technique in order to compare the effects of various herbicides on barley plants (Sauter et al., 1991). In the end of the 1990s, metabolite profiling was the basis of the development of a comprehensive GC–MS-based methodology for a simultaneous determination of a very large number of metabolites in a range of plant species by pioneers (Willmitzer, Trethewey, Kopka, Fiehn, Roessner) at the Max-Planck-Institute for Molecular Plant Physiology in Golm, Germany (Fiehn et al., 2000, Roessner et al., 2000). These scientists were also the first to apply mathematical tools for classification and visualization, such as principle component analysis (PCA) or hierarchical cluster analysis (HCA), onto large data sets accumulated from metabolite profiling (Fiehn et al., 2000, Roessner et al., 2001a, Roessner et al., 2001b). Another concept, first introduced by Steve Oliver in 1997, where he proposed the need for the measurement of the metabolic phenotype to access gene function in yeast (Oliver, 1997), was adopted for plant metabolism by the Max-Planck scientists. Using the metabolite profiling data sets, coresponse analysis between metabolites was carried out for further metabolic network establishments (Fiehn 2003, Weckwerth et al., 2004). Today, off-the-shelf instruments are able to rapidly and quantitatively detect up to 500 compounds simultaneously in crude plant extracts, depending on tissue and extraction procedure. In the last few years, GC–MS technology has been applied and optimized for simultaneous analyses of metabolites in many different plant species, such as *Arabidopsis thaliana* (Fiehn et al., 2000), *Solanum tuberosum* (Roessner et al., 2000), *Medicago truncatula* (Duran et al., 2003), *Lycopersicon esculentum* (Roessner-Tunali et al., 2003a), *Saccharum officinarum* (S. Bosch, personal commun.), *Lotus japonicus* (Colebatch et al., 2004), *Cucubita maxima* (Fiehn 2003), and *Hordeum vulgare* (Roessner et al., 2006).

It soon became obvious that GC–MS alone does not cover all of the chemical diversity of plant metabolites, and other complementary approaches had to be established. One of these was the application of liquid chromatography coupled to electrospray ionization mass spectrometry (LC–ESI–MS). The main advantages of LC–ESI–MS are twofold. First, compounds do not have to be chemically altered prior to analysis and secondly, highly polar, thermo-unstable, and high-molecular weight compounds, such as oligosaccharides or lipids, are to be separated and quantified. LC in combination with ultraviolet or visible light (UV/VIS) or diode-array detection (DAD) has been applied for many years in plant metabolite analyses. An enormous range of different columns and elution procedures exist for the separation and detection of many different classes of compounds. When coupled to MS, these provide further selectivity, unbiased detection, and most importantly, information about the structure of detected compounds. This multidimensional approach has been successfully applied for the analysis of a wide range of primary and secondary metabolites in plant tissues (Tolsitkov and Fiehn, 2002, Huhman and Sumner, 2002). Recently, the use of a monolithic column enabled the separation of several hundred chromatographic peaks derived from extracts of *Arabidopsis* (Tolstikov et al., 2003). Another research group has reported the detection of 1400 components (based on mass-to-charge ratios) by direct injection of *Arabidopsis* extracts into a quadrupole time-of-flight (QTOF) hybrid mass spectrometer (von Roepenack-Lahaye et al., 2004). The resolution and selectivity of mass detection can be dramatically increased up to 5000 signals from a single plant extract by application of Fourier-transform ion cyclotron resonance mass spectrometry (FT–ICR–MS) as shown by Aharoni et al. (2002).

An additional challenge in plant metabolite analyses is the development of technologies for the isolation and detection of metabolites from very small samples sizes in order to increase spatial resolution in single cell or tissue-specific investigations. These techniques have to be designed to combine high sensitivity with selectivity. First remarkable reports have been given on the determination of the distribution of IAA in *Arabidopsis* plants (Muller et al., 2002) or even the distribution of ATP in *Vicia faba* embryos (Borisjuk et al., 2003). Future research has now to face multi-parallel analyses of metabolites on a cell and organ level. One attractive technology to increase sensitivity is capillary electrophoresis in combination with laser-induced fluorescence (CE–LIF) or mass spectrometric detection (CE–MS), which has been already proven to give promising results. For example, CE–LIF allowed the separation and quantification of a large range of amino acids and sugars in approximately 50 picoliters of phloem sap or in five-pooled mesophyll cells of *Cucurbita maxima* (Arlt et al., 2001). By using CE–MS, more than 80 main metabolites belonging to glycolysis, photorespiration, or the oxidative pentose phosphate pathway could be analyzed in rice leaf extracts (Sato et al., 2004). It is worthwhile to note that in this study, the ability to analyze many unstable substances in parallel, which only occur in low concentrations in *planta*, such as fructose-1,6-bisphosphate or ribulose-1, 5-bisphosphate, was presented.

Another important technique, only very recently introduced in plant metabolomics, is nuclear magnetic resonance spectroscopy (NMR) (for review see Krishnan et al., 2005). Its major advantage is that the analysis is a noninvasive approach, meaning that samples could be used for extraction of other cell products following an NMR scan. In addition, NMR analysis covers a large range of compound classes simultaneously; it is fast and the resulting spectra can easily be accessed for post-multivariate analysis such as PCA.

Currently, scientists planning a metabolomics experiment on their plant system of interest will have to face a large number of different analytical techniques for the measurement of many different plant metabolite classes. Depending on experiences and resources, the most applicable extraction procedures and analytical techniques have to be chosen, but if the working definition for metabolomics means the analysis of all metabolites in a plant, it requires a platform of complementary analytical technologies for comprehensive selectivity and sensitivity.

8.3 PLANTS, THEIR METABOLISM AND METABOLOMICS

8.3.1 Plant Structures

Most seed-producing plants have the same three basic organs: leaves, stems, and roots. Various developmental adaptations of these organs have enabled plants to survive a large range of different environments and as plants are often immobile, they have to withstand temporary extreme conditions. Plant cells have unique structures compared to cells of other organisms; in addition they contain a central vacuole, plastids, and a thick, plasma membrane surrounding the cell wall.

In general, it can be said that plants are made of three types of cells which form four types of tissue. The most abundant type of cells in plants is parenchyma cells, which are the least structurally specialized, contain a very large central vacuole, and have thin and flexible cell walls. Parenchyma cells occur throughout the plants and fulfill many functions, including photosynthesis, storage product accumulation, and general metabolism. Other types of cells are collenchyma cells supporting the growing parts, and sclerenchyma cells, supporting the nongrowing parts of plants. The sclerenchyma cells have too thick cell walls that the cells die when matured, for example, fibers (cotton), and sclereids (walnut shell) are made from these type of cells. The three types of plant cells make up the four basic plant tissues: the vascular, the dermal, the ground, and the meristematic tissue, which themselves form into the organs leaves, roots, and stems.

Roots typically grow underground and are very important structures because they anchor the plant in the soil. They also absorb and transport water and nutrients from the soil to the upper parts of the plant. Interestingly, roots are selective about which mineral they absorb; some are even excluded. There are 13 minerals essential for all plants, including macronutrients, such as N and P, and micronutrients, such as Na, K, B, Mn, Fe, Ca, etc. Severe mineral deficiencies lead to dramatic growth retardations and can even kill the plants, but on the contrary excess amounts of some of the minerals can be toxic. In both cases, plant metabolism is dramatically affected; plants are able to develop mechanisms in order to cope with either deficiency or toxicity. Currently, metabolomics is used to follow metabolic responses to mineral deficiencies (e.g., P) and toxicities (e.g., Na^+ or B) to understand more about the mechanisms behind adaptation and tolerance to these types of stresses (Roessner et al., 2006, Roessner, personal commun.). In addition, roots of some plant species (legumes) are able to build symbiotic relationships with nitrogen fixating bacteria by the formation of nodules, which is an amazing metabolic process, and is in detailed studied using a metabolomics approach by Colebatch et al. (2004).

The stems have two major following functions: firstly, to hold up the leaves for best exposure to the sunlight, and secondly, to transport water, soluble carbon sources, and hormones between the roots and leaves. In some species, stems also function as storage organs, for example, potato tubers are underground stems storing large amounts of starch. To transport, two types of systems are developed in stems. The phloem moves the soluble carbon sources from the place of production (source—leaves) to places of need (sink—any heterotrophic, meaning nonphotosynthetic active tissue—roots, fruits). So far it was believed that the major transported food compound in plants is sucrose or other soluble carbohydrates, such as raffinose or sorbitol. By an in-depth metabolite analysis of phloem sap, it could be demonstrated that a large range of different metabolic compounds, including amino and organic acids, can be found in phloem sap of *Cucibta maxima* (Fiehn, 2003). Many of the detected substances were not identifiable, and therefore, this work has clearly demonstrated the potential of metabolomics for increasing our knowledge about plant physiology as well as identifying novel biosynthetic pathways. Water and minerals are transported through the xylem, which actually exists in all organs of a plant. As aerial parts of the plants lose large amounts of water by transpiration, replacement water has to be "pulled" from the roots via the xylem. Again, in literature it has been stated that xylem transports only water and nutrients, but when xylem sap was analyzed using GC–MS, many more primary and also secondary metabolites were detected (Roessner, personal commun.). The investigation of what the functions of these metabolites are and from where-to-where they are transported will be a major task in plant biology research.

The main function of leaves is to capture light energy during photosynthesis allowing them to produce glucose from carbon dioxide and water. In addition, leaves have important functions in defense mechanisms against animals, fungi, bacteria, or virus. Figure 8.2 shows a simplified scheme of a cross-section of a typical leaf. The epidermis of a leaf has two specialized structures developed as adaptations for photosynthesis; a waxy cuticle for water loss protection and strictly regulated stomata, allowing carbon dioxide to enter the leaf and water and oxygen to go out. These pores are formed by two kidney-shaped, so-called guard cells, which open and close the stomata depending on environmental condition and the needs of the plant. The middle region is called mesophyll. Mesophyll cells are packed with chloroplast, which are specialist compartments in plant cells where photosynthesis occurs.

The complex anatomy of plant tissues and organs has to be strongly considered for any metabolomics approach. Presently, most developed analytical methodologies need a certain amount of tissue to be extracted to be able to detect and quantify metabolite levels. Very often, parts of tissues, whole organs (e.g., leaves or roots), or even whole plants are homogenized and metabolites extracted. This may include many different cell types, which might be actually characterized by their specific metabolite profile. The development of instrumentation with highly increased

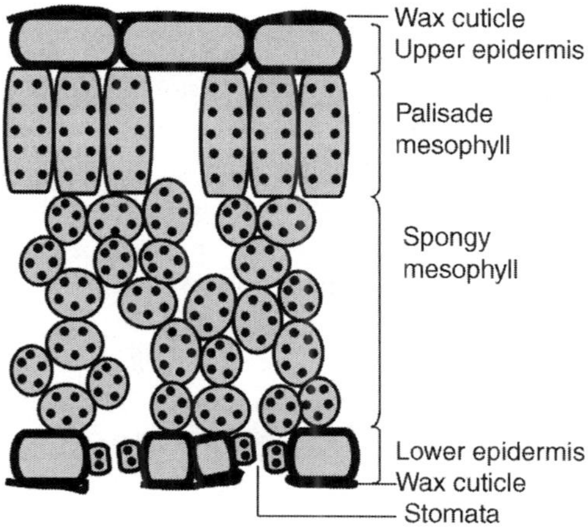

Figure 8.2 Schematic cross section of a photosynthetic active plant leaf showing the different types of tissues (epidermis, palisade, and spongy mesophyll) and cells (stomata).

sensitivity may help substantially, but the major issue is that it is very difficult or even sometimes impossible to separate and isolate single cells from plant tissues. First success on a single cell metabolomics approach has been reported by using cryo-sectioning to preserve cellular structures, specific cell types were cut and collected using laser micro-dissection to a sufficient amount of cells which allowed the detection of about 68 major metabolites in these cells by GC–MS (Schad et al., 2005). Another potential approach might be the production of cell-type specific protoplasts; these are wall-free cells, which can be cultured and therefore large amounts can be produced.

8.3.2 Plant Metabolism

Most plant primary metabolic pathways exist essentially in the same form as in all other organisms. But as plants are autotrophic certain unique features can be found in plant metabolism. Most known is the photosynthesis in which the plant produces ATP and reducing equivalents NADPH by using light as the energy source. This process is located in the chloroplasts of green tissues. In the second part of photosynthesis, which is a light-independent process, ATP and NADPH are used for the production of glucose from carbon dioxide. The overall reaction of photosynthesis is summarized as follows:

$$6\ CO_2 + 12\ H_2O + \text{light energy} \rightarrow C_6H_{12}O_6 + 6\ O_2 + 6\ H_2O$$

It is outside the scope of this book to go in much detail of the very interesting features and steps of the photosynthetic process and the reader is referred to any plant physiology book.

In addition to photosynthesis, there are other well-studied plant-specific metabolic pathways. Worthwhile to mention in this chapter is the photorespiration, which is a specialized mechanism of plants to survive with the situation where the CO_2 levels inside a leaf become too low for the photosynthesis process to operate. This happens on hot dry days when a plant is forced to close its stomata to prevent excessive water loss and therefore, sufficient CO_2 cannot be taken up efficiently. In this case, Rubsico accepts O_2 instead of CO_2 as substrate, producing the toxic compound phosphoglycolate and no ATP. The detoxification of phosphoglycolate by several enzymatic steps and involvement of different compartments lead to the production of serine and a consequent loss of carbon for the plant. Furthermore, plant mitochondria possess specific features; unlike those from animals, they have a specific transport system for NAD(P)H produced during glycolysis. Direct fixation of CO_2 into pyruvate in the cytosol using NADH or NADPH oxaloacetic acid is produced, which is then transported into the mitochondria, creating a shuttle system for reducing equivalents. The plant-specific carbohydrate storage product is starch, which is an important food component in most crops, fruits, and vegetables, but it is also of great importance for industrial application such as raw material for glue production. The biosynthetic pathway of starch has been a scientific target for many years (see Figure 2.4.) aiming for development of plants with increased starch levels or altered starch features. Unlike animal cells, those of plants are surrounded by a cell wall, which consists of different carbohydrate polymers, such as cellulose or hemicellulose. The biosynthesis of cell walls is very complex and involves the production of mainly UDP-activated sugar molecules for polymer extensions. As already mentioned in Chapter 2, plants are characterized by the ability to produce a vast diversity of secondary metabolites. Each plant species is able to produce a specific set of secondary metabolites depending on environmental conditions or ecological interactions with other organisms. Scientists have long been interested in the production of these phytochemicals and have investigated them extensively since the 1850s. The study of natural products has stimulated the development of separation techniques and methodologies for structure elucidation. Many of these compounds have been shown to play important adaptive roles in the protection against herbivory and microbial infection, as attractants for pollinators and seed-dispersing animals, as well as allelopathic agents that affect the plant's survival profoundly.

8.4 SPECIFIC CHALLENGES IN PLANT METABOLOMICS

8.4.1 Light Dependency of Plant Metabolism

Plant metabolism is highly light-dependent resulting in differential metabolite levels between day and night. During the day, when there is light, photosynthesis happens and carbon sources are produced and made available, e.g., many storage processes are functional, such as starch synthesis. During the night, on the contrary, photosynthesis is down regulated and storage products are degraded for energy availability through respiration. Many other metabolic pathways are dependent on carbon availability and therefore undergo diurnal rhythmus; depending on their function they are more active either during the day or during the dark phase (Figure 8.3, Urbanczyk-Wochniak et al., 2005a). Therefore, special care has to be taken about the time-point when plant tissue samples are harvested; in general, as a role, all samples should be taken at the same time-point or in a very small time frame. This may become difficult when a large set of plants are under investigation, then it can be of help to harvest in a randomized way (not one genotype after the other throughout the day) in order to capture day time differences in metabolite profiles in the variability throughout the data set.

Plant metabolism is dependent not only on availability of light, but also on the strength and wavelength of light. This especially affects leaf metabolism as in most plants each leaf is differently exposed to light, for example, upper leaves give shadow to lower leaves, leading to quite differential metabolite profiles for each leaf of one and the same plant. One way to overcome this is to grow again the set of plants under investigation in a randomized way and also select a similar exposed leaf always, either upper or lower.

As already described in Chapter 3, metabolic reactions can be extremely fast and therefore a rapid quenching of metabolism during tissue harvest is crucial. For plant tissues, this can be done either using freeze clamps or by shock freezing in liquid

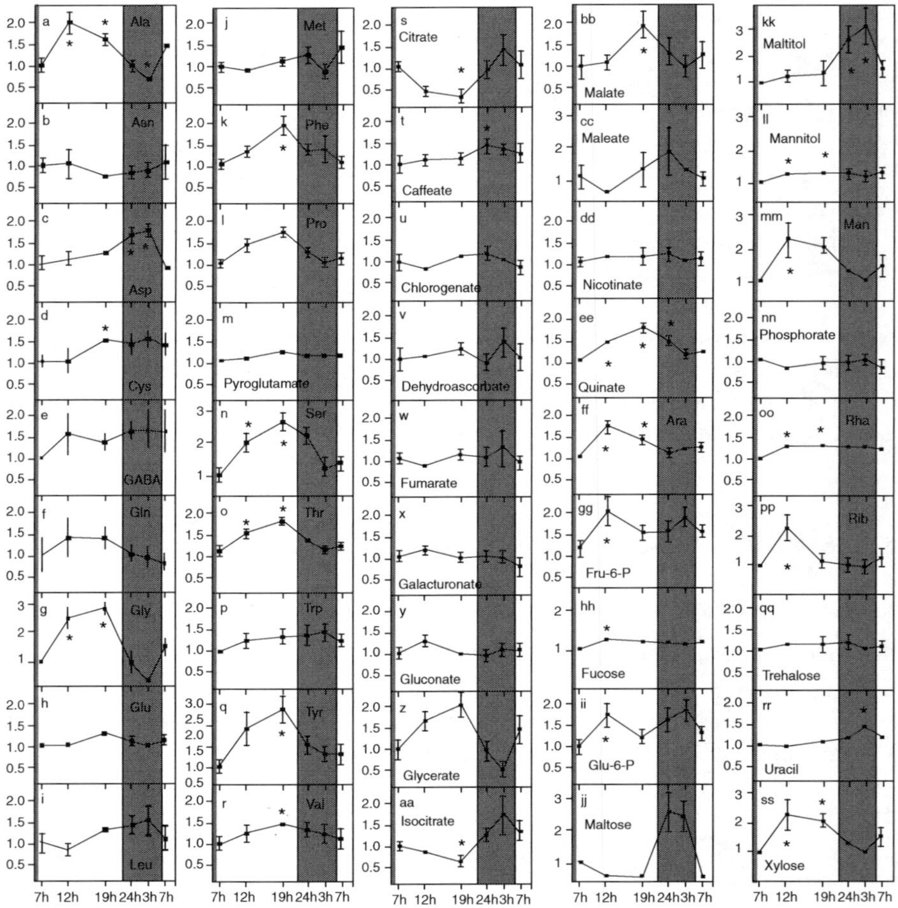

Figure 8.3 Diurnal changes in metabolite levels in tomato leaves: Ala (a), Asn (b), Asp (c), Cys (d), GABA (e), Gln (f), Gly (g), Glu (h), Leu (i), Met (j), Phe (k), Pro (l), Pyroglutamate (m) Ser (n), Thr (o), Trp (p), Tyr (q), Val (r), Citrate (s), Caffeate (t), Chlorogenate (u), Dehydroascorbate (v), Fumarate (w), Galacturonate (x), Gluconate (y), Glycerate (z), Isocitrate (aa), Malate (bb), Maleate (cc), Nicotinate (dd), Quinate (ee), Ara (ff), Fru-6-P (gg), Fucose (hh), Glu-6-P (ii), Maltose (jj), Maltitol (kk), Mannitol (ll), Mannose (mm), Phosphorate (nn), Rhamnose (oo), Ribose (pp), Trehalose (qq), Uracil (rr), Xylose (ss). At each timepoint, samples were taken from mature source leaves and the data represent the mean ±SE of measurements of six plants. The dark period is indicated by the grey box. Asterisks represent values that are significantly different from the first sampling point. With kind permission of Springer Science and Business Media. Figure 2 of Urbanczyk-Wochniak et al., 2005a.

nitrogen. The latter one has proven to be extremely efficient for many different plant tissues, but tissue pieces have to be small enough so that every part is frozen; if the piece is too large there will be a delay of freezing in the inner parts. Frozen plant tissue samples can be stored at −80°C until extraction.

8.4.2 Extraction of Plant Metabolites

Special care has to be taken for the extraction of metabolites from different plant species. Most crucial is the homogenization step and breakage of plant cells as they are often surrounded by very rigorous cell wall. Different homogenization procedures were introduced in Chapter 3, and the procedures most used for plant tissues are mortar and pestle or ball mills. It is extremely important that the homogenization process takes place under liquid nitrogen to prevent defrosting of tissue which, when happens, will dramatically alter the metabolite profile. Many plant enzymes survive freezing and will be quite active after defrosting. For example, the enzyme invertase, which cleaves sucrose to glucose and fructose very efficiently, not only survives freezing but also the extraction in a 1:1 mixture of chloroform and water at −20°C, therefore leading to a completely altered sugar profile (Roessner et al., 2006). To what extent other enzymes are stable throughout different extraction methods are to be confirmed for each tissue and procedure. As a role it is helpful to shorten the actual extraction step as much as possible and separate from insoluble components and dry the extract to prevent any enzymatic activity. An alternative is to extract in nonaqueous solution as most enzymes need water for their functionality.

It is then important to separate the small molecules from the insoluble components of the cell, such as protein, starch, cell wall, and other high-molecular weight carbohydrates. For many separation and detection techniques, the pigments contained in plant tissues, such as chlorophyll and carotenoids, disturb the analysis and should be separated from other metabolites (of course only if they are not the target of analysis).

8.4.3 Many Cell Types in One Tissue

As mentioned above, plant tissues are very heterogeneous, that means different cell types form a plant tissue. Each cell type may be characterized by a specific metabolic profile depending on their function, time of the day, environment, etc, which will not be seen when whole tissues are homogenized and extracted. For example, even a potato tuber, which grows in the dark and consists of the same cell types (apart from outer skin) and is therefore supposed to be very homogenous, is characterized by a gradient of metabolites driven by the supply of sucrose from leaves via the stolon. This also results in a light-dependent metabolism in potato tubers as the photosynthetic sucrose supply alters during the day (Roessner-Tunali et al., 2003b). Because of this tissue in-homogeny it is particularly important to take care that for comparative metabolomics always similar tissue parts, tissues or organs of each plant are sampled.

In addition, the developmental stage of a plant is another factor that affects its metabolite profile dramatically. Therefore each plant should be harvested in a similar developmental stage. This may become extremely difficult when, for example, mutants with growth retardations or developmental delays, compared with wild type, are to be analyzed. Specific developmental stages have to be defined, for example, appearance of first flowers or ripening of fruits.

8.4.4 The Dynamical Range of Plant Metabolites

Often, in plant extracts, only a small number of metabolites occur in extremely high concentrations, for example, hexoses (most leaves and tomato fruit), sucrose (potato tuber), citrate (tomato fruit), sorbitol (apple and peach trees and their fruits), and malate (barley leaf and apple fruit) (Roessner, personal commun.). In addition, certain environmental factors lead to the production of high amounts of specific metabolites (often referred as to osmolites or osmoprotectants), e.g., proline can increase several hundreds fold after a high salt or drought event. Water limitation also leads to the degradation of storage carbohydrates resulting in high concentrations of soluble sugars. On the contrary, many metabolites are present in very low amounts, especially pathway intermediates or signaling molecules, such as phytohormones. This variability of abundance, which has been estimated to exceed 6 orders of magnitude, represents an additional challenge for a metabolomics approach as most technologies, either the separation or detection, or both, cannot cover this high dynamic range. A separation of the high-abundant metabolites is often not feasible, as low- and high-abundant compounds may belong to the same compound class, and most prepurification procedures such as solid phase extraction, target-specific compound classes, for example, it is almost impossible to remove sucrose from the extract without losing other disaccharides and even mono- and trisaccharides. One potential approach would be to produce specific antibodies for single metabolites to be purified by affinity. Another possibility is to analyze different amounts of metabolite extract in order to cover larger dynamic ranges (Roessner et al., 2000; Roessner-Tunali et al., 2003a; Roessner et al., 2006). But care has to be taken to avoid column overloading or blocking of interacting sites, resulting in no separation at all.

8.4.5 Complexity of the Plant Metabolome

As mentioned in other chapters, the metabolome consist of a large range of compounds having many different chemical structures. This is particularly the case for plant metabolites. It is estimated that the whole plant kingdom is capable of producing between 200,000 and 400,000 different metabolic compounds, whereby a single species may be producing about 5000–10,000 compounds at one point of time in a certain environment. The new analytical approach of metabolomics, which is non-targeted metabolite detection, results in a large number of chromatographic peaks and mass spectra, which cannot be identified easily with respect to the chemical nature of the compound. It has been shown in many examples that up to 70% of all peaks in a typical GC–MS chromatogram of a plant extract still remains unidentified. Figure 8.4 shows a typical outcome of a deconvolution process of a plant GC–EI–MS chromatogram using AMDIS and the MSRI mass spectral library (see Section 6.4.6.). The software filtered more than 600 single metabolites of which about 220 could be assigned to a library spectra. These numbers also include artifacts like peaks resulting from solvents or the column but the ratio of the detected and the identified compound, is similar.

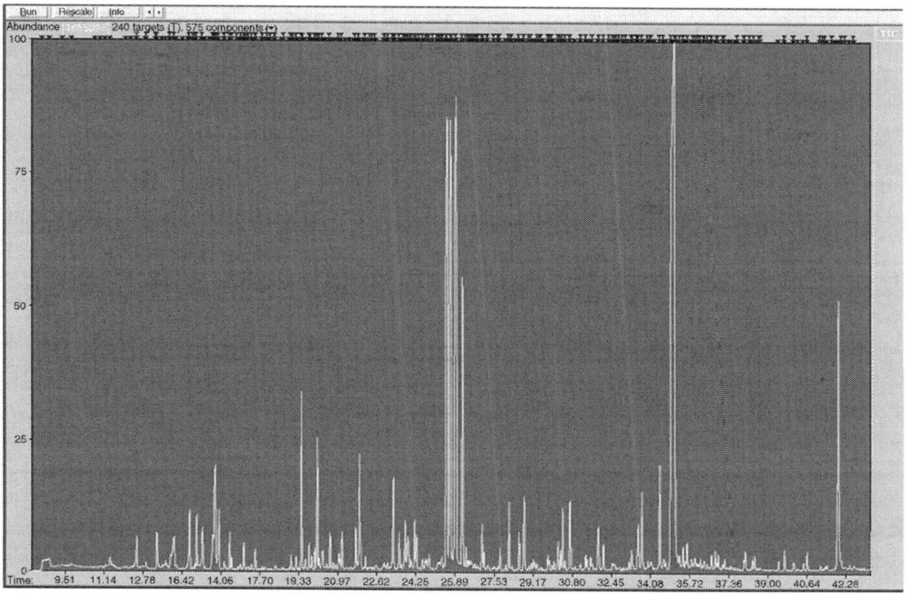

Figure 8.4 AMDIS deconvolution result of a GC–MS chromatogram of a wheat leaf extract. Deconvoluted mass spectra were matched against the MSRI mass spectral library (http://csbdb.mpimp-golm.mpg.de/gmd.html). Out of 575 deconvoluted mass spectra (components, indicated with triangles), 240 were found to match a library mass spectrum (targets, indicated with "T").

The interpretation of mass spectra following GC–EI–MS analysis is very difficult for two reasons. First, derivatization dramatically alters the chemical structure of the compounds. Secondly, the use of electron impact (EI) to ionize the compounds is a very harsh method that leads to complex fragmentation patterns. As a result, two strategies are used to identify the chemical nature of as many peaks as possible. First, the spectra of all resolved peaks are compared with commercially available EI mass spectrum libraries such as NIST (http://www.nist.gov/: National Institute of Standards and Technology, Gaithersburg, USA). However, although these libraries contain over 350,000 entries, the majority of these are nonbiological compounds. In the second approach, commercial standard compounds that are assumed to be present at detectable levels within plant tissues are analyzed. A reference library containing both the retention time of these compounds (as determined under the same conditions) and the corresponding mass spectrum can be created (Wagner et al., 2003). Identification by retention time is verified by co-chromatography of each standard substance obtained in the plant extract. A major problem with this approach is that most plant compounds are not commercially available, especially the enormous number of secondary metabolites. Very recently, the publication of the first "biological" public domain GC–EI–MS mass spectra library (MSRI; http://csbdb.mpimp-golm.mpg.de/gmd.html) was described (Kopka et al., 2005 and Schauer et al., 2005). This library contains a large number of identified and unknown, but repeatedly observed EI-mass spectra of many different plant species and organs. A feature of this library is its compatibility with the NIST software and GC–MS evaluation software packages such as AMDIS (see below).

For LC–MS signal identification, the situation is much more complex. Mass spectra generated by LC–MS are typically instrument dependent and therefore, standard reference LC–MS spectral libraries are of limited use. The minimum information acceptable for the identification of novel organic compounds or metabolites has been traditionally defined by the scientific literature criteria and often includes elemental analysis, NMR, and MS spectral data for the isolated compound. One method for preliminary identification of unknown compounds appears to be the use of multidimensional instrumental techniques (based on combinations of GC–MS, LC–MS, MS/MS, or MS/NMR), which enable both comparative profiling and structural elucidation. For example, LC–QTOF–MS/MS (liquid chromatographic quadrupole tandem time-of-flight mass spectroscopy) has the potential to provide accurate mass and product-ion information of chromatographically separated metabolites. Experimental mass data can then be used for the calculation of an elemental composition and be compared with available mass information in, for example, the NIST or KEGG database for possible structure suggestions. Further stepwise fragmentation by tandem MS (MS^n) leads to product-ion information, which can be used to determine/confirm structure. Although this gives much information about the potential structure of the compound, the final confirmation of the identity of the compound has to be done either by analysis of an authentic standards substance or by analysis of the purified sample using NMR.

The chosen method for unambiguous peak identification is NMR, which offers high chemical selectivity. In combination with LC and MS (LC–MS–NMR), it represents the ultimate technology for high-throughput peak identification and structure elucidation of unknown plant compounds (Wolfender et al., 2003), although the in-line version of this combination till date is still highly limited by the low sensitivity of the NMR instrument.

8.4.6 Development of Databases for Metabolomics-Derived Data in Plant Science

In the past, it has been noted by several scientists that the large data sets generated by postgenomics technologies have to be transmitted, stored safely, and be made available in convenient and accessible formats (Goodacre et al., 2004). The implementation of relational databases for data storage requires well-designed data standards. The DNA microarray community has agreed on the development of minimum information about a microarray experiment (MIAME, Brazma et al., 2001) and its structure has been widely accepted. Similar initiatives are underway for the proteomics community (PEDRo, Taylor et al., 2003). Although metabolic databases such as the KEGG system (Goto et al., 2002) or MetaCyc (Krieger et al., 2004) provide detailed information about metabolic pathways and enzymes of a variety of organisms, the development of a data standard equivalent to MIAME and PEDRo describing metabolomics data in their experimental context has been proposed only very recently (MIAMET, Bino et al., 2004, ArMet, Jenkins et al., 2004). On the contrary, it will be important not only to store metabolic profiling data but also to integrate these data with metabolic pathway information which will be the future source of knowledge discovery. Recently, a database has been developed that assembles information about all known *Arabidopsis thaliana* metabolic pathways (AraCyc) and provides diagrams showing metabolites and genes encoding the enzymes in each pathway (Mueller et al., 2003). For a holistic integration of numerous multiparallel genomic, proteomic, metabolomic, and metabolic flux analysis datasets with metabolic pathway information, the "Pathway Tools Omics Viewer", has been developed (http://www.arabidopsis.org:1555/expression.html), which in an easy and powerful manner paints experimental data onto the biochemical pathway map. Another example for such "mapping" tools is MapMan (Thimm et al., 2004), which allows users to visualize comparative metabolic and also transcriptional profiling datasets on existing metabolic templates. For a holistic integration of numeric multiparallel genomic, proteomic, and metabolomic datasets, a data managing system for editing and visualization of biological pathways was developed, which on a publicly available domain will be very important for data-mining in the functional genomics field (MetNetDB, Syrkin Wurtele et al., 2003, PaVESy, Luedemann et al., 2004). These software tools henceforth will become important in mapping novel findings onto metabolic pathways and fully understand the function of each gene, encoded protein, and metabolite.

8.5 APPLICATIONS OF METABOLOMICS APPROACHES IN PLANT RESEARCH

8.5.1 Phenotyping

Once a robust metabolite analysis platform has been established and reliable data have been produced, the range of plant research applications is enormous. These can vary from answering simple biological questions, that is, what are the metabolic differences between two cultivars, to investigations regarding complex metabolic networks. For example, a metabolomics approach can be used to determine the influence of transgenic and environmental manipulations on the metabolite profile as demonstrated by a detailed characterization of the metabolic complement of a number of transgenic potato tubers altered in their starch biosynthetic pathway and wild-type tubers incubated in different sugars using GC–MS (Figure 8.5, Roessner et al., 2001a, 2001b). As a result of this nontargeted approach, many unintended differences of transgenic tubers compared with wild type were detected (Roessner et al. 2001a, Figure 8.6). This study showed that using a metabolomics approach, it is possible to phenotype genetically and environmentally diverse plant systems easily. In addition, this work has demonstrated the importance of using metabolomics to monitor and evaluate effects (risk assessment) on metabolism in genetically modified organisms (GMO). In some cases, it was already shown that the introduction

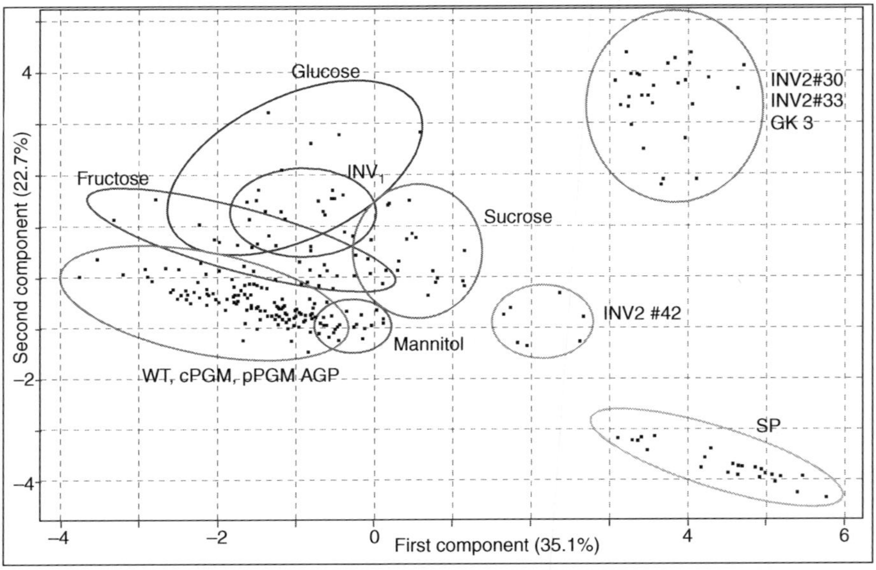

Figure 8.5 Principal component analysis (PCA) of metabolite profiles of environmentally and genetically modified potato tubers (Roessner et al. 2001b). Samples representing wild-type tubers and tubers incubated in buffer alone, plastidial (pPGM) and cytosolic (cPGM) phosphoglucomutase antisense tubers; ADP-glucose pyrophosphorylase (AGP) antisense tubers (dark green circle), mannitol-fed tubers (black circle), fructose-fed tubers (dark blue circle), sucrose-fed tubers (yellow circle), glucose-fed tubers (light red circle), apoplastic invertase (INV1) expressing tubers (light blue circle), cytosolic invertase (INV2) expressing tubers line #30; #33 and cytosolic invertase and glucokinase (GK3) expressing tubers (light green circle), cytosolic invertase (INV2) expressing tubers line #42 (dark red circle), and sucrose phosphorylase (SP) expressing tubers (lilac circle) are marked as described for ease of comparison. PCA Vectors 1 and 2 were chosen for best visualization of differences between experimental treatments and include 57.8% of the information derived from metabolic variances. © American Society of Plant Biologists. (See color plates.)

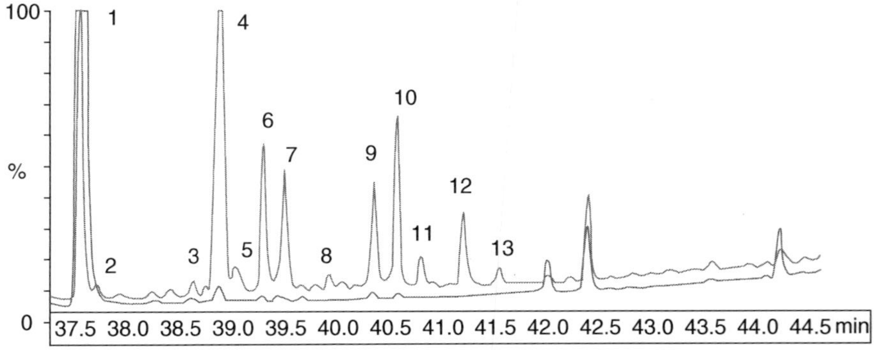

Figure 8.6 Comparison of a specific region of a GC–MS chromatogram of wild-type potato tuber (WT, lower line) compared to tubers expressing a yeast invertase in the cytosol (INV, upper line). 1: sucrose; 3: maltose TMS; 4: maltose MEOX1; 5: trehalose TMS; 6: maltose MEOX2; 7: maltitol TMS; 12: isomaltose MEOX1; 13: isomaltose MEOX2, 2, 8, 9, 10, 11, 14, 15 and 16 are not identified, mass spectra suggest they are sugars or sugar derivatives. (See color plates.)

or deletion of a gene in plants resulted in additional, not expected beforehand, alterations of the plant's metabolism, even when the altered gene activity was not involved directly in metabolic reactions but rather in cell or plant structure building. As shown in Figure 8.6, many additional metabolites were detectable in extracts of potato tubers expressing a yeast-derived gene encoding the sucrose cleaving enzyme invertase, but interestingly, only when the gene product was directed to the cytosol. This pattern was not seen in wild-type tubers or tubers expressing the same gene directed to the apoplast or vacuole. Most of these additional signals could be assigned as being disaccharides (on the basis of their retention time and mass spectra), which was somewhat surprising as invertase cleaves not only sucrose but also many other disaccharides. The reason for the occurrence of these additional sugars in the invertase expressing tubers in the cytosol could not be deciphered so far. In the recent past, metabolomics, due to its unbiased approach, has become a major tool in the analysis of direct transgenisis/mutation effects as well as for the investigation of indirect and potentially unknown alterations of plant metabolism.

8.5.2 Functional Genomics

One of the most useful application of metabolomics is on functional genomics studies, which aim to identify gene functions using high-throughput phenotyping technologies, for example, in the identification of responsible genes and their products on plant adaptations to different abiotic stresses. Often the role of certain metabolites in stress response could be assigned, for example, proline plays a major role in salt stress adjustments in rice. The detailed characterization of metabolic adaptations to low and high temperatures in *Arabidopsis thaliana* has already demonstrated the power of this approach (Kaplan et al., 2004; Cook et al., 2004). Interestingly, it could be shown that low temperatures have more profound effects than high temperatures, and novel findings of metabolic adaptations to temperature stress were identified (Kaplan et al., 2004). Another important report on using metabolomics as a tool in investigating metabolic responses of *Medicago truncatula* cell cultures to biotic and abiotic elicitors has revealed both elicitor-specific responses as well as more generic responses in which similar metabolites responded independently of the type of stress (Broeckling et al., 2005). Nutrient deficiencies and toxicities represent another example of common stress situations, e.g., it has been already demonstrated that the availability of inorganic nitrogen can reprogram carbohydrate metabolism (Stitt et al., 2002). This has been recently verified in more detail by a metabolomics investigation of the effects on tomato leaf metabolism grown in saturated, replete, and deficient nitrogen supplement conditions (Urbanczyk-Wochniak et al., 2005b), showing the impact of nitrogen levels in the growth solutions on a wide range of metabolites. Similar striking effects on metabolite levels have been found when barley plants were grown in conditions where other inorganic nutrients were unavailable, e.g., phosphate or zinc (Roessner, unpublished results). In future, this approach will lead to the determination of the role of both metabolites and genes in stress tolerance and thus provide new ideas for genetic engineering and breeding of novel stress-resistant crops.

8.5.3 Fluxomics

The measurement of steady-state levels of metabolites gives new insights into metabolic networks at a given time. But the real behavior of plant metabolism can be only understood by determination of the dynamics of metabolism. The basis of metabolic flux analysis (MFA) is a combination of stable isotope labeling under steady-state conditions and NMR or MS-based detection systems to follow the distribution of label. This technique has been applied in detail in microbial physiology but it will play an increasingly important role in plant research (for Review see Schwender et al., 2004). The application of a multiparallel detection method such as GC- or LC–MS allows the determination of isotope label in many metabolites in a single experiment and therefore gives the opportunity to calculate metabolic fluxes of many different pathways simultaneously (Schwender et al., 2003; Roessner-Tunali et al., 2004). The limitation of this method is the necessity of steady-state metabolite level determinations. In conclusion, metabolomics in combination with stable isotope metabolic flux analysis will provide important insights into plant functional genomics studies. Another obvious use of this information will be in more rational approaches in metabolic engineering of novel, valuable biotech-crops (Sweetlove et al., 2003).

8.5.4 Metabolic Trait Analysis

Another challenging application of metabolomics is in the identification of genetic loci involved in specific trait appearance. This can be done by comparison of the metabolite profiles of a set of lines derived from a cross between two parents differing in the desired trait, for example, tolerance level to a certain stress situation. Using the technique of QTL (quantitative trait locus) analysis, single metabolite QTLs can be identified and also loci that affect whole metabolic pathways or in an ideal situation the whole metabolite network. The first exciting example of this approach was presented very recently by Schauer et al. (2006). These authors have used a GC–MS – based metabolite profiling approach to metabolically phenotype a tomato introgression line (IL) population in which marker-defined regions of a cultivated tomato variety (*Solanum lycopersicon*) were substituted by a homologous region of a wild and nonripening tomato species (*Solanum penellii*). The initial aim of the work was to gain a greater understanding in fruit metabolism and ripening and to identify new genes being involved in these processes. Interestingly, this approach allowed the identification of a large number (almost 900) of single metabolite QTLs additional to many QTLs which affect a number of compounds in metabolic pathways (Figure 8.7). Most importantly, by integration of metabolite profiling data with other phenotypical observations, such as morphological traits, the whole plant phenotype—fruit metabolism networks could be established suggesting an important influence of plant phenotypes on the final metabolic composition of the fruit (Schauer et al., 2006). This work has opened a new dimension in the application of metabolomics to study genetic variation. In the past, the approximate positions of genetic loci controlling quantitative traits have been identified through associating marker and phenotype variation in a structured population. In

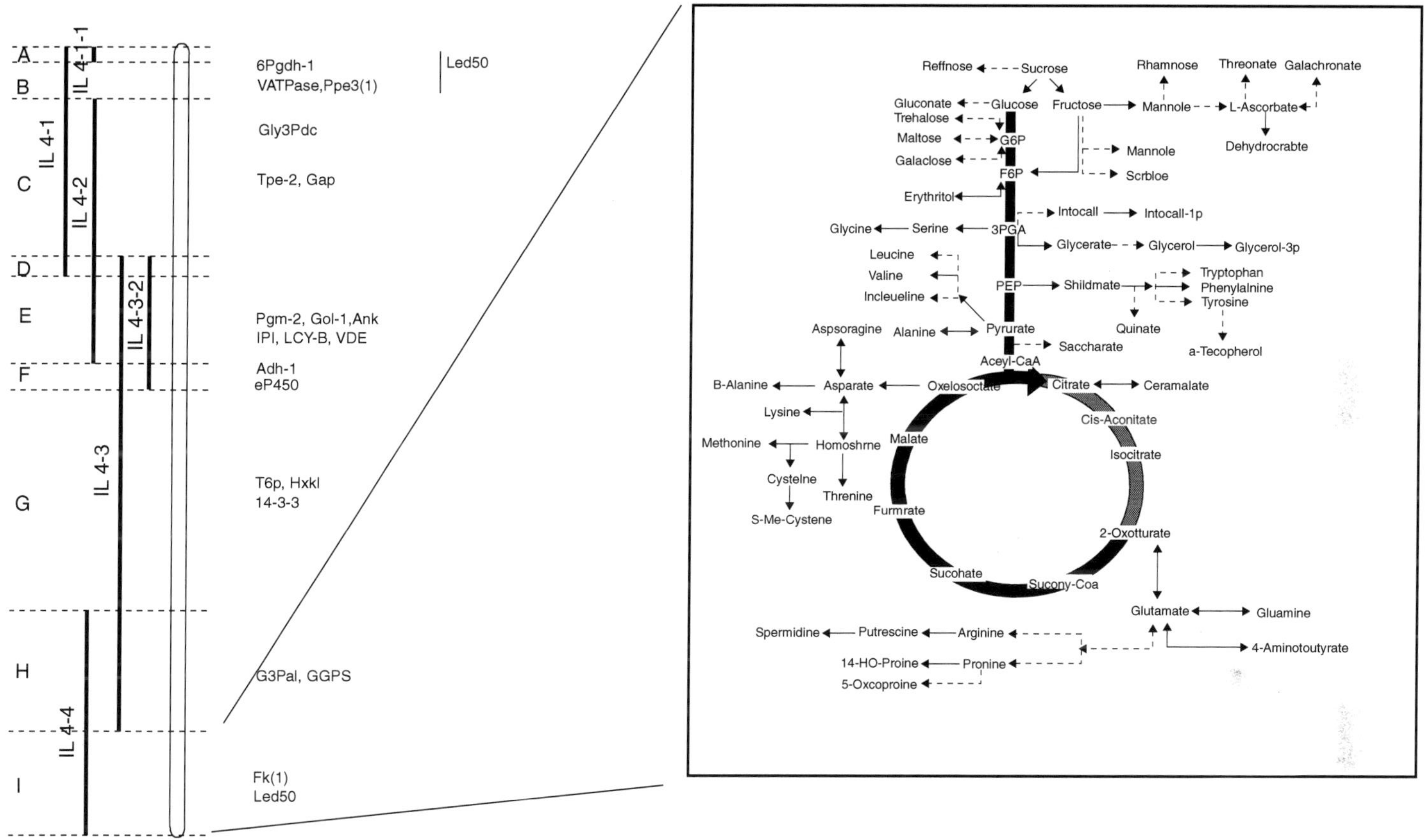

Figure 8.7 Correlation of metabolite accumulation assigned to metabolic pathways with fine maps of genomic regions established following an interspecific cross of two tomato cultivars (Schauer et al. 2006). Red colored metabolites were increased in the introgression line IL4-4 but not in IL 4-3 and therefore this pattern was related to Bin I of chromosome 4 of the tomato (*S. lycopersicum*) genome. Picture source: N. Schauer, Max-Planck-Institute for Plant Molecular Physiology, Germany.

the near future, the goal will be to utilize the new emerging high-throughput and highly parallel phenotyping technologies, such as transcriptomics, proteomics, and to an even greater extent metabolomics, to study genetic segregation and identify novel genes.

8.5.5 Systems Biology

The next step of interpretation of plant metabolomics datasets can be achieved when they are integrated with other "omics" data such as transcriptomics or proteomics data. First attempts to face this challenge have been presented by Urbanczyk-Wochniak and co-workers who combined data obtained from microarrays analysis and metabolite profiling of the same sample (Urbanczyk-Wochniak et al., 2003). A co-response analysis of both datasets has resulted in a large number of significant correlations between mRNA transcripts and metabolites. Some of these could be explained easily with existing biochemical knowledge but most were found to be novel, and thus highlighted the power of this integrated approach for gene and metabolite function identifications. A similar investigation simultaneously analyzed transcripts and metabolite levels in *Lotus japonicus* nodules to study symbiotic nitrogen fixation in detail (Colebatch et al., 2004).

This report has shown clear interrelationships between transcript and metabolite responses dependent on a physiological event.

Last but not least, it has to be noted that a detailed characterization of the metabolome of a biologic organism plays an integral role in a systems-biology approach. The aim of the emerging area of systems-biology is to investigate the dynamics of all genetic, regulatory, and metabolic processes in a cell and to understand the complexity of cellular networks (Kitano, 2002). Further, this will give the opportunity to investigate the behavior of biologic systems with respect to the environment.

8.6 FUTURE PERSPECTIVES

This chapter has hopefully given a short introduction about the potential metabolomics has to offer for plant research. In summary, metabolomics will become a major player in the investigation of plant metabolism and the phenotypic analysis of many different plant species following environmental and genetic perturbations. This will offer a number of approaches in which metabolomics will be of great use, such as functional genomics, metabolic and genetic engineering, or the development of novel biotech crop. It will also play an outstanding role in phenotyping and determination of novel pathways. In addition, when plant metabolomics will be linked to the field of nutrigenomics, in which scientists are studying the role of human metabolites in the development of modern-world diseases for example coronary heart diseases or diabetics, it will give the opportunity for selecting crops and food for novel bioactive plant compounds (phytochemicals) and provide invaluable tools for the investigation of the distribution of metabolite concentrations in crops and food and the relationship of those to diseases.

REFERENCES

Aharoni A, Ric de Vos CH, Verhoeven HA, Maliepaard CA, Kruppa G, Bino R, Goodenowe D. 2002. Nontargeted metabolome analysis by use of Fourier Transform Ion Cyclotron Mass Spectrometry. OMICS 6:217–234.

Arlt K, Brandt S, Kehr J. 2001. Amino acid analysis in five pooled single plant cell samples using capillary electrophoresis coupled to laser-induced fluorescence detection. J Chrom A 926:319–325.

Bino RJ, Hall RH, Fiehn O, Kopka J, Saito K, Draper J, Nikolau B, Mendes P, Roessner-Tunali U, Beale M, Trethewey RN, Lange BM, Syrkin Wurtele E, Sumner L. 2004. Opinion: Potential of metabolomics as a functional genomics tool. Trends Plant Sci 9:418–425.

Broeckling CD, Huhman DV, Farag MA, Smith JT, May GD, Mendes P, Dixon RA, Sumner LW. 2005. Metabolic profiling of *Medicago truncatula* cell cultures reveals the effects of biotic and abiotic elicitors on metabolism. J Exp Bot 56:323–336.

Borisjuk L, Rolletschek H, Walenta S, Panitz R, Wobus U, Weber H. 2003. Energy status and its control on embryogenesis of legumes: ATP distribution within Vicia faba embryos is developmentally regulated and correlated with photosynthetic capacity. Plant J 36:318–329.

Brazma A, Hingamp P, Quackenbush J, Sherlock G, Spellman P, Stoeckert C, Aach J, Ansorge W, Ball CA, Causton HC, Gaasterland T, Glenisson P, Holstege FC, Kim IF, Markowitz V, Matese JC, Parkinson H, Robinson A, Sarkans U, Schulze-Kremer S, Stewart J, Taylor R, Vilo J, Vingron M. 2001. Minimum information about a microarray experiment (MIAME)-toward standards for microarray data. Nat Genet 29:365–371.

Colebatch G, Desbrosses G, Ott T, Krusell L, Montanari O, Kloska S, Kopka J, Udvardi MK. 2004. Global changes in transcription orchestrate metabolic differentiation during symbiotic nitrogen fixation in *Lotus japonicus*. Plant J 39:487–512.

Cook D, Fowler S, Fiehn O, Thomashow MF. 2004. A prominent role for the CBF cold response pathway in configuring the low-temperature metabolomie of *Arabidopsis*. Proc Natl Acad Sci USA 101:15243–15248.

Duran AL, Yang J, Wang L, Sumner LW. 2003. Metabolomics spectral formatting, alignment and conversion tools (MSFACTs). Bioinformatics 19:2283–2293.

Fiehn O, Kopka J, Dörmann P, Altmann T, Trethewey RN, Willmitzer L. 2000. Metabolite profiling for plant functional genomics. Nature Biotechnol. 18:1157–1161.

Fiehn O. 2003. Metabolic networks of *Cucurbita maxima* phloem. Phytochem 62:875–86.

Goodacre R, Vaidyanathan S, Dunn WB, Harrigan GG, Kell DB. 2004. Metabolomics by numbers: Acquiring and understanding global metabolite data. Trends Biotechnol 22:245–252.

Goto S, Okuno Y, Hattori M, Nishioka T, Kanehisa M. 2002. LIGANS: Database of chemical compounds and reactions in biological pathways. Nucleic Acid Res 30:402–404.

Huhman DV, Sumner LW. 2002. Metabolic profiling of saponins in *Medicago sativa* and *Medicago truncatula* using HPLC coupled to an electrospray ion-trap mass spectrometer. Phytochem 59:347–360.

Jenkins H, Hardy N, Beckmann M, Draper J, Smith AR, Taylor J, Fiehn O, Goodacre R, Bino RJ, Hall R, Kopka J, Lane GA, Lange BM, Liu JR, Mendes P, Nikolau BJ, Oliver SG, Paton NW, Rhee S, Roessner-Tunali U, Saito K, Smedsgaard J, Sumner LW, Wang T, Walsh S, Syrkin Wurtele E, Kell DB. 2004. A proposed framework for the description of plant metabolomics experiments and their results. Nat Biotechnol 22:1601–1606.

Kaplan F, Kopka J, Haskell DW, Zhao W, Schiller KC, Gatzke N, Sung DY, Guy CL. 2004. Exploring the temperature-stress metabolomie of *Arabidopsis*. Plant Physiol 136:4159–4168.

Kitano H. 2002. Systems biology: A brief overview. Science 295:1662–1664.

Krieger CJ, Zhang P, Mueller LA, Wang A, Paley S, Arnaud M, Pick J, Rhee SY, Karp PD. 2004. MetaCyc: A multiorganism database of metabolic pathways and enzymes. Nucleic Acid Res 32: Database issue: D438–442.

Kopka J, Schauer N, Krueger S, Birkemeyer C, Usadel B, Bergmüller E, Dörmann P, Gibon Y, Stitt M, Willmitzer L, Fernie AR, and Steinhauser D. 2005. The Golm metabolome database. Bioinformatics 21:1635–16358.

Krishnan P, Kruger NJ, Ratcliffe RG. 2005. Metabolite fingerprinting and profiling in plants using NMR. J Exp Bot 56:255–265.

Luedemann A, Weicht D, Selbig J, Kopka J. 2004. PaVESy: Pathway visualization and editing system. Bioinformatics 20:2841–2844.

Muller A, Duchting P, Weiler EW. 2002. A multiplex GC-MS/MS technique for the sensitive and quantitative single-run analysis of acidic phytohormones and related compounds, and its application to Arabidopsis thaliana. Planta 216:44–56.

Mueller LA, Zhang P, Rhee SY. 2003. AraCyc: A biochemical pathway database for Arabidopsis. Plant Physiol 132:453–460.

Oliver S. 1997. Yeast as a navigational aid in genome analysis. Microbiol 143:1483–1487.

Roessner U, Wagner C, Kopka J, Trethewey RN, Willmitzer L. 2000. Simultaneous analysis of metabolites in potato tuber by gas chromatography-mass spectrometry. Plant J. 23:131–142.

Roessner U, Luedemann A, Brust D, Fiehn O, Linke T, Willmitzer L, Fernie AR. 2001a. Metabolic profiling allows comprehensive phenotyping of genetically or environmentally modified plant systems. Plant Cell 13:11–29.

Roessner U, Willmitzer L, Fernie A R. 2001b. High-resolution metabolic phenotyping of genetically and environmentally diverse plant systems—identification of phenocopies. Plant Physiol 127:749–764.

Roessner-Tunali U, Hegemann B, Lytovchenko A, Carrari F, Bruedigam C, Granot D, Fernie AR. 2003a. Metabolic profiling of transgenic tomato plants overexpressing hexokinase reveals that the influence of hexose phosphorylation diminishes during fruit development. Plant Physiol 133:84–99.

Roessner-Tunali U, Urbanczyk-Wochniak E, Czechowski T, Kolbe A, Willmitzer, Fernie AR. 2003b. *De novo* amino acid biosynthesis in plant storage tissues is regulated by sucrose levels. Plant Physiol 133:683–692.

Roessner-Tunali U, Lui J, Leisse A, Balbo I, Perez-Melis A, Willmitzer L, Fernie AR. 2004. Flux analysis of organic and amino acid metabolism in potato tubers by gas chromatography-mass spectrometry following incubation in ^{13}C labelled isotopes. Plant J 39:668–679.

Roessner U, Patterson J, Forbes MG, Fincher G, Langridge P, Bacic A. 2006. An investigation of boron toxicity in barley using metabolomics. Plant Physiol 142:1087–1101.

Sato S, Soga T, Nishioka T, Tomita M. 2004. Simultaneous determination of the main metabolites in rice leaves using capillary electrophoresis mass spectrometry and capillary electrophoresis diode array detection. Plant J 40:151–163.

Sauter H, Lauer M, Fritsch H. 1991. Metabolic profiling of plants: a new diagnostic technique. In: Baker DR, Fenyes JG, Moberg WK (Eds.), *American Chemical Society Symposium Series No. 443*, American Chemical Society, Washington DC, pp. 288–299.

Schad M, Mungur R, Fiehn O, Kehr J. 2005. Metabolic profiling of laser microdissected vascular bundles of *Arabidopsis thaliana*. Plant Methods 1: (doi: 10.1186/1746-4811-1-2).

Schauer N, Steinhauser D, Strelkov S, Schomburg D, Allison G, Moritz T, Lundgren K, Roessner-Tunali U, Forbes MG, Willmitzer L, Fernie AR, Kopka J. 2005. GC-MS libraries for the rapid identification of metabolites in complex biological samples. FEBS Lett 579:1332–1337.

Schauer N, Semel Y, Roessner U, Gurb A, Balbo I, Carrari F, Pleban T, Perez-Melisa A, Bruedigam C, Kopka J, Willmitzer L, Zamir D, Fernie AR. 2006. Quantitative genetics of metabolite accumulation in intraspecific introgressions of tomato. Nature Biotech 24:447–454.

Schwender J, Ohlrogge JB, Shachar-Hill Y. 2003. A flux model of glycolysis and the oxidative pentosephosphate pathway in developing *Brassica napus* embryos. J Biol Chem 278:29442–29453.

Schwender J, Ohlrogge J, Shachar-Hill Y. 2004. Understanding flux in plant metabolic networks. Curr Opin Plant Biol 7:309–317.

Stitt M, Muller C, Matt P, Gibon Y, Carillo P, Morcuende R, Scheible WR, Krapp A. 2002. Steps toward an integrated view of nitrogen metabolism. J Exp Bot 53:959–970.

Sweetlove LJ, Last RL, Fernie AR. 2003. Predictive metabolic engineering: A goal for systems biology. Plant Physiol 132:420–425.

Syrkin Wurtele E, Li J, Diao L, Zhang H, Foster CM, Fatland B, Dickerson J, Brown A, Cox Z, Cook D, Lee E-K, Hofmann H. 2003. MetNet: Software to build and model the biogenetic lattice of Arabidopsis. Comp Funct Genom 4:239–245.

Taylor CF, Paton NW, Garwood KL, Kirby PD, Stead DA, Yin Z, Deutsch EW, Selway L, Walker J, Riba-Garcia I, Mohammed S, Deery MJ, Howard JA, Dunkley T, Aebersold R, Kell DB, Lilley KS, Roepstorff P, Yates JR 3rd, Brass A, Brown AJ, Cash P, Gaskell SJ, Hubbard SJ, Oliver SG. 2003. A systematic approach to modeling, capturing, and disseminating proteomics experimental data. Nat Biotechnol 21:247–254.

Thimm O, Blasing O, Gibon Y, Nagel A, Meyer S, Kruger P, Selbig J, Muller LA, Rhee SY, Stitt M. 2004. MAPMAN: A user-driven tool to display genomics data sets onto diagrams of metabolic pathways and other biological processes. Plant J 37:914–939.

Tolstikov VV, Fiehn O. 2002. Analysis of highly polar compounds of plant origin: Combination of hydrophilic interaction chromatography and elctrospray ion mass trap spectrometry. Anal Biochem 301:298–307.

Tolstikov VV, Lommen A, Nakanishi K, Tanaka N, Fiehn O. 2003. Monolithic silica-based capillary reversed-phase liquid chromatography/electrospray mass spectrometry for plant metabolomics. Anal Chem 75:6737–6740.

Urbanczyk-Wochniak E, Luedemann A, Kopka J, Selbig J, Roessner-Tunali U, Willmitzer L, Fernie AR. 2003. Parallel analysis of transcript and metabolic profiles: A new approach in systems biology. EMBO Rep 4:989–992.

Urbanczyk-Wochniak E, Baxter C, Kolbe A, Kopka J, Sweetlove LJ, Fernie AR. 2005a. Profiling of diurnal patterns of metabolite and transcript abundance in potato (Solanum tuberosum) leaves. Planta 221:891–903.

Urbanczyk-Wochniak E, Fernie AR. 2005b. Metabolic profiling reveals altered nitrogen nutrient regimes have diverse effects on the metabolism of hydroponically-grown tomato (*Solanum lycopersicum*) plants. J Exp Bot 56:309–321.

von Roepenack-Lahaye E, Degenkolb T, Zerjeski M, Franz M, Roth U, Wessjohann L, Schmidt J, Scheel D, Clemens S. 2004. Profiling of *Arabidopsis* secondary metabolites by capillary liquid chromatography coupled to electrospray ionization quadrupole time-of-flight mass spectrometry. Plant Physiol 134:548–559.

Wagner C, Sefkow M, Kopka J. 2003. Construction and application of a mass spectral and retention time index database generated from plant GC/EI-TOF-MS metabolite profiles. Phytochem 62:887–900.

Weckwerth W, Loureiro ME, Wenzel K, Fiehn O. 2004. Differential metabolic networks unravel the effects of silent plant phenotypes. Proc Natl Acad Sci USA 18:7809–7814.

Wolfender JL, Ndjoko K, Hostettmann K. 2003. Liquid chromatography with ultraviolet absorbance-mass spectrometric detection and with nuclear magnetic resonance spectroscopy: A powerful combination for the on-line structural investigation of plant metabolites. J Chromatogr A 1000:437–455.

9

MASS PROFILING OF FUNGAL EXTRACT FROM *PENICILLIUM* SPECIES

By Jørn Smedsgaard

This chapter illustrates the use of direct infusion electrospray mass spectrometry (DiMS) as an efficient tool to study secondary metabolism in filamentous fungi. DiMS analysis can be used for a rapid chemical classification of samples, e.g., for taxonomy, to detect strain similarity and identify mutations, and it also gives an indication of metabolite production. To illustrate the potential of DiMS, a selected set of species from *Penicillium* subgenus *Penicillium* is analyzed by a rapid extraction method followed by DiMSometry. The data are analyzed by simple chemometrics and the results are related to known secondary metabolism of these species.

9.1 INTRODUCTION

The metabolome is used to describe the complete pool of metabolites in an organism in a given state as discussed in Chapters 1 and 2. Therefore, it comprises metabolites both from the central metabolism as well as from secondary metabolisms. While the central metabolism reflects nutritional and growth status, the secondary metabolism represents differentiation and complex responses to the environment as well as to other organisms. The secondary metabolism is much more complex and involves many dedicated genes for the production of the great variety of amazingly complex secondary metabolites (see Figure 9.1). Secondary metabolites can be uniquely found in one or a few species or are widespread in nature, and the same metabolites can even be found in organisms from different kingdoms. Among the organisms with a very active

Metabolome Analysis: An Introduction, by Silas G. Villas-Bôas, Ute Roessner, Michael A. E. Hansen, Jorn Smedsgaard and Jens Nielsen

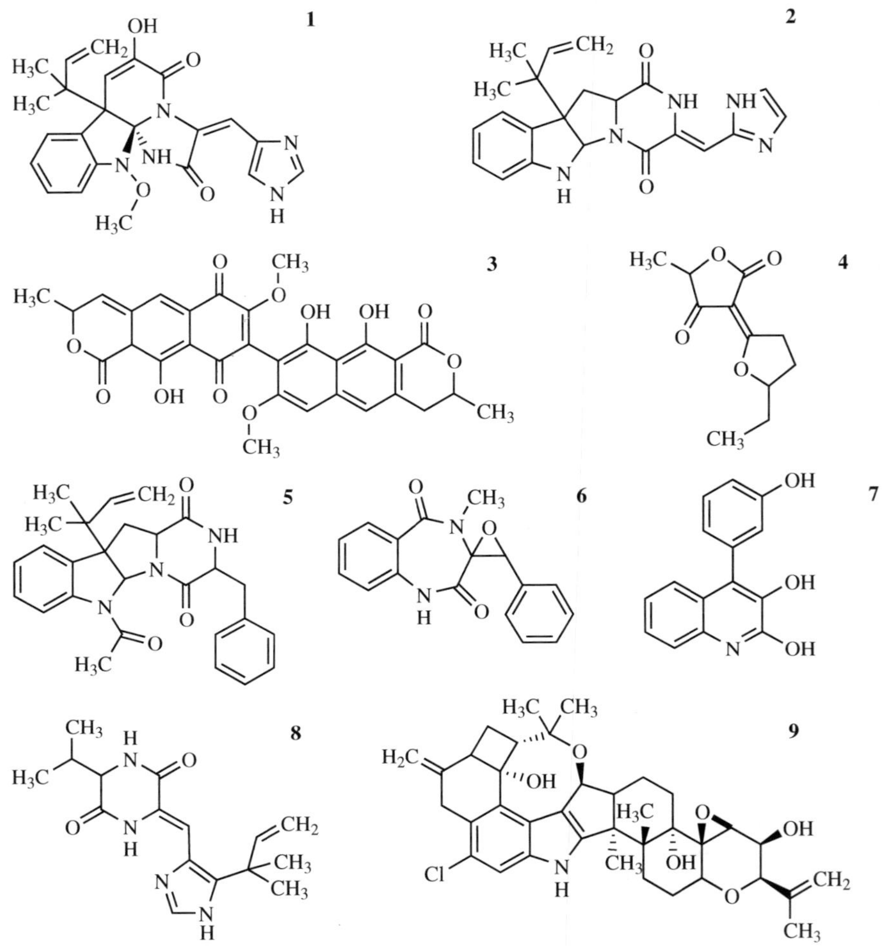

Figure 9.1 Structures of selected metabolites from Table 9.1 showing the fascinating chemical diversity found in even a small group of closely related *Penicillium* species. **1:** melagrin, **2:** roquefortine C, **3:** viomellin, **4:** terrestic acid, **5:** puberuline, **6:** cyclopenin, **7:** viridicatol, **8:** aurantiamine, **9:** penitrem A.

secondary metabolism are the filamentous fungi of which most genera and species are well known for their ability to produce a wide range of secondary metabolites. As the production of most secondary metabolites are coded by a few to many specialized genes, the secondary metabolites are today considered as a part of the species-specific phenotype, on the same level as cell differentiation and other phenotypic characters.

In the group of filamentous fungi *Penicillium* subgenus *Penicillium,* we find many fungi that are common in our environment either as contaminants in food and in our household or are used industrially for production of biotech products. Most of these fungi will produce a broad range of secondary metabolites where many are of unknown chemical structure and others are well-known mycotoxins. To illustrate the use of direct infusion electrospray mass spectrometry, a small subset of eight species (series *Viridicata* from *Penicillium* subgenus *Penicillium*) that are common contaminants in stored cereals in tempered zones have been selected to illustrate this case story. A more detailed study of these fungi can be found in further reading.

Table 9.1 lists some of the most important metabolites produced by these eight species but nowhere all metabolites are produced by every species. The structures of selected metabolites are shown in Figure 9.1.

TABLE 9.1 Metabolites Produced by the Species in the Series *Viridicata* from *Penicillium* Subgenus *Penicillium*. See Samson and Frisvad (2004) for Further Details.

Metabolite	Mass	M+H+	I	II	III	IV	V	VI	VII	VIII
Terrestric acid	210.0892	211.0970	X						X	
Puberulonic acid	223.9957	225.0035		X						
Viridicatin	237.0790	238.0868		X	X		X	X		
3-Methoxy-viridicatin	251.0946	252.1024		X	X		X	X		
Viridicatol	253.0739	254.0817		X	X		X	X		
Viridicatic acid	256.0947	257.1025	X						X	
Aspterric acid	266.1518	267.1596						X		
Dehydro-cyclopeptin	278.1055	279.1133		X	X		X	X		
Cyclopeptin	281.1212	281.1290		X	X		X	X		
Cyclopenin	294.1004	295.1082		X	X		X	X		
Aurantiamine	302.1743	303.1821	X		X		X			
Viridamine										X
Cyclopenol	310.0954	311.1032		X	X		X	X		
Auranthine	330.1117	331.1195	X							
Anacine	330.1692	331.1770	X					X		
Rugulosuvine	333.1477	334.1555		X				X	X	
Brevianamide A	365.1743	366.1817								X
Roquefortine C	389.1852	390.1930				X				
Normethyl-verrucosidin	400.1886	401.1964	X			X		X		
Verrucofortine	409.2365	410.2443		X				X	X	
Verrucosidin	414.2046	415.2120	X			X		X		
Asteltoxin	418.1992	419.2070							X	
Meleagrin	433.1750	434.1828				X				
Puberuline	443.2209	444.2287		X				X	X	
Xanthoviridicatin G	444.0845	445.0923								X
Viridic acid	454.2216	455.2294								X
Rubrosulphin	528.1056	529.1134								X
Viomellein	560.1319	561.1397		X	X	X		X	X	
Penitrem A	633.2857	634.2935				X				

I *Penicillium aurantiogriseum*, II *P. cyclopium*, III *P. freii*, IV *P. melanoconidium*, V *P. neoechinulatum*, VI *P. polonicum*, VII *P. tricolor*, VIII *P. viridicatum*. See Figure 9.1 for structures of selected metabolites.

As discussed in Section 4.5.3, electrospray ionization mass spectrometry (ESI–MS) has the advantage of being a soft and sensitive ionization technique that can be optimized mainly to produce protonated or sodiated ions (assuming positive ESI) from a very broad range of metabolites. Therefore, spectra obtained from injection of crude extracts from fungal culture can be considered as mass profile of the sample (or a fingerprint, see discussion in Section 4.1). The main advantage of mass profiling by direct infusion mass spectrometry (DiMS) is its high-throughput in obtaining profiles or fingerprints that are usually achieved in minutes and it contains both metabolite and chemical structure information. A further advantage is its easy storage of generated spectra in databases. However, when complex samples containing many components and with a wide concentration range are infused directly into the electrospray source, it may lead to serious discrimination due to what is known as matrix effects, see Section 4.5.3.

These matrix effects can seriously interfere with the metabolites seen in the spectra, e.g., some metabolites with high surface potential and proton affinity (or co-extracted media components, e.g., PEG and TWEEN) may "steal" more than their share of charge, thereby suppressing other metabolites. Also, not all metabolites are equally efficiently ionized, and the abundance seen in the spectra, therefore, does not reflect the quantitative composition of the sample. These effects can be reduced by keeping the concentration within a suitable (low) range, using nano-ESI techniques and careful selection of the solvent composition. The usability of DiMS for studying fungi was already demonstrated 10 years ago by Smedsgaard and Frisvad (1996) where they took advantage of direct infusion ESI–MS profiling to study a large group of fungal species (43 species and two growth media, approx. 293 stains). By chemometric analysis of these spectra (or mass profiles), they showed that it was possible to group more than 80% of these species into chemical classes that corresponded to the species as determined by classical phenotypic identification. Furthermore, it was shown that ions corresponding to the protonated mass of many well-known metabolites could be detected.

9.2 METHODOLOGY FOR SCREENING OF FUNGI BY DiMS

If the cultures are grown on solid media, as it is the common practice in classification and taxonomy, the overall workflow for profiling fungal cultures can be summarized as:

- Selection and retrieval of strains and phenotypic description (identification)
- Cultivation
- Extraction
- Analysis
- Data evaluation and processing.

9.2.1 Cultures

Selection of cultures is of course determined by the study and what is available (obtainable). In general, it is desirable to have a detailed description of the strains and preferably also a proper identification. The latter is far from trivial and many fungi can be identified properly by only experts in taxonomy. Unfortunately, there is a lot of misidentification in the literature, and one should, therefore, read literature critically and be aware that one cannot always rely on which metabolites are produced by what species. A full and detailed strain description is of utmost importance as is expert identification to compare results from different experiments. In the example discussed here, the isolates were selected from the study by Samson and Frisvad (2004), two leading experts in taxonomy, and were described and identified by using all available techniques.

Inoculation and cultivation. Although fungi may have the genes to produce a broad range of secondary metabolites, not all metabolites may be produced under all conditions or on all media. In general, the penicillia will show their full metabolic potential on a relatively few different growth media with Czapek yeast extract agar (CYA) and yeast extract sucrose agar (YES) being general and most popular. However, the cultivation temperature and atmospheric conditions do influence the growth and metabolite production. The penicillia from the series *viridicata* all grow well at 25°C and are normally cultivated in the dark for 7 days as is used in this case. See Samson et al., 2004 for details about isolation, cultivation, and identification of these fungi.

9.2.2 Extraction

Compared to the primary metabolism, the dynamics of the secondary metabolism is very slow, and therefore quenching and extraction is much simpler. Also, for screening purposes, the use of solid media will not only give a better differentiation (cellular and chemical), but it is also much easier to work with. As already discussed in Chapter 3 and illustrated in the other case stories, sample preparation can be anything from simple to daunting. In this case, screening of the fungal cultures is done in a simple HTS manner by the rapid plug extraction procedure (Smedsgaard, 1997) as illustrated in Figure 9.2.

By the plug extraction method, a few plugs are cut from the colony and transferred to a small vial. Extraction solvent is added and the sample is sonicated by ultrasound for about 45 min. The solvent phase is transferred to a clean vial and is evaporated to dryness. While the solvent is evaporated, the plugs may be re-extracted by a second solvent to ensure efficient extraction of a broader range of metabolites. The solvent phase from the second extraction may be combined with the first and evaporated to dryness. In this case, the first extraction solvent was 0.5 ml of ethyl acetate with 0.5% (v/v) formic acid and the second solvent was 0.5 ml 2-propanol. The combined residues were redissolved in 0.3 ml methanol, filtered, and are then ready for analysis. In general, extraction is not trivial and

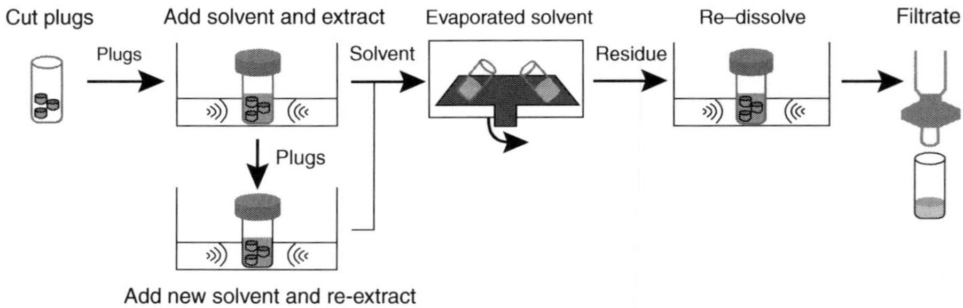

Figure 9.2 The simple plug extraction procedure used to prepare cultural extract from fungi on solid media. Although extraction by sonication requires time, many samples can be prepared in parallel.

consideration should be given not only to the discrimination between metabolites in the extraction procedure but also to ensure that minimal sample matrix is coextracted to minimize matrix effects and other interferences in the subsequent analyses.

9.2.3 Analysis by Direct Infusion Mass Spectrometry

The methanol extracts were analyzed by injection into positive electrospray mass spectrometry (di-ESMS) on a Micromass Q-Tof time-of-flight mass spectrometer with a 3.6 GHz time-to-digital detection. A modifier was added online by a syringe pump to facilitate a more efficient ionization. One μl extract was infused at a rate of 15 μl/min using methanol as carrier. Just prior to the source water containing 2% (v/v), formic acid was added at a rate of 5 μl/min to facilitate a more efficient ionization, giving a combined flow of 20 μl/min going into the source. The final composition was as follows: 75% (v/v) methanol with 0.5% (v/v) formic acid; continuum spectra were collected at a rate of 1 spectrum per second from m/z 150 to 1000 with 0.1 s interscan time; data were collected from 0 to 2 min after injection, and samples were injected at approximately 3 min interval to minimize cross contamination. The instrument was tuned to a resolution better than 8500 using a leucine-enkphaline solution (0.5 μg/ml in 50% (v/v) acetonitrile with 0.2% (v/v) formic acid) and calibrated on a solution of PEG giving a residual error of less than 2 mDa on more than 28 reference peaks by a 5th order calibration.

The data. The continuum data were stored and archived in the instrument format and processed either by the instrument software or by in-house written routines. These procedures are discussed more in details in Chapter 5, but a few examples are introduced below. Please note that each raw file from a high resolution instrument is about 20 Mb; thus, analyzing at a rate of 3 min per sample will produce about 400 Mb data per hour. Therefore, data archiving has to be taken into account while dealing with these kinds of experiments.

9.3 DISCUSSION

9.3.1 Initial Data Processing

Figure 9.3 illustrates the results and basic data processing of direct infusion mass profiles (DiMS data), in this case an extract of *Penicillium freii* cultivated on CYA (the same sample as shown in Figures 4.28 and 4.29).

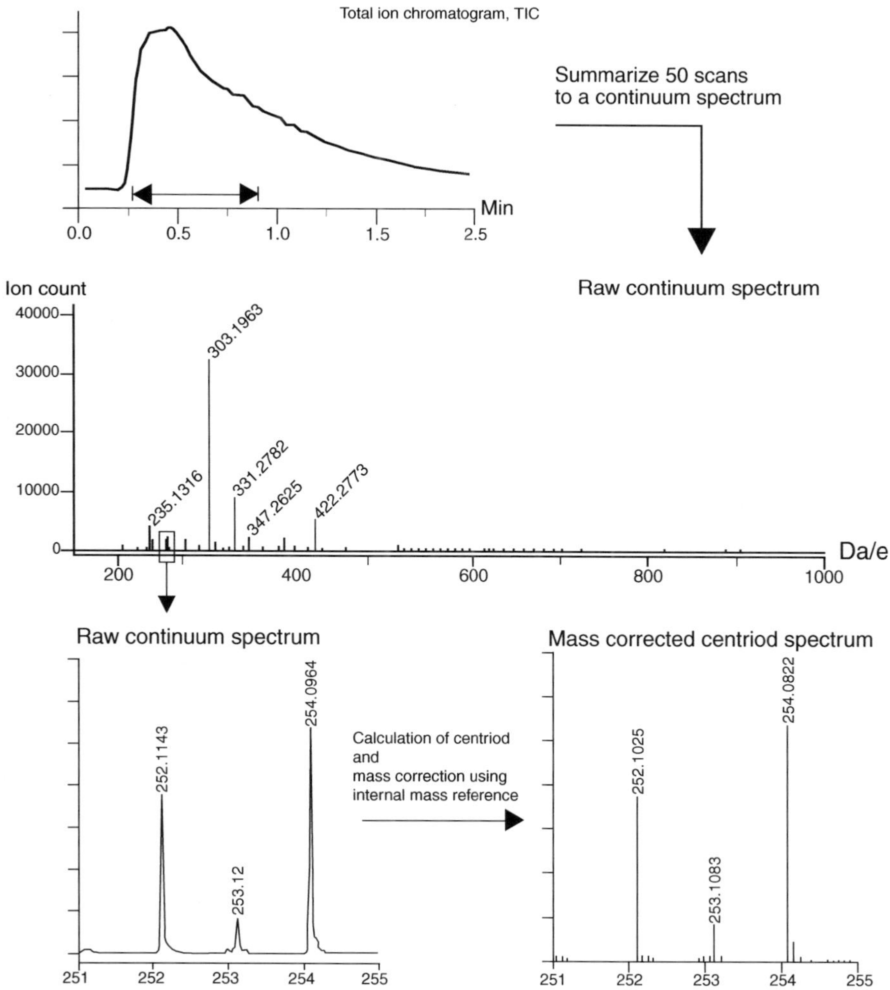

Figure 9.3 The standard data processing of raw spectra from direct infusion mass spectrometric analysis of crude extracts. A number of spectra are summarized to a single spectrum and then converted to a centroid spectrum. Internal mass calibration can be used in case of high-resolution mass spectra to get bet best mass accuracy. *Penicillium freii* (IBT 11273) cultivated on CYA, aurantiamine ($M + H^+$ at 303.1851 Da/e) was used for mass correction.

The sample reaches the source after about 15 s and the majority of the sample reaches the source during a following 40-s period as seen on the total ion profile on the top. Summarizing the continuum spectra collected during the elution of the sample results in a raw continuum mass spectrum with improved signal-to-noise ratio as shown in the middle. In this case, the high-resolution raw spectrum consists of approximately 115,000 data points. These combined raw spectra are the basis for all further processing. Note that if data are collected as centroid spectra, they cannot be combined in a similar fashion. Combining centroid spectra require binning, as discussed in Section 4.7, where it has to be decided which peaks belong to the same ions and which belong to different ions; thus, which to combine and which not to combine. Advanced chemometric processing can be applied directly to the raw continuum spectra same as that discussed in Chapter 5. However, the common procedure is to calculate a centroid spectrum. As these data are produced by a high-resolution TOF instrument, an internal mass reference can be used to improve the mass accuracy when calculating the centroid spectra. Rather than adding a reference compound to the sample, a metabolite produced by the fungus is used as internal mass reference. *P. freii* produce the metabolite aurantiamine ($[C_{16}H_{23}N_4O_2 + H]^+$ seen at 303.1821 Da/e), see Table 9.1, which is used as mass reference to improve the mass accuracy, as this metabolite is consistently produced and well ionized in positive elecrospray. The result is a centroid spectrum with very accurate masses as shown in part at the bottom (to the right) of Figure 9.3.

9.3.2 Metabolite Prediction

The accuracy of these high-resolution mass spectra is sufficient to limit possible elemental compositions for each ion to a relatively few formulas. If we assume a mass accuracy better than 5 ppm (typical for an average tof instrument) and that if all ions are composed of only the main isotopes of the common bioelements: carbon, hydrogen, nitrogen, and oxygen, then all possible compositions of each ion can be predicted. Figure 9.4 shows an elemental composition report calculated from the spectrum in Figure 9.3 limiting the calculation to ions above 5% base peak. For each ion, one or more elementary compositions fall within limits; however, some of these do not make sense in biology and can be rejected. Still, in most cases several formulas are possible. If the goal is to limit the number of candidates to just one, it requires very high accuracy (typically well below 1 ppm and resolution above 20,000 FWHM). The ion at 303.1821 Da/e is the internal mass reference used to correct the mass scale and should be ignored. The 304.1874 Da/e ions are actually the ^{13}C isotope (^{13}C was not included in this calculation) of aurantiamine (calculated 304.1854 Da/e). The elementary composition for the ions found at 238.0870 Da/e, 252.1025 Da/e, and 254.0822 Da/e all correspond to the protonated compositions of well-known metabolites produced by *P. freii* (viridicatin, 3-methoxy-viridicatin, virirdicatol), see Table 9.1, whereas most other ions listed are unknown. These findings can be confirmed by looking at the results from LC–MS analysis of exactly the same sample as shown in Figure 4.29. Ion traces from these two metabolites are shown and are confirmed by the UV-spectra shown in Figure 4.28. However, other

Elemental composition report

Multiple mass analysis: 13 mass(es) processed - displaying only valid results
Tolerance = 5.0 PPM / DBE: min = -1.5, max = 50.0
Isotope cluster parameters: Separation = 1.0 Abundance = 1.0%
Monoisotopic mass, Odd and even electron ions
5383 formula(e) evaluated with 41 results within limits
(up to 50 closest results for each mass)
Relative Abundance = min = 5.00% max = 100%

Mass	RA	Calc. Mass	mDa	PPM	DBE	Score	Formula
235.1197	12.72	235.1195	0.2	0.8	6.5	2	C11 H15 N4 O2
		235.1208	-1.1	-4.9	6.0	1	C13 H17 N O3
238.0870	5.76	238.0868	0.2	0.8	10.5	1	C15 H12 N O2
252.1025	5.78	252.1025	0.0	0.2	10.5	1	C16 H14 N O2
		252.1016	0.9	3.5	-1.5	2	H14 N9 O7
254.0822	8.30	254.0817	0.5	1.9	10.5	1	C15 H12 N O3
263.2381	5.38	263.2375	0.6	2.3	3.5	1	C18 H31 O
276.0618	5.27	276.0620	-0.2	-0.9	15.0	1	C11 H4 N10
		276.0620	-0.2	-0.9	9.5	2	C12 H10 N3 O5
		276.0607	1.1	4.0	10.0	3	C10 H8 N6 O4
303.1821	100.00	303.1821	0.0	0.0	7.5	2	C16 H23 N4 O2
		303.1808	1.3	4.4	2.5	4	C15 H27 O6
		303.1808	1.3	4.4	8.0	3	C14 H21 N7 O
		303.1834	-1.3	-4.4	7.0	1	C18 H25 N O3
304.1874	25.87	304.1872	0.2	0.5	2.5	4	C13 H26 N3 O5
		304.1872	0.2	0.5	8.0	3	C12 H20 N10
		304.1886	-1.2	-3.9	7.5	1	C14 H22 N7 O
		304.1886	-1.2	-3.9	2.0	2	C15 H28 O6
		304.1859	1.5	4.9	3.0	5	C11 H24 N6 O4
331.2628	27.85	331.2624	0.4	1.3	6.0	2	C20 H33 N3 O
		331.2637	-0.9	-2.7	5.5	1	C22 H35 O2
332.2667	6.67	332.2662	0.5	1.6	1.5	2	C15 H34 N5 O3
		332.2675	-0.8	-2.4	1.0	1	C17 H36 N2 O4
347.2423	7.56	347.2420	0.3	0.8	2.0	3	C16 H33 N3 O5
		347.2420	0.3	0.8	7.5	4	C15 H27 N10
		347.2434	-1.1	-3.0	7.0	2	C17 H29 N7 O
		347.2434	-1.1	-3.1	1.5	1	C18 H35 O6
		347.2407	1.6	4.7	2.5	5	C14 H31 N6 O4
387.1476	6.45	387.1476	0.0	0.1	2.5	2	C10 H23 N6 O10
		387.1471	0.5	1.4	15.0	6	C24 H21 N O4
		387.1484	-0.8	-2.1	20.0	7	C25 H17 N5
		387.1489	-1.3	-3.4	7.5	3	C11 H19 N10 O6
		387.1489	-1.3	-3.4	2.0	1	C12 H25 N3 O11
		387.1462	1.4	3.6	3.0	4	C8 H21 N9 O9
		387.1457	1.9	4.9	15.5	5	C22 H19 N4 O3
422.2555	16.12	422.2556	-0.1	-0.2	11.5	3	C24 H32 N5 O2
		422.2561	-0.6	-1.4	-1.0	6	C10 H34 N10 O8
		422.2543	1.2	2.9	6.5	1	C23 H36 N O6
		422.2543	1.2	2.9	12.0	2	C22 H30 N8 O
		422.2569	-1.4	-3.4	11.0	4	C26 H34 N2 O3
		422.2575	-2.0	-4.6	-1.5	5	C12 H36 N7 O9

Figure 9.4 Elemental compositions of all ions above 5% of base peak height. The columns shown form the left: measured mass, relative abundance (RA) in pct of base peak, calculated mass, error in mDa and ppm, double bond equivalents (DBE), and internal score and formula. Conditions: hydrogen less than 1000, carbon less than 500, oxygen less than 12, nitrogen less than 10, error maximal 5 ppm, less than 50 DBE.

ions, clusters, and fragments as those listed in Table 4.2 should be considered. Other elements, e.g., S, P, Cl, and Na are of course relevant and should be considered in the analysis of biological samples. However, the more the elements included, the more the formulas within limits will be returned.

To obtain the highest mass precision, the instrument has to be operated and maintained carefully, and most importantly a good tuning and calibration has to be maintained. In case of MCP–TDC detectors, the ion count is within the detector limit to avoid dead time problems.

9.3.3 Chemical Diversity and Similarity

These eight closely related fungi from the series *Viridicata* from *Penicillium* subgenus *Penicillium* show a remarkable diversity as illustrated in Figure 9.5 where mass profiles

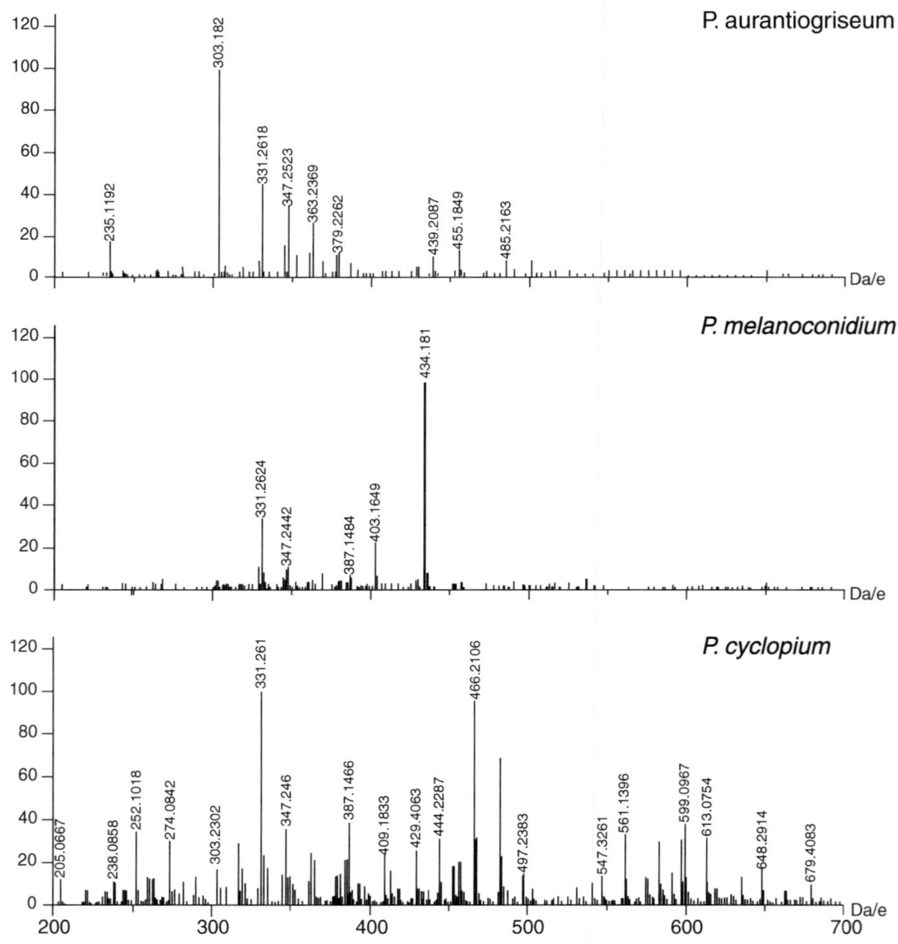

Figure 9.5 Mass profiles from three different *Penicillium* species all grown on CYA media, extracted and analyzed by direct infusion electrospray mass spectrometry. Only the mass range from *m*/*z* 200–700 is shown. Aurantiamine is used for internal mass correction in *P. augantiogriseum* (IBT collection no 21519), roquefortine C for *P. melanoconidium* (IBT collection no 21534) and verrucofortine for *P. cyclopium* (IBT collection no 21542).

from three different species grown under the same conditions are shown. However, similarities can also be an important feature as obvious from these three spectra.

It can be seen that all the spectra contain ions corresponding to the protonated mass of many of the metabolites listed in Table 9.1, but they also contain a lot of ions of unknown structure. Similarly, a remarkable consistency is observed within a species even over longer period of time; these data are not shown; however, it should be considered that changes in the analytical approach may seriously influence the mass profiles recorded. This diversity between species and similarity within species seen in mass profiles are, therefore, an efficient tool for classification/identification of the samples.

Eight to ten strains of each of the eight major *Penicillium* species associated with cereals (*Penicillium* subgenus *Penicillium* series *Viridicata*) were cultivated and analyzed as described above, and from these cultures 73 DiMS mass profiles were produced (including those showed in the figures above). The spectra were binned using an intelligent binning approach. Ions in each spectrum were binned into 0.5 *m*/*z* wide bins placed from −0.1 Da/e to + 0.4 Da/e and +0.4 Da/e to +0.9 Da/e around each nominal mass. If more than one ion fell into a bin, the most intense ion was selected; empty bins and those with ion count below threshold were removed. The result was aligned spectra that could be represented as vectors (bin, ion-count) representing each sample. These vectors were organized in a matrix and submitted to chemometric analyses as described in Chapter 5.

A cluster analysis was done on the aligned data matrix (after centering and scaling) using the correlation distances and clustering by WPGMA (weighted average distance) linkage. The result is shown in the dendrogram in Figure 9.6. Here, it can be seen that all samples are classified into the correct species as determined by classical phenotypic classification done by an expert taxonomist, thereby confirming that the mass profile contain sufficient information for species identification. In the study

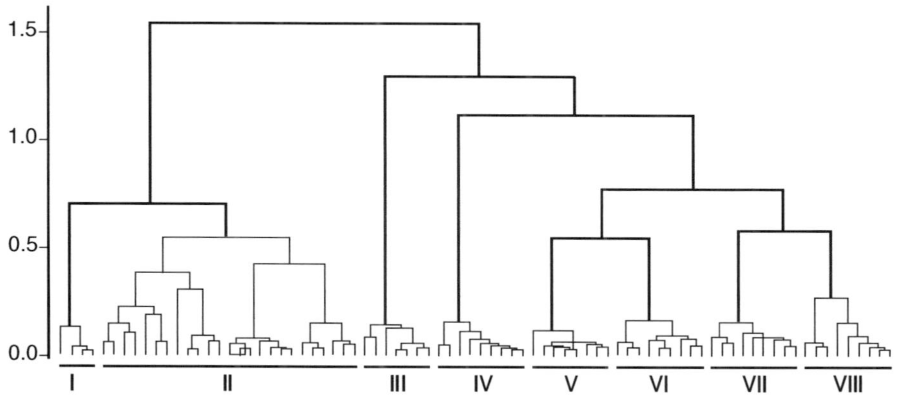

Figure 9.6 Classification of 73 mass profiles (spectra) from eight species selected from *Penicillium* subgenus *Penicillium* series *Viridicata*. All strains included in the study is classified into cluster in full agreement with identification by expert taxonomists. Based on intelligent binning using 0.5 mDa bins, see text. The species are: I *Penicillium aurantiogriseum*, II *P. cyclopium*, III *P. freii*, IV *P. melanoconidium*, V *P. neoechinulatum*, VI *P. polonicum*, VII *P. tricolor*, VIII *P. viridicatum*.

by Samson and Frisvad (2004), it was shown that approximately 60% of 57 species can be classified into species from mass profiles.

With this knowledge, it is logical to use the data base facility built into most instrument software packages. As an extension of the study by Smedsgaard and Frisvad (1996), a database of quadrupole mass profiles (spectra) from 43 *Penicillium* subgenus *Penicillium* species on two different media was build, in which 629 spectra (about 300 strains) were included. When this database is searched with the modern TOF spectrum as shown in Figure 9.3, a search report as shown in Figure 9.7 can

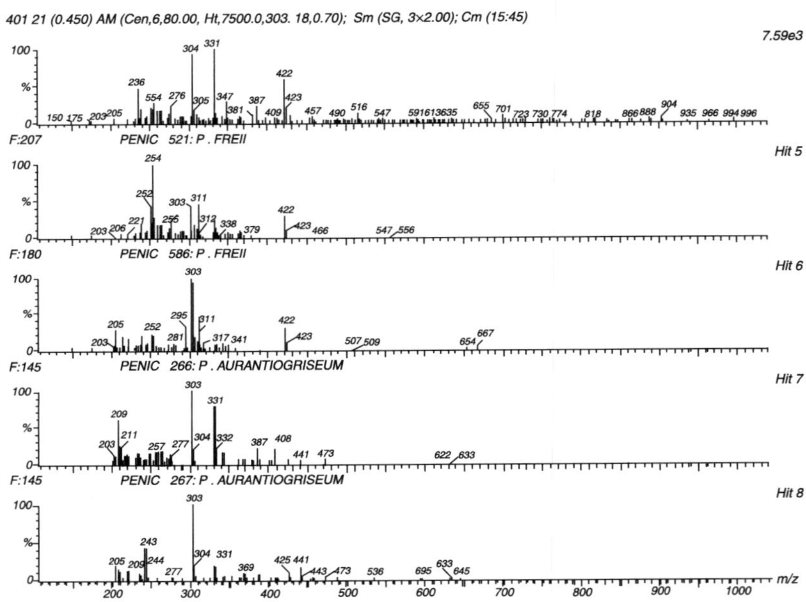

Hit	Compound name	CAS	Rev	For
1	P. FREII	10004-10-0	432	253
2	P. FREII	15783-10-0	357	241
3	P. FREII	15162-10-0	414	226
4	P. FREII	15374-10-0	395	216
5	P. FREII	15783-11-0	415	207
6	P. FREII	16692-10-0	352	180
7	P. AURANTIOGRISEUM	12957-11-0	312	145
8	P. AURANTIOGRISEUM	12957-10-0	282	145
9	P. AURANTIOGRISEUM	6689-11-0	281	127
10	P. AURANTIOGRISEUM	6689-10-0	262	125
11	P. AURANTIOGRISEUM	14264-11-0	201	86
12	P. PANEUM	13321-11-0	158	49

Figure 9.7 Most mass spectrometric software can be used to build libraries of spectra. Although not intended for complex mixtures they can easily be used for sample identification. An unknown high resolution mass profile (the one from Figure 9.3, *P. freii*) is search in a library of nominal spectra (approx 629 spectra) from most species in *Penicilium* subgenus *Penicillium*. The CAS number is used for strain collection number and a media code (10 is CYA).

be produced. The report shows *P. freii* spectra in the top six hits (only five different *P. freii* are included in the database), and the strain collection numbers can be read from the CAS number. The middle number, e.g., 10, indicates that the media used was CYA, the same as used for the spectrum showed in Figure 9.3 for the first four hits. Using the instrument database software like this was of course not the intention of the manufacturer; therefore, the search routines are not always optimal for this type of query. Furthermore, the scores will be much lower than usually seen from searches of EI–MS spectra of pure compounds. Finally, it is important to remember that on searching a database without limiting the criteria, the search will always return something, which may be without relevance to the sample.

Principle component analysis can also be used to find similarities in the data as discussed in Chapter 5. However, PCA will also reveal which of the variables, in this case which ions, are the main factors for sample discrimination or grouping seen in a scores plot (not shown). By plotting the first three loadings as a function of the mass from a PCA analysis of the binned data matrix, we get the plot as shown in Figure 9.8. Ions with a numerical high loading (highest or lowest values) are those contributing most to the segregation between species and to the grouping cluster formation. By comparing the *m*/*z* of these high loadings with Table 9.1, we can see that they correspond to the protonated or sodiated mass of many of the well-known metabolites.

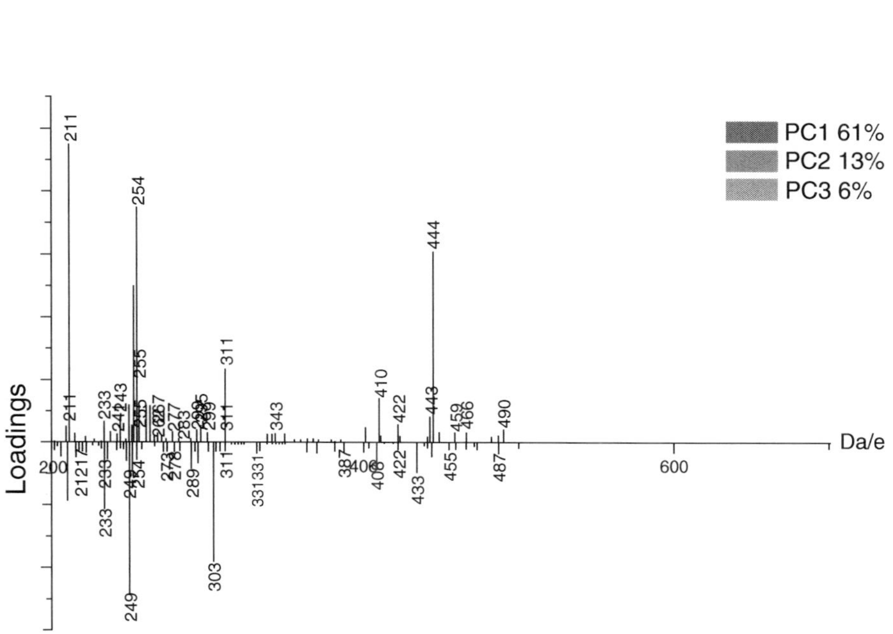

Figure 9.8 The loadings from principal component analysis (PCA) can tell how much each variable or mass contribute to the grouping or spreading of the samples along the principal component. Here, the three first loading are shown accounting for about 50% of the variation. Most of the masses with a high or a low contribution to the loading corresponds to the protonated (or sodiated) mass of known metabolites, compare to Table 9.1 or distinct ions in the spectra.

9.4 CONCLUSION

As seen from these few results analysis of crude extracts of fungal cultures by direct infusion, electrospray mass spectrometry is a very efficient tool for both indication of occurrence of a metabolite and for classification (or sample identification). However, one should be aware that matrix effect might hide important metabolites. On the contrary, the ability to efficiently group samples based on chemistry presents an efficient tool to limit the number of samples for the more complex analyses, e.g., LC–MS. This is of particular advantage in the search for organisms with capabilities of producing new or unexpected metabolites or to deselect chemically similar organisms so that further studies can focus on maximal diversity. Similarly, DiMS can be used as an efficient and rapid tool to examine mutant libraries in particular for the production of secondary metabolites.

REFERENCES

Samson RA, Frisvad JC. 2004. *Penicillium* subgenus *Penicilium*: New taxonomic schemes, mycotoxins and other extrolites. Studies in Mycology **49**, Centraalbuteau voor Schimmelcultures, P.O. box 85167, 3508 AD Utrecht The Netherlands ISBN 90-70351-53-6.

Samson RA, Hoekstra ES, Frisvad JC. 2004. Introduction to food- and airborne fungi. 7th edition. Centraalbuteau voor Schimmelcultures, P.O. box 85167, 3508 AD Utrecht The Netherlands.

Smedsgaard J, Frisvad JC. 1996. Using direct electrospray mass spectrometry in taxonomy and secondary metabolite profiling of crude fungal extracts. J Microbiol Met 25:5–17.

Smedsgaard J. 1997. Micro-scale extraction procedure for standardized screening of fungal metabolite production in cultures. J Chromatogr A 760:264–270.

10

METABOLOMICS IN HUMANS AND OTHER MAMMALS

By Dr. David Wishart

This chapter describes the preparation of samples and measurement of metabolites from mammals, specifically humans, rats, and mice. A brief review of mammalian metabolomics is provided along with a more detailed description of how mammalian biofluid and tissue samples can be obtained, extracted, and processed for metabolite analysis. This chapter also describes a number of metabolic profiling techniques that are somewhat unique to mammalian metabolomics. Finally, a brief description of a specific application of metabolomics for humans (metabolic disease diagnosis) is provided.

10.1 INTRODUCTION

The mammalian metabolome is very different from that of either microbes or plants. Unlike plants or most microbes, mammals are auxotrophs. In other words, mammals cannot synthesize all the nutrients or metabolites they need to stay alive. As a result, mammals must consume a variety of foreign plants, animals, and microbial products to fulfill their dietary requirements. Therefore, by definition, the mammalian metabolome consists of both endogenous and exogenous metabolites. Endogenous metabolites are those small molecules that are synthesized by the enzymes encoded by the host's genome, whereas exogenous metabolites are "foreign" chemicals consumed as food or generated by host-specific microbes. As a general rule, the concentration of most endogenous metabolites in mammals is much greater than the concentration of any given exogenous metabolite. While mammalian cells are much

Metabolome Analysis: An Introduction, by Silas G. Villas-Bôas, Ute Roessner, Michael A. E. Hansen, Jorn Smedsgaard and Jens Nielsen

larger, more specialized, and generally more complex than microbial cells, it appears that the mammalian metabolome is probably not much larger than that of any given microbe. Current estimates put the mammalian metabolome at about 1500 different compounds (www.hmdb.ca) whereas the yeast and *E. coli* metabolomes are believed to consist of between 600 and 800 compounds (Forster et al., 2003; Keseler et al., 2005). Unlike microbes, however, it appears that the endogenous metabolome of mammals varies little among species – with rats, mice, and humans having essentially identical constituents and exhibiting only modest variations in interspecies concentrations. The interspecies uniformity and relatively small size of the mammalian metabolome stands in stark contrast to the number and variety of metabolites found in plants. In fact, it is estimated that the plant kingdom may encode more than 200,000 different metabolites, with any given plant species capable of synthesizing between 5000 and 10,000 different compounds (Trethewey, 2004; Hall et al., 2002). This enormous difference in metabolic complexity can be rationalized by the fundamental differences in mobility between plants and animals (and microbes). Because mammals are able to run, walk, or fly, they require a much smaller arsenal of defensive chemical agents than plants, which must "stand and fight" when attacked by a predator or parasite.

While the endogenous metabolome in mammals is relatively small, their exogenous metabolome is probably very large (>10,000 compounds). Humans, like most mammals, have a highly varied diet, and ingest a wide spectrum of plant, animal, and microbial (cheese, yogurt, wine, beer) products. These foods, many of which provide essential vitamins, fats, and amino acids (Table 10.1), also contain many other nonessential nutrients that must be broken down, processed, or secreted. Many foods consumed today are also supplemented with a growing number of synthetic additives

TABLE 10.1. Essential Minerals and Nutrients in Mammals.

Fatty Acids and amino acids	Vitamins and cofactors	Minerals and ions
Linoleic acid	Biotin	Chromium
Alpha-linolenic acid	Folate	Cobalt
Phenylalanine	Niacin	Copper
Valine	Pantothenic acid	Iodine
Threonine	Riboflavin	Iron
Tryptophan	Thiamin	Magnesium
Isoleucine	Vitamin A	Manganese
Methionine	Vitamin B6	Molybdenum
Histidine (children)	Vitamin B12	Potassium
Alanine (children)	Vitamin C (primates & guinea pigs)	Selenium
Leucine	Vitamin D	Zinc
Lysine	Vitamin E	Calcium
Taurine (cats)	Vitamin K	Phosphorus
Carnitine (conditional)	Pyrroloquinoline quinone (mice)	Sodium

(coloring, texture, and flavor enhancers). Of course, foods are not the only source of exogenous metabolites in mammals. Drugs, nutraceuticals, and other xenobiotics constitute an equally large and complex source of exogenous metabolites. Currently, there are more than 1200 FDA approved drugs and nutraceuticals in the market (Wishart et al., 2006). Furthermore, many of these drug molecules are subsequently modified via cytochrome P450s, glucuronidases, esterases, and other detoxifying enzymes to yield an even larger collection of metabolic by-products.

Foods, drugs, and nutritional supplements certainly contribute significantly to the size of the exogenous metabolome. However, another important and oft-neglected source of exogenous metabolites comes from the nearly 400 different microbial species that live in the mammalian gut (Eckburg et al., 2005). In humans, the gut microflora weigh between 1 and 2 kg and constitute a metabolically essential, albeit highly distributed, multicellular organ (Eckburg et al., 2005; Guarner and Malagelada, 2003). In ungulates and other herbivores, the gut microflora are even more important and represent an even larger portion of the organism's metabolic infrastructure. It is thought that these symbiotic microbes may contribute several hundred additional compounds to the exogenous metabolome of mammals, including at least 2 dozen essential nutrients. (Nicholson et al., 2005).

The issue of exogenous versus endogenous metabolites is not the only complication associated with describing the mammalian metabolome. Mammals have more than 200 different cell types, several dozen different organs, and many highly compartmentalized biofluid systems. Each of these cell types, tissues, or organs is metabolically specialized in some fashion or another, often producing a handful of unique metabolites that are not found in other cells or organs. The same metabolic specialization is true for many biofluids as well. These biofluids include blood, milk, cerebrospinal fluid, bile, saliva, mucus, lung exudates, lachrymal secretions, semen, lymph, and more. Perhaps the only places where the entire collection of all endogenous and exogenous metabolites might be found is in the urine (for water soluble molecules) and feces (for fat soluble molecules).

Cell, tissue, and organ variations make a "single" mammalian metabolome hard to define. So too, does the wide range of metabolite concentrations found in mammals. These concentrations, which can range from as low as picomolar levels (i.e., exogenous chemicals, certain hormones, and many signaling molecules) to as high as molar concentrations (urea), are a function of diet, gender, time of day, age, health, and genetic background. They are also a function of the solubility, size, toxicity, and physiological role of the chemical itself. So, while the genome of mammals can be formally defined (3.272 billion base pairs and 23,300 genes in the human) and is uniformly the same between different cells and tissues, the mammalian metabolome can only be approximated. Furthermore, it appears that the mammalian metabolome varies tremendously between different cells, tissues, and biofluids. Therefore, the metabolome is actually defined by where and how it is measured (i.e., instrument sensitivity). Certainly, if we had infinite sensitivity, the human metabolome might easily exceed 100,000 chemicals. However, given that most analytical instruments have a detection limit of ~1 micromolar, it appears that the readily accessible metabolome is probably less than 1000 compounds. This is minimum estimate only.

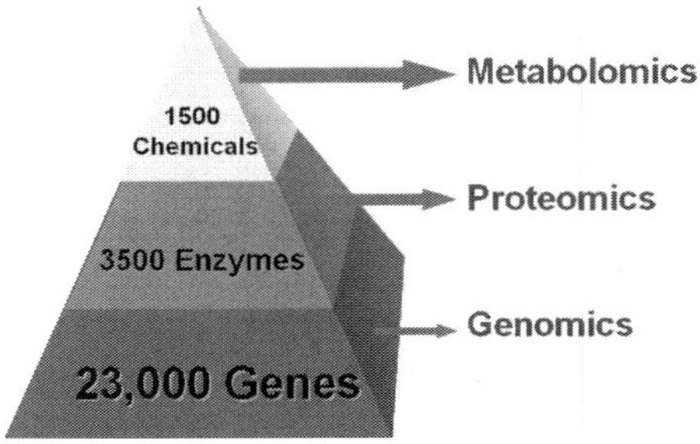

Figure 10.1 The "pyramid of life" illustrating the relationship between genes (genomics), enzymes (proteomics) and metabolites (metabolomics). Metabolites, which require an enormous proteomic and genomic infrastructure to be processed, exhibit the least diversity of all biological molecules. They are also the most sensitive to changes or mutations at the bottom of the pyramid.

Obviously, with pooling, extraction, sample concentration, and other targeted approaches, this lower limit can be readily extended.

While we have spent a good deal of time trying to define the mammalian metabolome, it is important to remember that whatever the metabolome is, it is a tremendously important part of biochemistry and physiology. Indeed, the power of metabolomics comes from the fact that small molecule metabolites effectively lie at the top of the genomic pyramid (Figure 10.1). An imperceptibly small genomic change, such as single base transition or a noncoding polymorphism in a gene, can be amplified many 1000s of times when the effect is measured at the metabolite level. This is because metabolites are essentially the end-products of dozens of interdependent macromolecular interactions. Indeed, small molecule metabolites could be considered to be the "canaries" of the genome. They are the body's advance warning system that something is wrong or about to go wrong. The fact that metabolomics measures the "downsteam products" of multiple protein, gene, and environmental interactions, makes it a particularly good reporter of an organism's phenotype or physiology. Indeed, metabolomics essentially offers researchers and physicians the capacity to generate a quantitative *molecular phenotype*. Because metabolic responses are often measured in seconds or minutes (whereas genetic responses are typically measured in days or weeks), metabolomics measurements can potentially yield important physiological information that is not normally accessible with genomic or proteomic analyses.

This chapter focuses on describing the techniques and technologies used to characterize the mammalian metabolome, with a particular emphasis on the applications toward mouse, rat, and human systems. Unlike plant and microbial metabolomics, many of the applications in mammalian metabolomics are health related, and many of the technologies emerged from the health sciences. This difference in focus and difference in origin partly explains the somewhat different technologies and analytical techniques used in studying the mammalian metabolome. In this chapter, we will describe and critically assess some of these techniques with the aim of helping the reader to select the best analytical techniques and the best sample preparation methods for their given purpose or chosen interest.

10.2 A BRIEF HISTORY OF MAMMALIAN METABOLOMICS

Metabolic profiling, in one form or another, has been a part of medical practice for thousands of years. As far back as the fifth century BC, both Hippocrates and Hermogenes described the diagnosis and detection of diseases through the sensory analysis of urine (color, taste, smell). The analysis of biofluids eventually becomes more quantitative with the development of clinical chemistry in the mid-19th century (Coley, 2004). Largely through the works and writings of a number of British scientists (William Prout, Henry Bence Jones, John Bostock, and Richard Bright), clinicians began to identify and quantify biofluid constituents and associate them with various medical conditions. However, it was not until the early 20th century through the systematic and wide ranging studies of the US chemists, Otto Folin (1867–1934) and Donald Van Slyke (1883–1971) that clinical chemistry and metabolic profiling became a part of routine medical practice (Rosenfeld, 2002). These two visionary scientists helped to develop many of the colorimetric tests, and early instrumentation used to quantify metabolites in blood and urine (Fandek et al., 1995; Rosenfeld, 2002). Nowadays, blood and urine tests, which offer from 5 to 50 different chemical readouts (Table 10.2), are routinely performed by multicomponent clinical analyzers or by simple paper strip tests (Fandek et al., 1995; Tietz, 1995). These semiquantitative tests typically depend on colorimetric assays where specific reagents are added to a sample and reactions are monitored spectrophotometrically to identify or quantify a targeted metabolite. In the nomenclature of chemical chemists, these metabolite-specific tests are called "point analyses," meaning that only one compound is monitored or detected in any given test (Matsumoto and Kuhara, 1996).

By the 1970s, a new generation of clinical chemistry instrumentation was appearing which permitted the identification of not just a single compound but a whole class of compounds. Gas chromatographic (GC) columns started being coupled to mass spectrometers (MS) to create GC–MS systems, which could detect organic acids from blood and urine. Indeed, the birth of metabolomics (or metabolic profiling as it was called then) could probably be traced to a seminal GC–MS paper written in 1974 (Sweeley et al., 1974). These authors used GC–MS to develop quantitative metabolic profiles of dozens of urinary organic acids. The MS spectra of the metabolites in combination with their chromatographic retention times were monitored against known standards to uniquely identify each compound. Many other studies have since been followed (Gates and Sweeley, 1978; Tanaka and Hine, 1982) and GC–MS continues to be the method of choice in organic acid profiling especially for genetic disease testing and monitoring (Matsumoto and Kuhara, 1996; Kuhara, 2005). Among clinical chemists, these class-specific tests are called "line analyses," meaning that they characterize or target a specific group of metabolites (i.e., organic acids). In metabolomics, line analysis is also called targeted analysis.

TABLE 10.2. List of Compounds Identifiable via Standard Clinical Chemistry Tests.

Clinical + electrolyte analyzers + immunoassays	GC–MS (organic acids)	Amino acid analyzer (HPLC)
Sodium	Methylmalonic acid	Alanine
Potassium	Ethylmalonic acid	Cysteine
Chloride	Methylsuccinic acid	Aspartic acid
Calcium	Lactic acid	Glutamic acid
Magnesium	Adipic acid	Phenylalanine
Iron	Methyladipic acid	Glycine
Bicarbonate	Suberic acid	Histidine
Phosphate	Homovanillic acid	Isoleucine
Ammonia	Azelaic acid	Lysine
Urea	Hippuric acid	Methionine
Urate	Citric acid	Asparagine
Creatinine	Sebacic acid	Proline
Glucose	Vanillylmandelic acid	Glutamine
Beta hydroxybutyrate	Stearic acid	Arginine
Bilirubin		Serine
Cortisol		Threonine
Thyroid hormone T3, T4		Valine
Triglyceride		Tryptophan
Testosterone		Tyrosine
Vitamin B12		Ornithine
Lactate		Taurine
Cholesterol		Homocysteine
Fructosamine		Citrulline

In the 1990s, tandem mass spectrometry (MS/MS) emerged as a powerful, new approach for the nontargeted detection and identification of a wide range of metabolites. This kind of nontargeted analysis is sometimes called "planar analysis" in the field of clinical chemistry (Matsumoto and Kahura, 1996). MS/MS permits very rapid (1–2 min), sensitive (femtomole detection limits from dried blood spots) and, with appropriate internal standards, the accurate quantification of up to 20 different types of metabolites with relatively minimal sample preparation and without prior chromatographic separation (Pitt et al., 2002). Because of these appealing features, MS/MS or direct injection mass spectrometry (DIMS) is being increasingly used in newborn screening programs in the USA, Canada, Australia, and elsewhere, with a particular focus on identifying amino acid, nucleic acid, and acylcarnitine markers for inborn errors of metabolism or IEMs (Mueller et al., 2003). Other metabolite profiling developments in the 1990s include the introduction of capillary electrophoresis (CE) methods for more precise and rapid metabolite separation (Terabe et al., 2001), the use of UPLC (ultrahigh pressure liquid chromatography) and two-dimensional HPLC methods for improved compound partitioning (Wilson et al., 2005; Guttman et al., 2004), and the debut of Fourier transform MS

(FT-MS) methods for large-scale metabolite screening (Leavell et al., 2002; Brown et al., 2005). More recently, infrared spectroscopy (FTIR) and NMR spectroscopy have entered the fray (Wevers et al., 1994; Jackson et al., 1999; Moolenaar et al., 2003). Indeed, it is not unusual to see metabolomics studies of mammals being done with robotically linked combinations of HPLC, CE, NMR, and/or MS instruments (Shockor et al., 1996).

The trend toward using NMR, FT-MS, and FTIR in metabolomics studies of humans and other mammals during the 1990s was paralleled by a trend toward using chemometric or multivariate statistical methods to analyze the spectra obtained from these instruments (Holmes et al., 2000; Smith and Baert, 2003). Rather than attempting to identify and quantify the individual chemical components of the biofluid being analyzed, the spectra were treated as uniquely classifiable metabolic fingerprints. Machine learning (ML) methods, principal component analysis (PCA), clustering, self-organizing feature maps, genetic algorithms (GA), or neural networks (NN) have all been used to interpret NMR, MS/MS, and FTIR spectral patterns (Holmes et al., 2000; Smith and Baert, 2003; Wilson et al., 2005). The intent of using this type of pattern classification software is not to identify any specific compound but, rather, to look at the spectral profiles of blood, tissue, or urine and to classify them in specific categories, conditions, or disease states. This trend to pattern classification represents a significant break from the classical methods of clinical chemistry, which traditionally depend on identifying and quantifying specific compounds. With these new chemometric profiling methods, one is not so interested in quantifying known metabolites, but rather in trying to look at all the metabolites (known and unknown) at once (Nicholson et al., 1999; Nicholson et al., 2002). The strength of this holistic approach lies in the fact that one is not selectively ignoring or including key metabolic data in making a disease classification or diagnosis. These pattern classification methods can perform quite impressively, and a number of groups have reported success in diagnosing certain diseases such as colon cancer (Smith and Baert, 2003) and breast cancer (Jackson et al., 1999), in identifying inborn errors of metabolism (Bamforth et al., 1999), in sorting out the location of toxic-substance injuries (Holmes et al., 2000), in tracking the time dependencies of drug toxicity (Nicholson et al., 2002), in monitoring organ rejection (Wishart 2005), in measuring HDL and LDL ratios (Cromwell and Otvos, 2004), and in classifying different strains of mice and rats (Wilson et al., 2005; Robosky et al., 2005).

Whether you call it clinical chemistry, metabolic profiling, or metabolomics, the study of mammalian metabolites has been an important part of medicine and physiology for hundreds of years. The close connection between health and metabolism has been a strong technology driver for new developments in metabolic profiling. As a result, many of the new technologies are applied first to mammalian systems, and then later migrated to the study of plants and microbes. In other words, if you want to see where metabolomics is going, it is often best to monitor what is going on in the study of mammalian systems. Certainly, the trends in mammalian metabolomics over the past 10 years have been toward the adoption of newer, more expensive technologies (FT–MS, NMR, MRI); a greater reliance on chemometric and multivariate statistical analyses; a greater focus on drug and xenobiotic interactions, and even the emergence of an alternative name (i.e., metabonomics) for metabolic profiling (Nicholson et al., 1999; Dunn et al., 2005). Many of these same technology trends and nomenclature preferences are now showing up in the literature describing metabolic studies of plants and microbes. Curiously, though, while most of the technology and analysis trends in metabolomics are first tested on mammals, many of the sample preparation techniques are first tested on plants and microbes.

10.3 SAMPLE PREPARATION FOR MAMMALIAN METABOLOMICS STUDIES

Key to any successful effort in a metabolomics experiment is having a high-quality biological sample. The choice of the sample (fluid, tissue, etc.) is dictated by the questions being asked, the sensitivity of the instrument, and the kind of metabolites being studied. One thing that distinguishes metabolomics studies of mammals from plants and microbes is the variety of samples or sample types that are available. Metabolomics studies in mammals have been reported on intact organs (van der Graaf et al., 2004), extracted tissues or biopsies (Smith and Baert, 2003), fine needle aspirates (Mountford et al., 2001), dried blood spots (Mueller et al., 2003), plasma or serum (Andreasen and Blennow, 2005; Daykin et al., 2002), urine (Matsumoto and Kuhara, 1996; Zuppi et al., 1997; Nicholson et al., 2002), cerebrospinal fluid (Lutz et al., 1998), bile (Paczkowska et al., 2003), seminal fluid (Hamamah et al., 1998), feces (Smith and Baert, 2003), saliva (Silwood et al., 2002), and many other biofluids. Overall, the clear majority of metabolomics measurements are performed on biofluids, not tissues. The choice of fluids over tissues is done with the assumption that the chemicals found in most biofluids are largely reflective of the physiological state of the organ that produces, or is bathed in, that fluid. Hence, urine reflects processes going in the kidney, bile—the liver, CSF—the brain, and so on. The blood is a special biofluid as it potentially reflects all processes going on in all organs. This can be both a blessing and a curse as metabolite perturbations in the blood, while easily detectable, cannot be easily traced to a specific organ or a specific cause. In metabolomics, the choice of biofluids over tissues is also dictated by the fact that fluids are far easier to process and analyze with today's NMR, MS, or HPLC instruments. Likewise, the collection of biofluids is generally much less invasive than the collection of tissues.

Regardless of whether the sample of interest is a biofluid or tissue, sample uniformity is a particular challenge in mammalian metabolomics. When it comes to rats, mice, and other laboratory mammals, care must be taken to ensure that sampling is reproducible in terms of sampling time, strain, breed, developmental stage, estrus cycle, age, and gender (Bollard et al., 2001; Stanley et al., 2005; Robosky et al., 2005). Likewise, sufficiently large sample sizes, either longitudinally (many samples from one individual over time) or cross-sectionally (many samples from multiple individuals at one time point), must be acquired in order to do the statistics needed to confidently report metabolite levels, responses, or trends. In other words, sufficient numbers of physiologically similar animals (biological replicates) must be available to provide multiple fluid/tissue samples. Likewise, a sufficient number or quantity of samples from each animal (technical replicates) must also be available in order to perform a well-validated metabolomics study. Depending on the questions being asked, the instrumentation and method of analysis as few as 2–3 biological and 2–3 technical replicates may be needed. For chemometric analyses, several dozens are typically needed to draw conclusions. In all metabolomics studies, a sufficient number of reference or control animals (or tissues or biofluids) must be available. Fortunately, for humans there are a number of books containing reference metabolite values that make the need for human controls a little less onerous (Tietz, 1995).

For lab animals, metabolic cages under controlled environmental conditions (sterile housing, uniform temperature, humidity, filtered air, controlled light/dark periods, identical diets) are frequently used to facilitate the collection of biofluids and to eliminate many unwanted variables. These cages, with only one rodent per cage, allow the controlled feeding and watering of the animals and the collection of urine in external graduated tubes without cross contamination by feces, food, or fur (Dickman, 1953).

When it comes to human metabolomics studies, it is essentially impossible to achieve the same level of environmental and dietary control as seen in lab animals housed in metabolic cages. Certainly, humans tend to be more conscientious than lab rats when it comes to sanitation and much more amenable to following instructions. However, humans are intrinsically more variable and free-willed. Nevertheless, variations in diet, behavior, and drug intake can be partially controlled or monitored by having patients maintain diaries of activities as well as food, drink, and drug consumption. Alternately, collecting samples after fasting can help eliminate some of these dietary issues as well. As with lab animals, age, gender, disease state, diurnal changes, menstrual cycle status, level of activity, and lifestyle choices among humans can all affect metabolite readings (Tietz, 1995; Kaiser et al., 2005). These need to be controlled, matched, or accounted for as best as possible, given the resources available.

An additional challenge to working with animal samples is the need for proper protection and handling because of the risk of disease transmission. Human tissues, blood, and CSF are typically treated as level-2 biohazards requiring level-2 containment. This is because improper handling of these substances can lead to the transmission of hepatitis A, B, and C; HIV; and various prion diseases (CJD, vCJD). Human urine, being remarkably sterile, can typically be treated as a nonhazardous material requiring only level-1 biohazard certification or level-1 containment. Most animal (i.e., rodent) biofluids and tissues are also rated as level-1 biohazards requiring only level-1 containment. However, work with primates or animals infected with human pathogens may require higher containment levels (level-2 or -3) and greater attention to safety. Many biofluids can be "decontaminated" or extracted with organic solvents (see below), making them harmless and suitable for work in standard, level-1 lab space. Different jurisdictions may require different containment practices as well as different certification or vaccination requirements for lab personnel. Obviously, it is critical that lab supervisors and researchers be well-versed in safe laboratory practices and that all parties be made aware of any hazards associated with any biological material being analyzed. Given that many metabolomics specialists are analytical chemists having little formal experience with biohazardous materials, this issue is likely to be an ongoing concern.

10.3.1 Working with Blood

Because of the strong influence of clinical chemistry and current medical practices, the analysis of blood, serum, or plasma has always been held in high esteem for metabolic studies. Certainly, a key advantage of blood is that it a remarkably uniform and highly homeostatic biofluid. Indeed, blood is largely unaffected by such confounding factors as age, gender, diet, fluid consumption, diurnal cycles, and stress. However, a disadvantage of blood is that, in addition to small molecule metabolites, it contains many cellular components (red blood cells, white blood cells) and macromolecules such as proteins (albumin and immunoglobulins), lipids, and lipoproteins (HDL, LDL, VLDL). Furthermore, many of the small molecules of interest are tightly bound to the circulating proteins and lipoprotein particles. Given the problems of working with raw blood, there is a general preference by most specialists to work with serum or plasma instead. Serum and plasma are derivatives of blood products. Blood plasma is the liquid, straw-colored component of blood consisting primarily of water, blood proteins, inorganic electrolytes, and small molecule metabolites. Plasma is prepared by adding an anticoagulant (heparin, EDTA, citrate) to the blood specimen immediately after it has been obtained. The sample is then centrifuged to separate the plasma (top layer) from the blood cells (bottom layer). The top layer is typically removed and then stored at −80°C. Serum is the same as blood plasma, except that clotting factors, such as fibrin, have been removed.

The abundance of proteins (and potential pathogens) that still remain in either serum or plasma still make these fluids problematic for routine metabolomics analysis. As a result, most protocols for the analysis of blood, serum, or plasma, call for the extraction or deproteinization of the material. This process eliminates large macromolecules and pathogens, releases bound metabolites from proteins, and makes chromatographic separation, MS analysis, or NMR data collection much easier. Different analytical techniques, such as GC–MS, DIMS, or FTIR require different approaches for analyzing blood (Mueller et al., 2003; Smith and Baert, 2003). However, one approach based on studies performed by Daykin et al. (2002) seems to work particularly well for both LC–MS and NMR studies. In this simple protocol, fresh plasma is mixed with an equivalent volume of acetonitrile (AcN) and shaken for 30 s. The mixture is then sonicated for 15 min to insure good mixing. The sample is then centrifuged at 7000 rpm and 4°C for 25 min to remove the precipitates. The supernatant is then removed and placed in a separate tube. A second extraction step is then performed on the remaining protein pellet wherein an equivalent volume of aqueous methanol (1:1 MeOH/H_2O, v/v) is added to the pellet, shaken for 30 s, and then sonicated for 15 min. The sample is then centrifuged to remove the remaining precipitates and the MeOH supernatant combined with the AcN supernatant. The AcN and MeOH are removed using a rotary evaporator, and the sample is concentrated to dryness using a freeze-dryer. In this dried state, the sample may be reconstituted in a more concentrated form and placed into an NMR tube or injected directly into an HPLC or LC–MS system. Obviously, this process, with its many drying steps, tends to remove volatile substances such as ethanol, trimethylamine, and acetone. However, NMR studies comparing the extracted material to whole plasma indicate that most metabolites are preserved and present in the same amounts as in unprocessed serum (Daykin et al., 2002).

10.3.2 Working with Urine

Urine is the by-product or waste fluid secreted by the kidneys and transported to the bladder where it is stored and later secreted. It is composed of 95% water, 2% urea, 2% salts, and 1% small molecule metabolites. In mammals, urine serves as a means for flushing waste molecules collected from the blood, for homeostasis of body fluids, and (except for humans) for olfactory communication. While long despised by clinicians as a medically useful biofluid, urine is perhaps the ideal fluid for metabolomics analysis. This is because urine contains and concentrates essentially all the exogenous and endogenous metabolites found in the body. Furthermore, unlike most biofluids, urine is abundant, sterile, easily and non-invasively obtained, safe to handle, and usually devoid of proteins or other macromolecules. This latter fact makes the chromatographic separation, MS analysis, or NMR spectral collection of urine relatively easier and trouble-free. There are, however, some drawbacks of working with urine. First, the collection of urine from rodents and other small mammals is often difficult and frequently leads to cross contamination with other unwanted material. Likewise, the collection of urine from human infants is also difficult as similar cross contamination issues can arise. Secondly, urine is subject to considerable variations in dilution, making the reporting, and comparison of metabolite concentrations difficult or inconsistent. Indeed urinary metabolites are significantly affected by such factors as age, gender, diet, fluid consumption, diurnal cycles, and stress (Lenz et al., 2004; Bollard et al., 2005). Thirdly, urine is not a biofluid that can be sampled continuously such as blood or saliva. Rather urine is only an indicator of metabolic or physiological processes that happened hours or even days before collection. Fourthly, because urine is a waste product, it is over enriched with exogenous metabolites or xenobiotics that have little to do with the organism's essential metabolism. Most of these problems are not insurmountable, and the primary issue concerning urinary metabolite concentrations has long been dealt with by reporting concentrations relative to urinary creatinine. This abundant breakdown product of muscle metabolism is secreted at a remarkably constant rate and is easily measured. In some cases, these potential problems are actually benefits. For instance, because urine concentrates waste products or toxins, it is particularly a good indicator for hundreds of metabolic disorders (Matsumoto and Kuhara, 1996; Wishart et al., 2001; Moolenar, 2003), many different kinds of infections (Gupta et al., 2005), and certain kinds of cancers (Fauler et al., 1997). It is also particularly good for monitoring food consumption, nutritional balance, and illicit drug consumption.

The metabolomics analysis of urine is relatively easier. In most cases, it can be placed directly into chromatographic equipment, MS instruments, amino acid analyzers, and NMR spectrometers with little or no sample preparation. In some cases, particularly if there is a concern about the presence of possible human pathogens, blood, or high levels of protein, urine can be extracted, decontaminated, or deproteinized using the following simple protocol. In this method, urine is mixed with an equivalent volume of acetonitrile (AcN) and then allowed to sit on ice for a minimum of 5 min. The sample is then centrifuged at 7000 rpm and 4°C for 20 min to remove any precipitates. The supernatant is then removed and stored separately. A second extraction of the pellet is then performed using aqueous methanol (1:1 $MeOH/H_2O$, v/v). This mixture is allowed to sit on ice for a minimum of 5 min followed by centrifugation to remove any precipitates or particulates. The MeOH supernatant is then removed and combined with the AcN supernatant. The sample is then concentrated by removing the MeOH and AcN by rotary evaporation or speed-vac evaporation. NMR studies comparing the extracted material (solubilized in an H_2O buffer) with raw urine indicate that most nonvolatile metabolites are preserved and present in the same amounts as in unprocessed urine.

10.3.3 Working with Cerebrospinal Fluid

Cerebrospinal fluid (CSF) is a clear biofluid found around the cortex, the ventricular system of the brain, and the spinal cord. The total amount of CSF in humans at any given time is about 150 ml, although about 500 ml is produced each day. CSF is important for cushioning the brain (mechanical protection), for distribution of neuroendocrine hormones, and for facilitation of cerebral blood flow. CSF is not easily obtained. It must be acquired through a medical procedure called a lumbar puncture or spinal tap. A spinal tap may yield 5–15 ml of CSF at any given time. Generally rodents are too small for lumbar punctures, so CSF is usually acquired from larger lab animals, such as cats and dogs. Because the CSF bathes the neural system, it can be used for the detection, diagnosis, and monitoring of a number of neurological conditions. These include meningitis, subarachnoid hemorrhage, Alzheimer's disease, multiple sclerosis, and numerous neurometabolic disorders (Hoffmann et al., 1998; Andreasen and Blennow, 2005). Like blood, CSF is highly regulated and exhibits very little variation because of age, gender, diet, fluid consumption, diurnal cycles, or stress. However, in certain metabolic disorders such as Canavan's disease, some metabolites—such as *N*-acetylaspartic acid – may be greatly elevated (Wevers et al., 1995; Hoffmann et al., 1998). Relative to blood and urine, which typically have thousands of metabolites (many of which are still to be identified), CSF is quite limited in its metabolic repertoire having less than 70 compounds—most of which appear to be known (Table 10.3).

Like urine, CSF is largely protein free making metabolomics analysis of this biofluid relatively easier. In most cases, CSF can be placed directly into analytical instrument of choice with little or no sample preparation. In some cases, particularly if there is a concern about the presence of possible human pathogens, prions, blood, or high levels of protein, CSF can be extracted, decontaminated, or deproteinized using the same protocol described earlier for urine. Handling human CSF generally requires level-2 containment procedures.

TABLE 10.3. Table of ~65 Metabolites, Concentrations Ranges and Disease Conditions for Normal and Abnormal Human Cerebrospinal Fluid (CSF).

Metabolite	Normal concentration range (μmol/L)	Abnormal concentration range (μmol/L)	Condition associated with abnormal concentration range
3-methoxy-4-hydroxyphenylglycol	0.0500 (0.00580–0.0942)		
5-hydroxylindoleacetic acid	0.093 (0.059–0.127)	0.124 (0.081–0.167)	Depression
5-methyltetrahydrofolate	0.0746	0.0536	Rett syndrome
Acetic acid	2280 (50–4500)		
Acetoacetate	284 (161–407)	322 (240–404)	Bacterial meningitis
Acetone	67.1 (43.1–91.2)		
Adenosine	>10 (NMR)		
Adrenaline	0.0346 (0.00633–0.0628)		
Alanine	27 (10–44)	192 (161–223)	Tuberculous
Alpha-Aminobutyric acid	3.33 (1.8–4.86)		
Alpha-hydroxy-n-butyrate	>10 (NMR)		
Alpha-oxoalutarate	>10 (NMR)		
Arginine	20.5 (15.7–25.3)		
Aspartate	219 (0–482)		
Beta-galactose	>10 (NMR)		
Beta-Hydroxybutyrate	286 (207–365)	430 (359–501)	Bacterial meningitis
Bilirubin	>10 (NMR)		
Cholesterol	8.32 (7.88–8.76)		
Choline	1.82 (0.28–3.36)		
Citric acid	370 (110–630)	2400	Canavan disease
Citrulline	2.62 (1.35–3.89)		
Creatine	127 (108–146)	166 (156–176)	Bacterial meningitis
Cystine	29 (2–56)		
Dimethyl amine	>10 (NMR)		
Dimethyl sulfone	11.3 (5.1–17.5)		
Dimethylamine	>10 (NMR)		
Dopamine	0.00209 (0.0010–0.0043)	0.00797–0.0118	Parkinsons disease
Ethanolamine	0.843 (0.262–1.424)		
Formate	>10 (NMR)		
Fumarate	>10 (NMR)		
Gamma-aminobutyric acid	>10 (NMR)		
Gamma-aminobutyric acid (GABA)	>10 (NMR)		
Glucose	1720 (1560–1880)		
Glutamate	<150		

(*Continued*)

TABLE 10.3. *(Continued)*

Metabolite	Normal concentration range (μmol/L)	Abnormal concentration range (μmol/L)	Condition associated with abnormal concentration range
Glutamine	627 (482–772)	Increased	Aneurysmal subarachnoid haemorrhage
Glycerol	>10 (NMR)		
Glycerophosphocholine	3.94 (1.60–6.28)	6.95 (3.70–10.2)	Alzheimers disease
Glycine	8.3 (5.9–10.7)		
Histidine	<30	130	Histidinemia
Homovanillic acid	0.20 (0.047–0.35)		
Indoxyl sulphate	>10 (NMR)		
Isoleucine	5.8 (3.3–8.2)		
Kynurenic acid	0.0019 (0.0017–0.0021)		
Lactic acid	3000 (1850–4150)	Increased	Subarachnoid haemorrhage
Leucine	13.5 (8.82–18.2)		
Lysine	23.9 (18.1–29.7)		
Methionine	4.07 (3.00–5.14)		
Myo-inositol	>0.01 (NMR Spectroscopy)		
N-Acetylaspartic acid	0.00	0.380	Canavan disease
Noradrenaline	0.0205 (0.00963–0.0314)		
Ornithine	4.87 (3.48–6.26)		
Oxaloacetate	>0.01 (NMR Spectroscopy)		
Phenylalanine	10.4 (7.58–13.2)		
Phosphocholine	1.42 (0.71–2.13)	2.16 (1.32–3.0)	Alzheimers disease
Pyruvate	153 (121–185)	195 (175–215)	Bacterial meningitis
Serine	28.9 (20.7–37.0)		
Serotonin	0.00125 (0.000624–0.00187)	0.00448–0.00933	Parkinsons disease
Succinic Acid	2.5 (0–5.0)	19.0	Canavan disease
Taurine	8.24 (5.48–11.0)	6.49 (4.6–8.38)	Parkinsons disease
Threonine	32 (4–60)		
Trimethyl amine	>10 (NMR)		
Trimethylamine-*N*-oxide	>10 (NMR)		
Tyrosine	10.1 (6.37–13.8)		
Uracil	>10 (NMR)		
Urea	1060 (820–1300)	1800 (1500–2100)	Tuberculosis
Valine	20 (10–30)		

A more complete version of this table, with references is available at www.hmdb.ca.

10.3.4 Working with Cells and Tissues

A particular challenge in mammalian metabolomics is the analysis or characterization of the intracellular metabolome. As a rule it is not as easy to get tissues from an animal as from a plant or a microbe. Certainly the acquisition of tissues from living humans is difficult and must be done in close coordination with surgeons doing "biopsies for cause" or surgical removal of tumors. As with any human body substance, ethics approval must be applied for and received, and appropriate containment (level-2) procedures must be in place. For non-human or non-primate tissues, the requirements are obviously not so rigorous, and the containment requirements are usually only at level-1. Nevertheless, even for animals, surgical procedures are still required, and appropriate ethics approvals must be obtained. An alternative, noninvasive approach to mammalian metabolomics is to analyze metabolites from mammalian cell cultures (Takesada et al., 2000; Farkas and Tannenbaum, 2005). This approach certainly avoids the problems of tissue extravisation and preservation. It also simplifies the extraction of metabolites by eliminating the presence of adipose tissues, connective tissue, and cartilage that make tissue extraction so difficult. However, cell cultures are neither organisms nor organs, and it is likely that the metabolism of clonal, immortalized cells is somewhat different from what goes on in most mammals. Likewise, metabolite contaminants from the growth media can confound the interpretation of cell culture results. As a result, the metabolomics of cell cultures can only serve as a proxy of what really goes on in a living animal.

Regardless of whether one uses cell cultures or biopsied tissue, a critical component of working with these samples is finding ways to rapidly quench metabolic processes after isolation or extravisation. The removal of tissues from living animals or the extraction of cells from an incubator induces considerable metabolic stress, leading to the rapid appearance of potentially confounding stress metabolites (lactate, acetate, creatinine, TMAO). The best way to rapidly quench metabolism is to snap-freeze the material in liquid nitrogen—typically within a minute or two of removal or isolation. Once frozen, the material can then be processed or extracted using a variety of mechanical or solvent-based techniques. Frozen tissues or cells can be processed by quickly grinding them into a powder using a mortar and pestle. Once the tissue or cell sample is powdered, the metabolites may be extracted into polar (methanol, water) and nonpolar (chloroform, hexane, ethyl acetate) solvents followed by removal of the cellular residue by centrifugation.

The key requirements of a solvent extraction technique are that it is efficient, it produces a high total tissue metabolite yield, and it does so with low variability. Perchloric acid extraction (cold 12% perchloric acid, sonication, centrifugation, and neutralization with NaOH) has long been used in tissue work as it seems to fulfill these criteria, at least for water-soluble metabolites (Le Belle et al., 2002). Methanol/chloroform (M/C) extractions are largely reserved for extracting hydrophobic metabolites. Recently, it has been shown that a single M/C extraction can be performed on mammalian cells that yield better results for both lipid and water soluble metabolites than perchloric acid (PCA) extraction (Le Belle et al., 2002).

In this protocol, methanol and chloroform (4°C) in a ratio of 2:1 (v/v) are added to either frozen ground tissue or frozen cell pellets. After the solvent – tissue mixture is allowed to thaw, it is sonicated (30 s). After approximately 15 min in contact with the first solvents, chloroform and distilled water (1:1 v/v) are added to the samples, thereby forming an emulsion. The samples are then centrifuged (13,000 rpm for 20 min) and the upper phase (methanol/water) separated from the lower (organic) phase. The protein pellet can be re-extracted using methanol/chloroform (1:1) to pull off any remaining metabolites. The water-soluble fractions are pooled separately from the organic fractions and dried by speed-vac, rotoray evaporation, or via dry nitrogen passage. NMR studies of the water-soluble and lipid-soluble metabolites generated in this way show that this simple method is superior to both PCA extraction alone, and PCA extraction followed by lipid extraction, with metabolite yields being 50–100% greater and sample-to-sample variations being 2–3 times smaller.

Of course, not all tissues or cell samples need to be extracted. Some analytical techniques such as NMR, MRI (magnetic resonance imaging), and MRM (magnetic resonance microscopy) allow metabolites to be identified and quantified directly from whole animals, organs, or cell cultures without the need for dissection, or any further tissue processing (Takesada et al., 2000; van der Graaf et al, 2004; Kaiser et al., 2005). Furthermore, very high-resolution NMR spectra of solid tissues and organs can be obtained using magic angle sample spinning (MAS). In conventional NMR, liquids are the preferred substrate as the analysis of a solid or semisolid sample (such as an organ or tissue) results in very broad lines and loss of spectral resolution due to sample inhomogeneity and dipolar coupling. In MAS–NMR, the sample is spun very quickly (600,000 rpm) at an angle of 54.7° (the so-called "magic angle") relative to the magnetic field. This rapid spinning at this precise angle has the effect of reducing dipolar coupling effects and narrowing the broad lines found in these samples. MAS–NMR has been used to metabolically characterize tumors and has permitted the identification of fucose as an important cancer biomarker (Smith and Baert, 2003).

10.4 SAMPLE ANALYSIS

In the previous section, we highlighted some of the key issues associated with working on biological samples obtained from mammals. We also described a number of techniques or protocols that permit the extraction or "matrix simplification" of blood, urine, CSF, and tissues. These extraction processes are relatively generic, at least for mammalian systems, and often serve as a necessary first step before most biological samples can be analyzed further. In the following section, we will describe additional sample processing steps that are more specific to certain types of instrumentation. We will also describe some of the associated data processing methodologies as well as the strengths and limitations of these technologies with reference to analyzing three important biofluids: urine, plasma, and CSF. While there are many analytical technologies now used in mammalian metabolomics (CE, FTIR,

IMS, electrochemistry), this section is limited to describing GC–MS, LC–MS, and NMR methods only.

10.4.1 GC–MS Analysis of Urine, Plasma, and CSF

The application of GC–MS to human metabolic characterization dates back to 1966, with the discovery of a case of valeric academia, an inborn error of organic acid metabolism, by Dr. Kay Tanaka (Tanaka et al., 1966). Since then, GC–MS has become a mainstay of many clinical chemistry and metabolic laboratories studying metabolic disorders of organic acids (Matsumoto and Kuhara, 1996; Kuhara, 2005). While primarily restricted to characterizing organic acids in blood and urine, GC–MS has recently been shown to be amenable to monitoring amino acids, nucleic acids, sugars, amines, and alcohols (Matsumoto and Kuhara, 1996). Relative to other separation techniques, gas chromatography is almost unmatched in its separation resolution (as measured by plate count) and reproducibility. In gas chromatography, chemically modified analytes are separated in the gas phase at temperatures of up to 300°C and detected by a mass spectrometer. The combination of the time taken by the analyte to travel the GC column (called retention time or RI) and the molecular weight information acquired from the mass spectrometer allows many compounds to be uniquely and rapidly identified. Specifically, in GC–MS, metabolite identification is performed by comparing GC retention times with known compounds or by comparing against pregenerated retention index/mass spectral library databases. The identification process can be facilitated by the use of freely available GC deconvolution software such as AMDIS (http://chemdata.nist.gov/mass-spc/amdis/), or commercial tools such as ChromaToF that support GC peak detection, peak area calculation, and mass spectral deconvolution. In gas chromatography, metabolites can be classified into two groups—volatile metabolites not requiring chemical derivatization and nonvolatile metabolites requiring chemical derivatization. Volatile metabolites include small organic amines (trimethyl, dimethyl, and methylamine) and small alcohols and ketones (ethanol, acetone). However, the majority of metabolites of interest are nonvolatile, including most organic acids, amino acids, and sugars. Chemical derivatization of these compounds is used to induce volatility and enhance thermal stability. The typical limit of sensitivity for GC–MS is in the high nM to low μM range.

The most widespread use of GC–MS in mammalian metabolomics continues to be the measurement of organic acids in blood, CSF, or urine. When measuring these acids in urine, plasma, or cerebrospinal fluid, either solvent extraction or ion-exchange chromatography should be used prior to GC–MS analysis. In solvent extraction, the biofluid is made acidic ($pH < 1$) through the addition of concentrated HCl (1:10 ratio of 6 M HCl to biofluid). To facilitate extraction, sodium chloride (in a 1:1 ratio) is usually added to the acidified solution. The organic acids can then be extracted by mixing the solution with ethyl acetate (using a 2:1 ratio of ethyl acetate to biofluid) for 5 to 10 min. After centrifugation, the organic layer, which contains the organic acids, can be separated and the ethyl acetate evaporated under reduced pressure. Solvent extraction is quick and easy, but quantification is often inaccurate because of interference from numerous endogenous components (urea, amino acids, creatinine) at acidic pH. Typically better results are obtained using ion-exchange methods, followed by solvent extraction (Verhaeghe et al., 1988). This gives better specific isolation from urinary components than solvent extraction. Both anionic- and cation-exchange methods can be used; however, a disadvantage of the anion-exchange method is that certain amino acids, which are co-eluted, tend to mask a number of important organic acids on GC chromatograms. Generally, cation-exchange columns using preconditioned Dowex resin (a strong cation exchanger) appear to offer the best results (Suh et al., 1997). Once the cation-exchange column step is completed, the sample is pH adjusted (to pH 3) to neutralize the negative charges of any anions and is typically solvent extracted and dried down as described above.

Once the dried material is obtained, it is derivatized by trimethylsilylation. This process volatilizes the compounds by replacing the hydrogens on polar functional groups with less polar trimethylsilyl (TMS) groups. This chemical substitution greatly reduces the dipole–dipole interactions allowing greater thermal volatility of the compounds. Typically, derivatization proceeds by dissolving the material of interest in a small amount (typically 50 μl) of a TMS reagent mixture consisting of *N*-methyl-*N*-trimethylsilyltrifluoroacetamide (MSTFA) and 1% trimethylsilyl chloride (TMS-Cl). By heating the mixture to 60°C for 15 min, the derivatization reaction is completed and the sample can be readily injected into the GC–MS system. Quantification of the organic acids is performed by comparing the signal intensities to internal standards, including isotopic analogs.

Recently, several GC–MS approaches have been described which permit "planar" or nontargeted analysis of a wide range of metabolites including organic acids, amino acids, nucleic acids, and sugars from either urine (Matsumoto and Kuhara, 1996) or blood (Andreasen and Blennow, 2005). Briefly, GC–MS metabolome analysis of urine involves four basic steps: urease treatment, ethanolic deproteinization, evaporation, and trimethylsilylation. The method is sensitive enough, such that dried urine specimens spotted on filter paper may be used. In this method, urine samples (100 μl) are incubated with urease for 10 min to remove urea. Because urea is, by far and away, the most abundant compound in urine, its presence can frequently mask the presence of other compounds. After urease treatment, the sample is then spiked with small amounts of isotopically labeled (deuterated) amino acids and organic acids, and then deproteinized with ethanol (added in a 9:1 ratio). The sample is centrifuged to remove any precipitate and evaporated to dryness. Once dried, the residue can be trimethylsilylated with 0.1 ml of BSTFA and TMCS (10:1) for 30 min at 80°C. This method permits the routine detection of more than 50 different metabolites from urine including many organic acids, most amino acids, sugars (galactose, galactitol), and some bases (uracil).

The use of GC–MS in the nontargeted or "planar" analysis of plasma samples is a little more complicated than for CSF and urine. Several protocols have been described, with the following being perhaps the simplest (Andreasen and Blennow, 2005). In this process, blood plasma is obtained by centrifuging EDTA anticoagulated blood at 1600 g for 10 min at 4°C. The blood plasma is then extracted or deproteinized using a mixture of plasma:organic solvent in a ratio of 1:9. The organic solvent is a mixture of methanol and water (8:1 v/v) containing all the internal (isotopic) standards. This organic extraction step precipitates the serum proteins, which may be separated by centrifugation. A 200 μl aliquot of the supernatant is then transferred to a GC/MS vial and evaporated to dryness. Prior to GC/MS analysis, the samples are methoxymated at room temperature for 16 h (with 30 μl of 15 mg/mL methoxyamine in pyridine) and trimethylsilylated with 30 μl of MSTFA with 1% TMS–Cl for 1 h. The method allows the resolution of up to 500 different components in blood plasma with concentrations as low as 100 nM. The method has been used to identify more than 80 compounds in serum including most amino acids, several sugars (glucose, fructose, sucrose), many organic acids, phosphorylated compounds (phyrophsophate, glycerophosphate), fatty acids (stearate, oleate), and even cholesterol.

GC–MS is still very popular in many clinical chemistry applications and metabolite profiling efforts. However, GC–MS is limited in its mass range (i.e., higher molecular weight compounds cannot be analyzed) and it is not easily applied to nonvolatile, nonderivatizable, thermo-labile metabolites such as sugars, vitamins, hormones, or phosphoylated metabolites. This introduces a selective bias in the metabolites typically reported by GC–MS analyses. The requirement for sample derivatization also makes the process time consuming as some reactions require up to 3 h to complete. Likewise, the stability of derivatized samples can be an issue as silylation can be easily reversed in the presence of water. Ideally, samples should be well dried and analyzed rapidly after derivatization. Even when these steps are carefully followed, there is always some sample degradation which is typically manifested by extra peaks in the ion current chromatogram. GC–MS is also limited in its scope for metabolite discovery. The identification of new or previously unexpected metabolites is difficult by conventional GC–MS because of the requirement for chemical modification, leading to unknown or unknowable chemical derivatives of the parent compound.

10.4.2 LC–MS Analysis of Urine, Blood, and CSF

Given the limitations of GC–MS and the rapid technological improvements occurring in LC–MS, there is a growing interest in using LC–MS or LC–MS/MS in both clinical chemistry and mammalian metabolome analysis (Dunn et al., 2005; Wilson et al., 2005). While liquid chromatography (LC) or high pressure liquid chromatography (HPLC) does not offer the resolution of gas chromatography, a key advantage of LC is the fact that chemical derivatization is not required making sample preparation and analysis relatively simpler. Furthermore, with LC systems nonvolatile as well as thermolabile metabolites can be directly detected and measured. The principles of metabolite identification for LC–MS are similar to those of GC–MS, with identifications being made on the basis of comparisons against elution time and molecular weight to libraries of known reference compounds. Generally, lower resolution spectrometers (single quadrupole or ion trap instruments) may not provide sufficient mass precision to positively identify many compounds from their parent ion masses. However, higher resolution MS analyzers such as TOF and Fourier transform (FT–MS) instruments can allow exact masses to be determined and permit the calculation of definitive molecular formulae (Brown et al., 2005). Further, the use of MS/MS, FT–MS, or certain kinds of ion trap mass spectrometers allows metabolites to be more firmly identified on the basis of their chemical structure as derived from their parent ion fragmentation patterns. MS/MS is also able to distinguish between chemical isomers because most isomers follow different fragmentation pathways yielding different product ions with different product intensities.

Till date, most LC–MS studies have been limited to somewhat targeted analyses, as opposed to nontargeted analyses of metabolites. This is because the chromatographic resolution of most unprocessed biofluids by HPLC is not particularly good, leading to analyte coelution, ion suppression, in-source fragmentation, and adduct formation. The relatively poor reproducibility of HPLC retention times (due to column, solvent, and instrument variations) relative to GC retention times also makes the use of reference HPLC retention indices for metabolite identification difficult or impractical. In short, the key limitation in LC–MS for metabolomics is not the MS component, but the liquid chromatography component.

Today, most metabolite separations are performed on C_{18} reversed-phase columns with volatile carrier solvents such as acetonitrile, methanol, or water. C_{18} columns, although offering excellent resolution for hydrophobic metabolites, are not particularly good for the separation of hydrophilic metabolites which typically come off in the void volume. Other studies have shown that the use of weak ion exchange columns or mixed mode "metabonomics" columns can permit the separation of sugars, nucleosides, and hydrophilic amino acids (Dunn et al., 2005; Wilson et al., 2005). Given the good separation of hydrophobic components with reversed-phase columns and the moderately good separation seen with ion exchange or mixed-mode columns, it stands to reason that the tandem coupling of two or more different column types together would lead to much better separations. Indeed, over the past few years several papers have been published showing the efficacy of multidimensional or 2D-HPLC separations for both urine and plasma (Guttman et al., 2004; Wilson et al., 2005). The quality and resolution of LC separations of complex metabolite mixtures can be further improved if the column-internal diameter and particle size can be decreased. Hence, the use of microbore or capillary HPLC columns can significantly enhance the resolution (up to 3X) and increase the sensitivity (Wilson et al., 2005). These columns limit diffusive band broadening which, in turn, increases signal-to-noise ratio. More recently, the introduction of ultrahigh pressure liquid chromatography (UPLC) that uses much smaller particle sizes than HPLC columns has been shown to improve resolution even further and shorten the separation time by a factor of 5 or 10. In fact, it is possible to generate UPLC chromatograms with up to 10,000 MS detectable peaks from urine or serum samples (Plumb et al., 2005; Wilson et al., 2005; Dunn et al., 2005).

While many different HPLC separation protocols exist for targeted metabolite separation, it is unlikely that any single protocol or single column will emerge which can be applied to nontargeted metabolite separation. Following is an example of a typical HPLC–MS protocol that would be applied to urinalysis. In this procedure

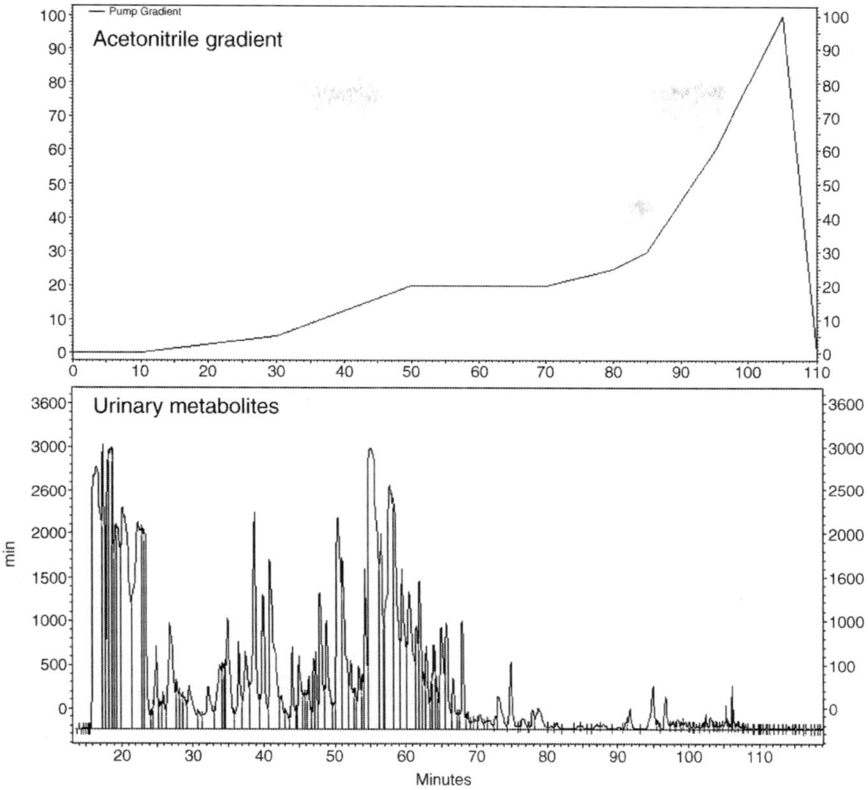

Figure 10.2 An example of an HPLC chromatogram showing the separation of urine on a 250 × 10 mm, 5 μm, Gemini C_{18} column, using a complex AcN gradient (mobile phases: A, 0.1% TFA in water, B, 0.1% TFA in acetonitrile).

0.1% formic acid is added to both the aqueous and organic (acetonitrile) mobile phases prior to separation. Typically, a 10 μl aliquot of urine is injected into an analytical C_{18} HPLC column. A linear gradient of 0.1% aqueous formic acid to 20% AcN is run over a period of 0.5–4 min followed by an increase in the AcN content to 95% over the period of 4–8 min. The 95% AcN level is run for an additional minute and then the column returned to its starting conditions. The separation achieved with this protocol may lead to 20–30 distinct peaks, with similar results expected for deproteinized serum or CSF. A more complex protocol for urinary compound separation is shown in Figure 2. This method uses several more gradient changes over a longer period of time, yielding a much better separation.

In LC–MS, the eluent from these LC runs must then be analyzed using both positive and negative ion modes on a conventional soft ionization (electrospray) mass spectrometer. Typically amino acids, amines, sugars, and nucleotide bases are detected in the positive ion mode whereas organic acids are detected in the negative ion mode. The best results are achieved on higher resolution models such as

MS–TOF instruments which permit continuous ion sampling. The total ion current (TIC) from these LC–MS runs will typically show 20–30 resolvable peaks, with each peak containing 30–40 different parent ions having mass ranges between 50 and 850 amu (Wilson et al., 2005). In other words, HPLC–MS methods can yield 1500–2000 unique peaks (not all of which are metabolites) from serum or urine. With continuous sampling of MS/MS instruments, these parent ions may be further fragmented to help positively identify selected metabolites.

After an LC–MS run has been completed, users have two options: either they can attempt to identify and quantify the peaks as is typically done by GC–MS or they can analyze the resulting spectra using chemometric or multivariate statistical methods (Wilson et al., 2005; Idborg-Bjorkman et al., 2003). The difficulty in identifying small molecules by LC–MS or LC–MS/MS lies in the fact that currently there are far fewer and far smaller MS/MS libraries than GC–MS libraries. Furthermore, these MS/MS libraries are somewhat instrument dependent (triple quad vs. ion trap vs. FT–MS). While several such libraries are being built (including one containing 300+ common mammalian metabolites – Liang Li, personal communication), this continues to be a key limitation for mammalian metabolome analysis. Given the current state of affairs, most LC–MS metabolomics studies reported till date rely on chemometric methods (principal component analysis) to assess differences or similarities between control and diseased animals (Wilson et al., 2005; Idborg-Bjorkman et al., 2003; Plumb et al., 2005). These methods do not require identification or quantification of metabolites. However, they do require extremely well controlled sample collection, preparation, and comparison for being effective.

10.4.3 NMR Analysis of CSF, Urine, and Blood

NMR is a high-resolution spectroscopic technique that measures the absorbance of radio frequency radiation by receptive nuclear spins exposed to high magnetic fields. Only certain elements or certain isotopes are NMR sensitive, including hydrogen (^{1}H), carbon (^{13}C), and nitrogen (^{15}N). ^{1}H NMR spectra are characterized by sharp peaks located at different positions (chemical shifts) of differing intensities (representing the number of chemically identical atoms), split into various multiplet patterns (via J-couplings). Each chemical compound has a unique or nearly unique spectral fingerprint defined by the number, intensity, and location of its NMR peaks. This NMR spectra fingerprint is analogous to an MS/MS fingerprint or GC–MS fingerprint. The application of NMR toward metabolic profiling in mammals is not new. Stable isotope tracer work using NMR has been used since the 1970s to determine metabolic fates, fluxes, and pathways of key metabolites (Cohen et al., 1979). More recently, NMR spectroscopy has been used to identify a number of inborn errors of metabolism (Wevers et al., 1994; Hoffmann et al., 1998; Moolenar et al., 2003), to measure lipoprotein (HDL, LDL) content in plasma (Freedman et al., 1998), to classify tumors from cell homogenates (Mountford et al., 2001), and to identify the location and extent of drug-induced organ damage (Nicholson et al., 1999; 2002). Magnetic resonance imaging (MRI) has also been used to map, identify, and monitor the concentration of key metabolites in the brain and muscles (Takanashi et al, 2002).

Among the advantages of NMR over MS-based methods are the fact that it is nondestructive, nonbiased (any compound with protons is detectable), easily quantifiable, requires little or no separation, permits the identification of novel compounds, and needs no chemical derivatization. A key disadvantage of NMR, relative to MS, is the fact that it is about 10–50X less sensitive, with a lower limit of detection of about 1–5 μM and a minimum sample size of ~500 μl. However, with the recent introduction of higher field magnets (900 MHz), cryogenically cooled probes (that reduce thermal noise and increase signal by a factor of three) as well as microprobes equipped to handle very small samples (60 μl), some of these issues of sensitivity are beginning to become less of a concern. Nevertheless, the aforementioned positives and negatives about NMR simply reinforce the view held by many that MS and NMR are complementary technologies, and that both techniques should be used in metabolomics studies.

As noted earlier, one of the key strengths of NMR in metabolomics is that samples from most complex biological fluids do not require chromatographic separation prior to analysis. This is because the chemical shifts of the constituent components effectively separate the metabolites into identifiable peaks. This phenomenon is sometimes called "chemical shift chromatography" (Figure 10.3). As a result, many biological samples, such as urine and CSF can be studied in their raw form, direct from the animal or patient. If necessary, CSF and urine can be extracted or decontaminated using the extraction protocols described earlier (see Sections 10.3.2 and 10.3.3). When using serum or plasma, the sample can be either deproteinized (see Section 10.3.1) or analyzed directly without any extraction. In the latter case, special NMR pulse sequences (CPMG or diffusion editing) can be applied which eliminate the broad resonances arising from the protein and lipoprotein constituents (Daykin et al., 2002; Van et al., 2003). Unfortunately, these spectral editing methods do not permit the level of quantitation accuracy that can be attained using extracted or deproteinized samples.

Normally NMR samples are spiked with 5% D_2O (to serve as a frequently lock signal) and a small amount of a chemical shift reference standard (DSS or TSP, 0.1 mM) that can also serve as a quantitation standard. Occasionally a small amount of imidazole (10 mM) is added to serve as a pH reference and as a second quantitation standard. The NMR spectra of urine, CSF, and plasma are heavily dominated by the water resonance or any contaminating extraction solvents (methanol, chloroform, ethyl acetate, acetonitrile). Normally, the water resonance can be greatly suppressed by the use of simple presaturation methods or more sophisticated WATERGATE or 1D-NOE pulse sequences (Sklenar, 1990; Piotto et al., 1992). The elimination of any contaminating organic solvent peak is usually best done during sample preparation by making sure that the sample is well dried before aqueous reconstitution. However, selective saturation techniques are also available to eliminate organic solvent peaks on the spectrometer (Simpson and Brown, 2005; Prost et al. 2002).

NMR spectra of biofluids can be very complex, with up to 5000 resonances being detectable in certain biofluids such as urine. This spectral complexity has led to the development of two very distinct schools of thought for collecting, processing, and interpreting metabolomics NMR data. In one version (the chemometric or

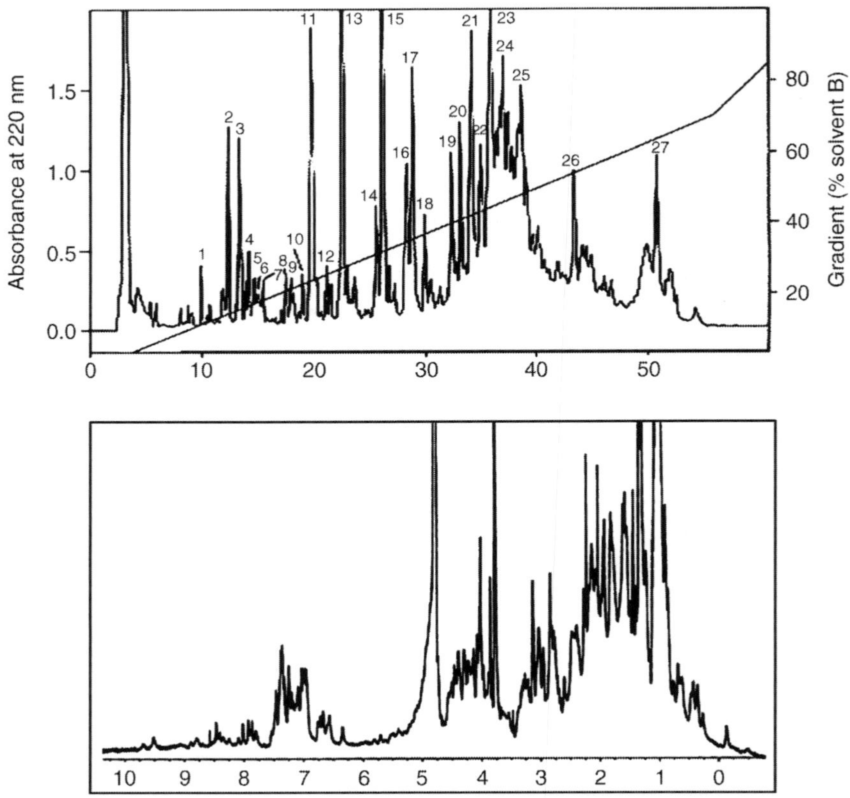

Figure 10.3 The concept of chemical shift chromatography. Just as analytes are separated by retention time on an HPLC chromatogram (top), analytes in NMR can be separated by their chemical shift in an NMR spectrum. The amino acid mixture separated in the HPLC chromatogram above is the same as the amino acid mixture separated in the NMR spectrum below.

metabonomics approach), the compounds are not formally identified—only their spectral patterns and intensities are recorded, compared, and used to make diagnoses or draw conclusions. The chemometric approach is based on computer-aided pattern recognition and sophisticated statistical techniques like principal component analysis (PCA). This method requires that the organisms (rats, mice) or cells be genetically identical and that they be grown, fed, and treated identically for long periods of time to facilitate direct spectral comparison and analysis (Nicholson et al., 1999; Nicholson et al., 2002; Robosky et al., 2005).

In the other approach to NMR-based metabolomics analysis, compounds are actually identified and quantified by comparing the biofluid spectrum of interest with a library of reference spectra of pure compounds (Wishart et al., 2001). This is somewhat similar to the approach historically taken by GC–MS methods and to a much more limited extent, LC–MS methods. For NMR, this particular approach requires

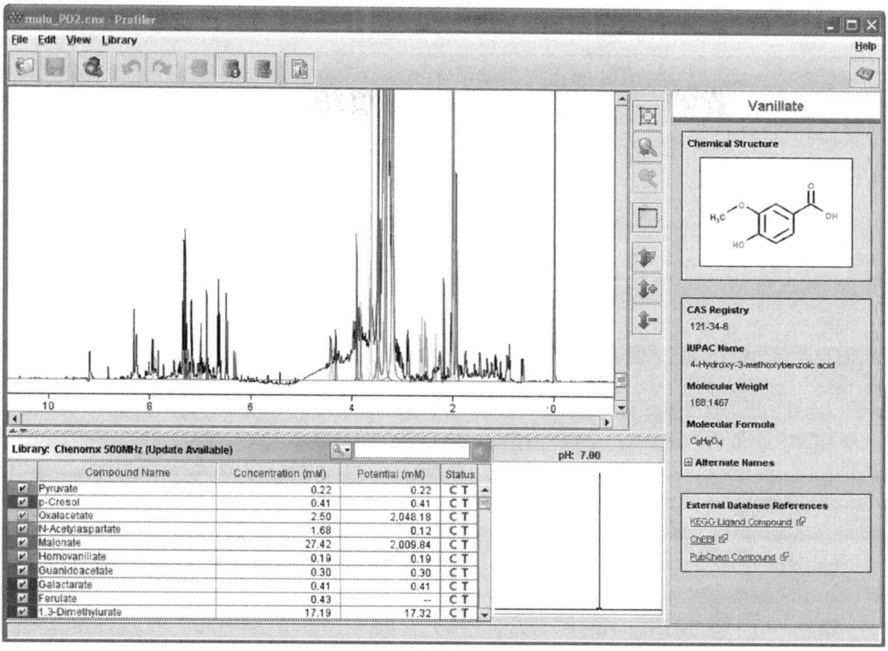

Figure 10.4 Screen shot of a urine NMR spectrum being analyzed by a type of "chemonomic" software, which permits the identification and quantification of metabolites on the basis of comparisons between their chemical shifts and those found in a library of compounds.

that the sample pH be precisely known or precisely controlled. It also requires the use of sophisticated curve-fitting software and specially prepared databases of NMR spectra collected at different pH values and different spectrometer frequencies (400, 500, 600, 700, and 800 MHz). An example of a biofluid spectrum analyzed using this kind of strategy is shown in Figure 4. A key advantage of this "chemonomic" approach is that it does not require the collection of identical sets of cells, tissues, or lab animals, and so it is more amenable to human studies. A key disadvantage of this approach is the relatively limited size of the spectral library (~300 compounds). Such a small library of identifiable compounds may bias metabolite identification and interpretation. Both the chemonomic and chemometric approaches have their advocates. However, it appears that there is a growing trend toward combining the best features of both methods.

10.5 APPLICATIONS

Metabolomics (or metabolic profiling) has been used in many ways to characterize mammalian physiology, genetics, and nutrition. Some of these applications include disease diagnosis, biomarker identification, mutation identification, metabolic state monitoring, organ transplantation, and drug testing (Dunn et al., 2005; Nicholson et al., 2002; Smith and Baert 2003). Describing all of these would easily fill an entire textbook. Nevertheless, of all the applications mentioned so far, perhaps the one that best describes the utility of metabolomics in mammals relates to the characterization of metabolic diseases. Indeed, most of the motivation leads to the establishment of such fields as clinical chemistry, biochemistry, human genetics, and now metabolomics can be traced back to the desire by physicians and scientists to understand metabolic diseases and disorders.

10.5.1 Identification and Classification of Metabolic Disorders

Strictly speaking, metabolic disorders refer to diseases or disorders of the internal body chemistry affecting metabolism or catabolism of lipids, nucleosides, sugars, and amino acids. Metabolic disorders can be either acquired or inherited. Some can be both. Classically, most inherited metabolic disorders are identified as inborn errors of metabolism or IEMs. IEMs are normally defined as diseases of amino acids, organic acids, the urea cycle, galactosemia, primary lactic acidoses, glycogen storage diseases, lysosomal storage diseases, and diseases involving peroxisomal and mitochondrial respiratory chain dysfunction. Some IEMs (such as cystinuria) are relatively milder, and many individuals with these disorders live normal, relatively asymptomatic lives. Certain other IEMs, such as lysosomal and peroxisomal storage diseases are only present in later childhood (Burton, 1998). Still other IEMs such as organic acidemias, urea cycle defects, and certain disorders of amino acid metabolism are typically present with acute life-threatening symptoms in infants within the first 2 weeks of life (Burton, 1998). Although individually rare, IEMs are collectively quite common, with about 0.5–1% of all newborns having some kind of disorder (Applegarth et al., 2000). These patients account for up to 10% of all pediatric admissions to hospitals. Many of these disorders are untreatable, but for those that are for a lifetime, therapy, monitoring, or palliative care can cost upwards of $3 million per patient (Braddock, 2002). The number, complexity, and varied clinical presentation of IEMs have often presented a formidable challenge to practicing physicians. Yet, in many cases, prevention of death or permanent neurologic sequelae in patients with these disorders is dependent on early diagnosis and implementation of appropriate therapy.

IEMs are not the only metabolic disorders of importance. Scientists are increasingly including such acquired metabolic conditions as obesity, diabetes, insulin resistance, Fanconi's syndrome, and malabsorption (celiac disease, lactose intolerance) as metabolic diseases. These acquired or induced metabolic disorders are much more frequent among adults (up to one third of the population) and the incidence of some (especially obesity and diabetes) is growing alarmingly. Indeed, acquired disorders of carbohydrate metabolism are perhaps the most common metabolic disorder in humans. These include diabetes, hypoglycemia, hyperinsulinemia, diabetic ketoacidosis, and hyperosmolar coma. Despite their frequency, the presentation and origin of some of these acquired disorders can often be just as confounding to the physician as are some of the most obscure IEMs.

Many of the clinical chemistry tests shown in Table 1 were developed to help identify and monitor metabolic diseases. However, the small number of compounds routinely scanned in clinical tests (column 1) or measured much less frequently in clinical GC–MS or amino acid analyzers (columns 2 and 3) only cover a tiny fraction of the metabolites that are known to be associated with metabolic disorders (Table 4). This means that only a tiny percentage of known metabolic disorders are capable of being properly diagnosed or monitored using conventional (or targeted) clinical chemistry tests. By contrast, nontargeted metabolomics methods have been shown to be capable of detecting and diagnosing nearly 200 different metabolic disorders (Moolenar et al., 2003; Mueller et al., 2003; Rinaldo et al., 2004; Kuhara et al., 2005). Furthermore, it has been shown that by increasing the number and type of detectable metabolites, the rate of IEM detection can be increased by 2–3X (Rinaldo et al., 2004). This has had a profound, positive effect on the treatment and prognosis of patients with these disorders.

Another positive consequence to nontargeted metabolite detection is the substantial improvement in IEM diagnostic specificity. Many metabolic disorders are present with diffuse and, or nonspecific symptoms, making diagnoses difficult. Single metabolite or point analyses certainly allow some disorders (PKU) to be detected, but as can be seen in Table 10.4, most metabolic disorders are characterized by a complex metabolic profile with several metabolites either significantly reduced or increased relative to normal levels. Obviously, single test analyses could not detect such profiles nor could they offer much more than a qualitative "yes/no" answer about a metabolite's presence. Using such techniques as NMR-based metabolomics, it is now possible to quantify these metabolite levels and provide a much more definitive assessment of the severity or potential severity of a given IEM. As is also evident from Table 10.4, it is not uncommon for very different disorders to share at least one metabolite in common. For instance, both homocystinuria and citrullinemia II share the amino acid methionine as a disease marker. This means that a simple test restricted to the detection of methionine would not be able to distinguish these two disorders. On the contrary, a nontargeted metabolomics approach (using NMR or GC–MS) would easily be able to detect all the necessary metabolites to positively identify the disorder.

The improved sensitivity of many of the newer metabolomics instruments (tandem MS, high field NMR, FT–MS, capillary electrophoresis) along with continuing improvements in the sensitivity of the more traditional instruments (GC–MS, amino acid analyzers) also has an important benefit in the study of metabolic disorders. These improvements are permitting the identification of new IEMs, the detection of asymptomatic IEMs, and the improved characterization of many well known IEMs. Indeed, most new IEMs are being identified by clinical research and testing laboratories employing the latest metabolomics technologies. Unfortunately, the adoption of new technologies in most commercial or medical testing labs often tends to be quite slow. Furthermore, the reimbursement schemes for laboratory testing and the requirement for directed, targeted testing by physicians means that targeted (i.e., point analysis) testing is well entrenched in the medical community. So, while the potential of nontargeted (i.e., metabolomics) testing is enormous and the benefits

TABLE 10.4. Metabolic Disorders (IEMs) and Their Associated Metabolites.

Metabolic disorder	Abnormal metabolites	Reference
2-Hydroxyglutaric aciduria	2-Hydroxyglutaric acid	Moolenar (2003)
2-Ketoadipic 2-aminoadipic aciduria	2-Oxoadipic acid; 2-Hydroxyadipic acid; 2-Aminoadipic acid; 2-Oxoadipic acid	Moolenar (2003)
2-Methyl-3-hydroxybutyryl CoA dehydrogenase deficiency	Tiglyglycine; 2-Methyl-3-hydroxybutyric acid	Moolenar (2003)
3-HMG-CoA lyase deficiency	3-Hydroxy-3-methylglutaric acid; 3-Methyglutaconi acid; 3-Hydroxyisovaleric acid	Matsumoto (1996)
3-Ketothiolase deficiency	2-Methyl-3-hydroxybutyric acid; 2-Methylacetoacetic acid; Tiglyglycine	Matsumoto (1996)
3-Methylcrotonylglcinuria	3-Hydroxyisovaleric acid; 3-Methylcrotonylglycine	Matsumoto (1996)
4-Hydroxybutyric aciduria	4-Hydroxybutyric acid	Moolenar (2003)
Adenosine deaminase deficiency	Deoxyadenosine	Moolenar (2003)
Adenylosuccinate lyase deficiency	SAICA-riboside; S-Adenosine	Moolenar (2003)
Alkaptonuria	Homogentisic acid	Moolenar (2003)
Arginosuccinic aciduria	Arinosuccinic acid; Orotic acid; Orotidine; Uracil	Moolenar (2003)
Aspartylglycosaminuria	N-Aspartylglucosamine	Moolenar (2003)
Beta-mannosidosis	Mannosyl(1-4)-N-acetyglucosamine	Moolenar (2003)
Canavan disease	N-Acetylaspartic acid	Moolenar (2003)
Citrullinemia	N-Acetylcitrulline; Citrulline; Orotic acid; Orotidine; Uracil	Moolenar (2003)
Citrullinemia type II	Methionine; Phenylalanine; Galactose	Rinaldo (2004)
Congenital adrenal hyperplasia	17-hydroxyprogesterone; adrostenedione; cortisol	Rinaldo (2004)
Cystathionine Beta-synthase deficiency	Methionine sulfoxide	Moolenar (2003)
Cystinuria	Cystine; Lysine; Ornithine	Matsumoto (1996)
Dihydropyrimidinase deficiency	5,6-Dihydro-uracil; 5,6-Dihydro-thymine; Thymine; Uracil	Moolenar (2003)
Dihydropyrimidine dehydrogenase deficiency	Thymine; Uracil	Moolenar (2003)
Dimethylglycine dehydrogenase deficiency	*N,N*-Dimethylglycine; Betaine	Moolenar (2003)
Ethylmalonic encephalopathy	Lactic acid; Ethylmalonic acid; C4 and C5 acylcarnitines	Rinaldo (2004)
Galactosemia	Galactose; Galactitol; Galactonic acid	Matsumoto (1996)

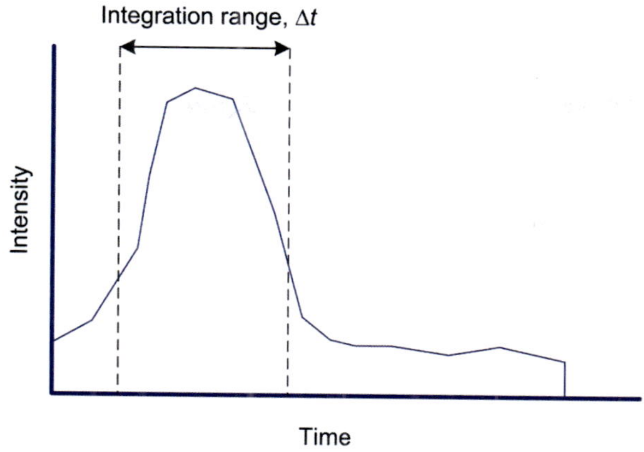

Figure 5.2 Elution profile for the direct infusion ESI-MS. In order to improve the signal to noise ratio only scans within the time injection interval, Δt, is used to calculate a spectrum representing the sample.

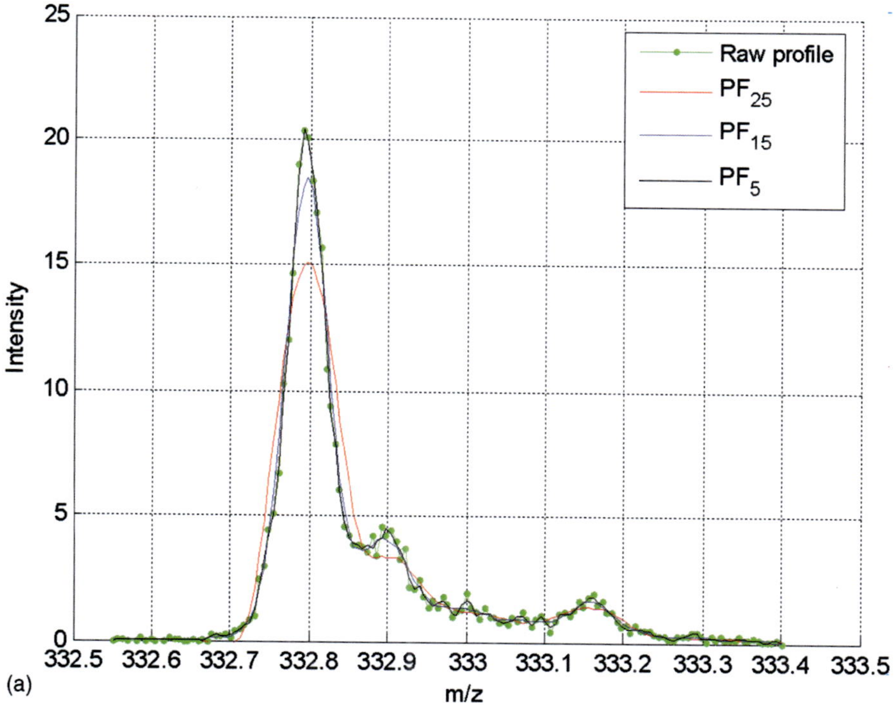

Figure 5.4 (a) Results of a moving average filter for different widths 25, 15, and 5 points. We see that the intensity of the peak is reduced even when a small size window is applied, and skewed when applying larger kernels no matter what size window is used. (b) Results of a polynomial filter of the same MS profile for different widths 25, 15, and 5 points and the polynomial of the order 3. With this filter the filtered profile maintains its shape almost all window sizes except from 25. This indicates that in this example the optimal size of window lies between 15 and 25.

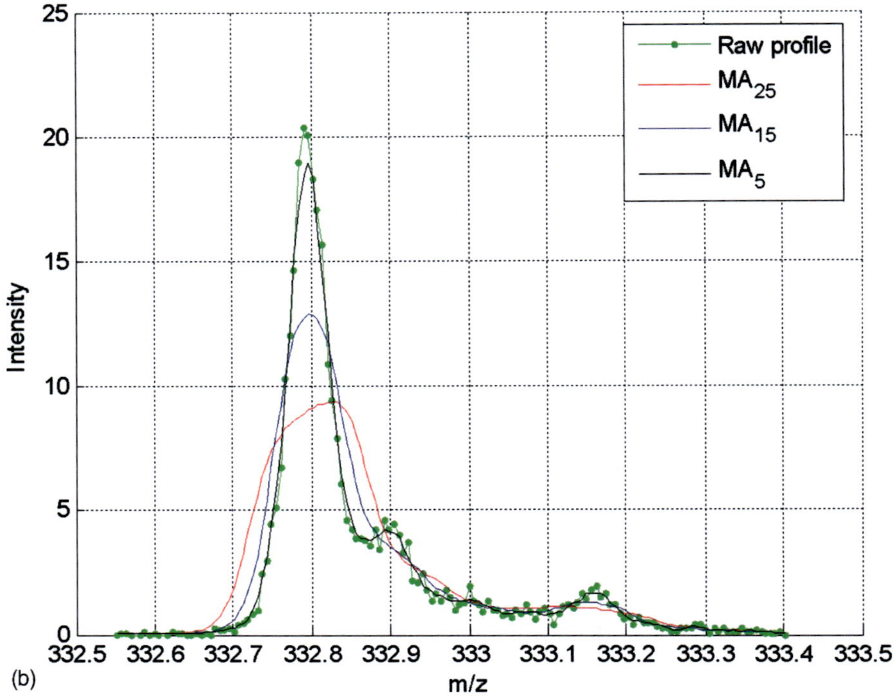

Figure 5.4 (*Continued*)

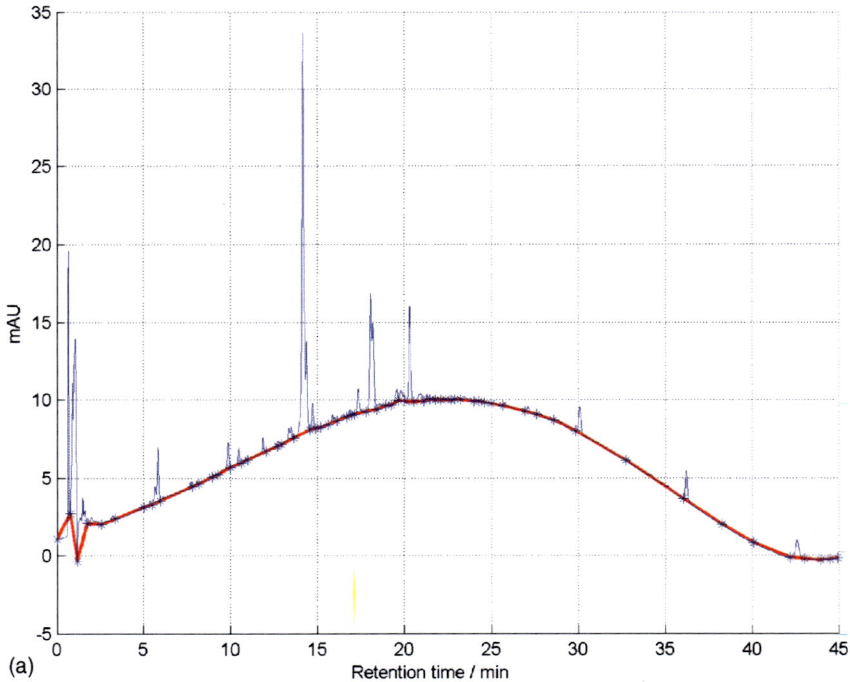

Figure 5.7 Illustration of the piecewise linear baseline correction. Figure (a) shows the chromatogram (blue line) and estimated local minima (marked with '*'). Between the segments defined by these local minima the background is estimated as lines (red line). Figure (b) shows the result after having subtracted the background from the chromatogram.

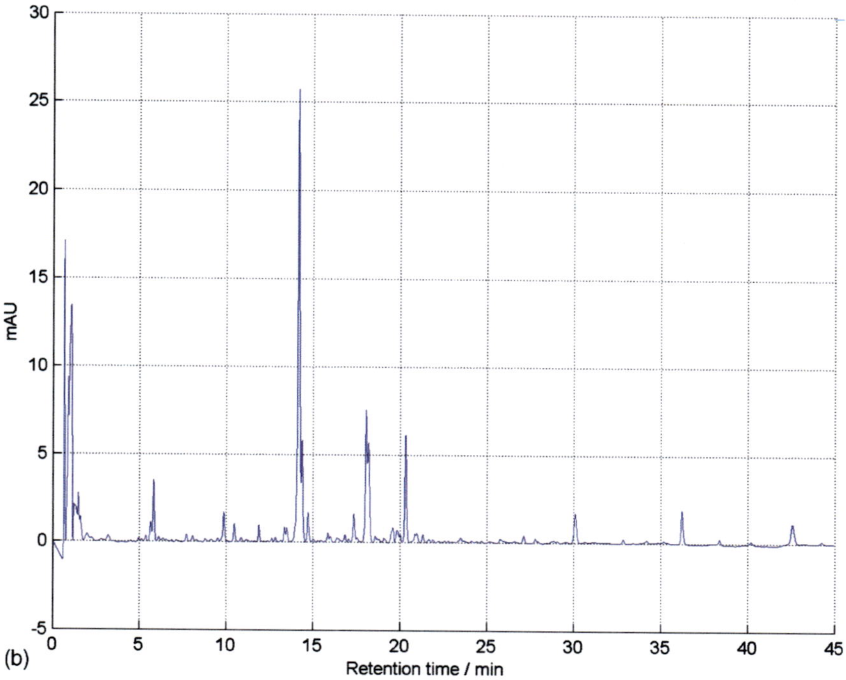

(b)

Figure 5.7 (*Continued*)

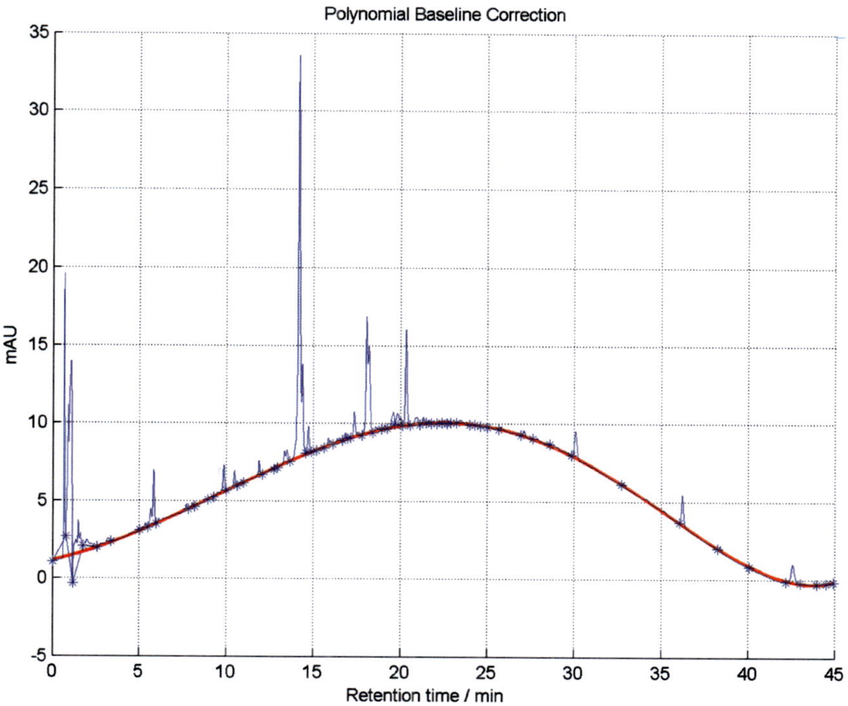

Figure 5.8 Illustration of the polynomial baseline estimation. The figure shows the chromatogram (blue line) and estimated local minima (marked with '*'). The background is estimated from these points as a 5th order ($P = 5$) polynomial.

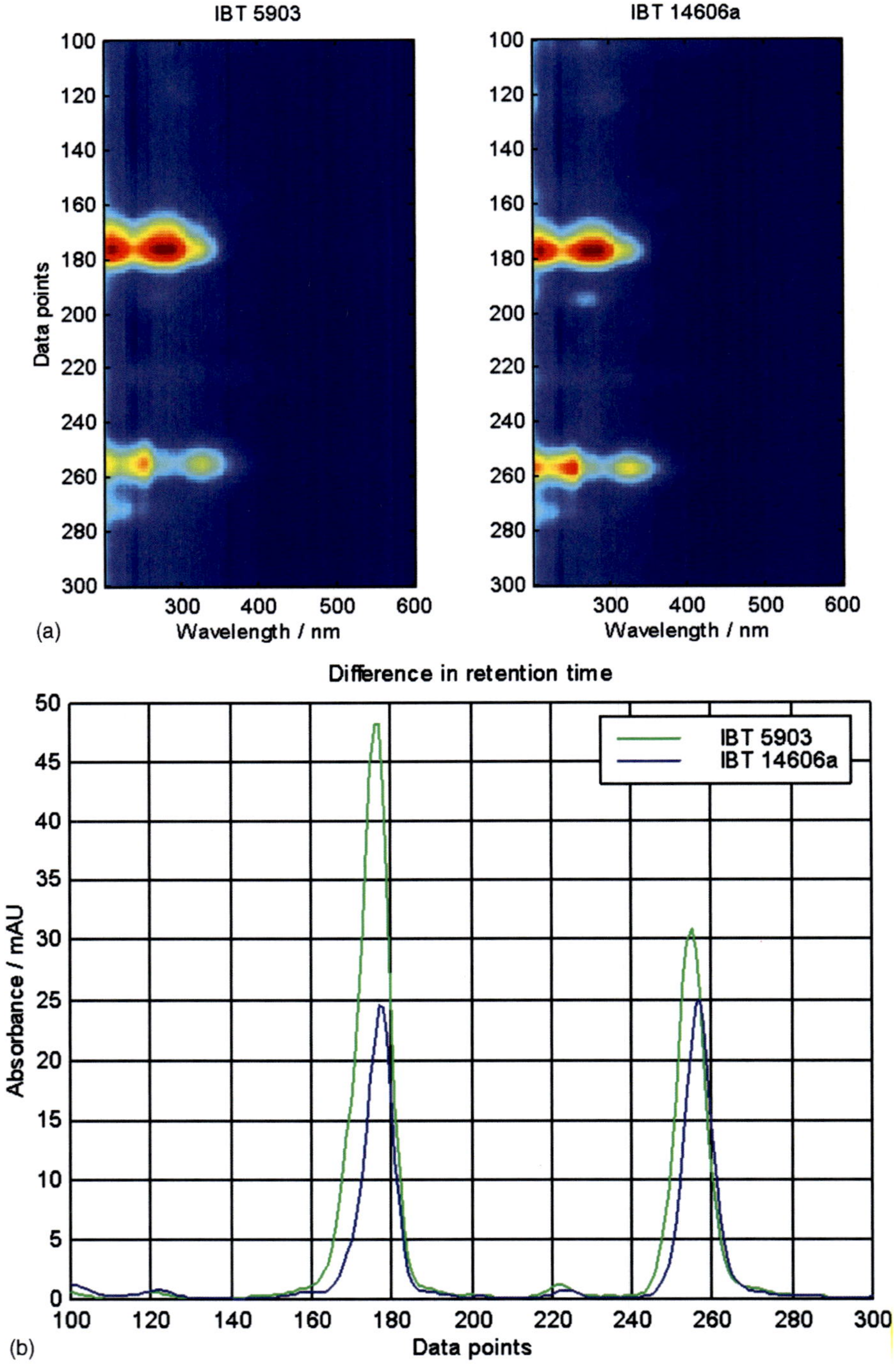

Figure 5.9 Illustration of the problem with shifts in retention time between two HPLC runs. Figure (a) shows a section of the UV absorbance of two complex fungal metabolite extracts containing two peaks. The color illustrate the amount of absorbed light going from low absorbance (blue) to higher absorbance (red). In Figure (b) the two traces along 230 nm are plotted. From the figures we see that there is a significant difference between the peak maxima for the two profiles. It is the aim of the aligning algorithm to correct for these shifts in retention time.

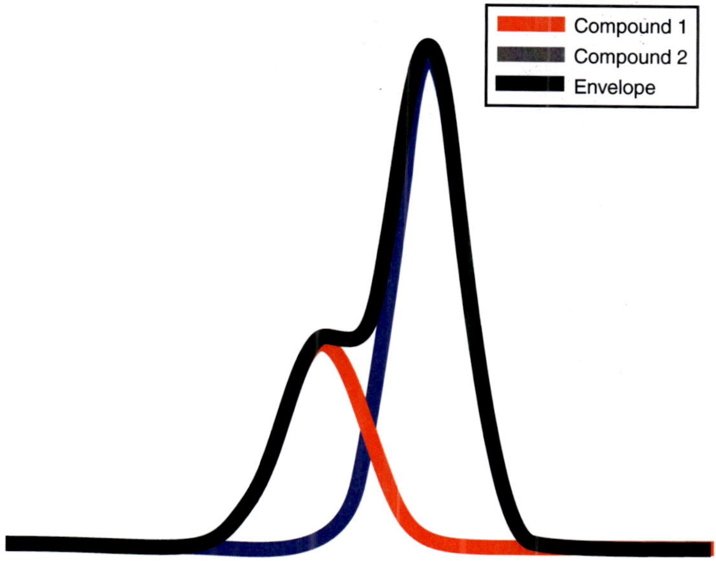

Figure 5.11 Schematic illustration of the deconvolution problem. If two compounds elute at the approximately same time they will overlap and give rise to an "artificial" spectrum being a sum of the two.

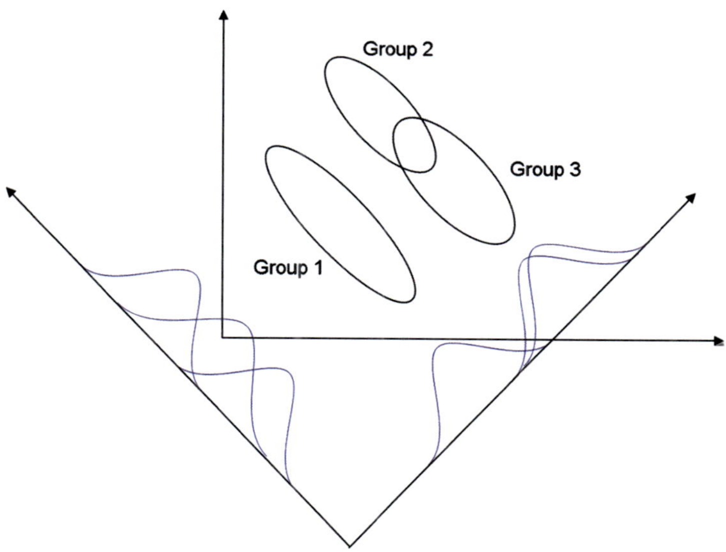

Figure 5.13 Stylized scatter plot for three-group discriminant analysis problem.

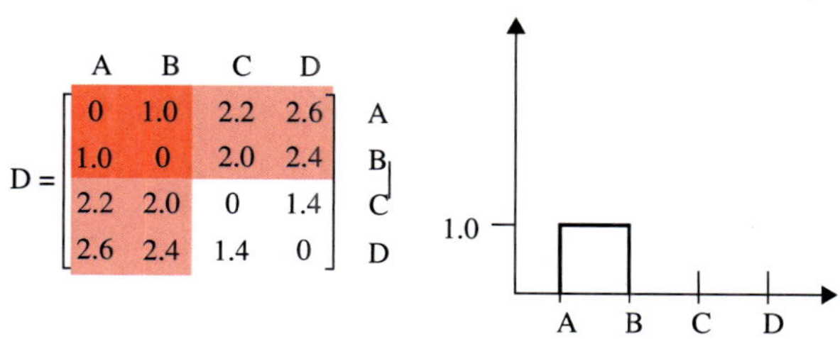

(a)

Figure 5.16 Illustration of the hierarchical clustering method using single linkage.

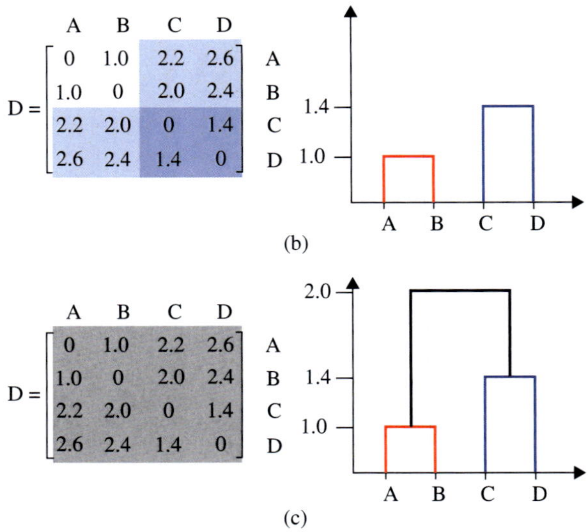

Figure 5.16 (*Continued*)

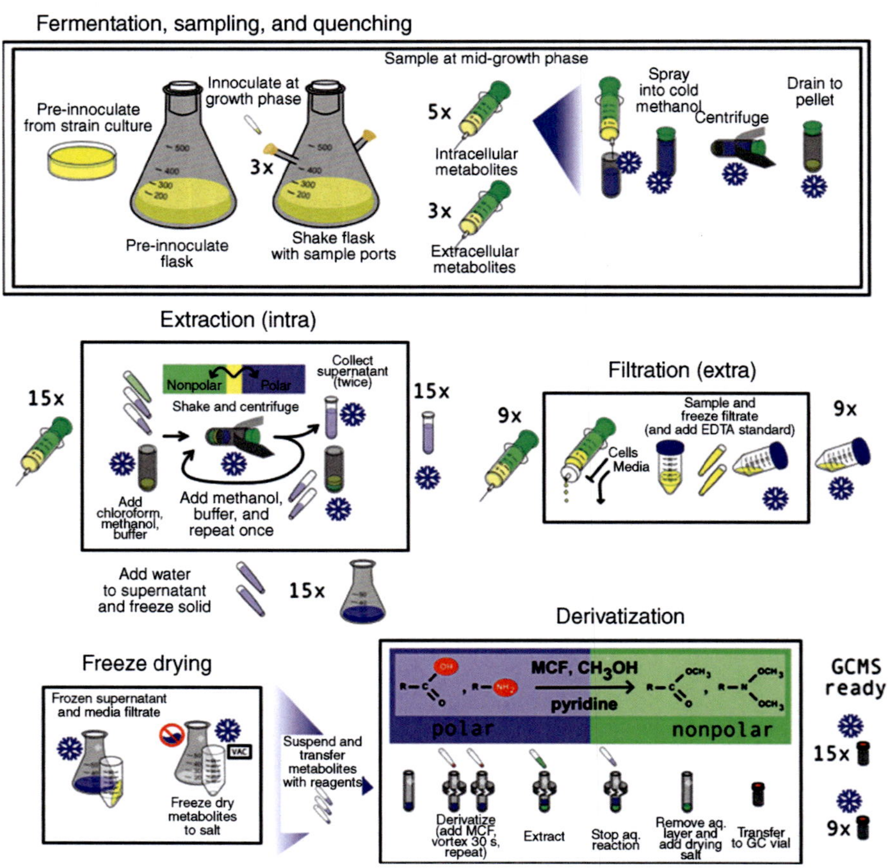

Figure 6.1 Summary of the methodology applied for the analysis of intra- and extracellular metabolites of yeasts according to Villas-Bôas et al. (2005a). Shake flasks were inoculated

Figure 6.1 (*Continued*) from the same pre-inoculum's culture at exponential growth phase. Samples were harvested at mid-exponential phase (O.D.600 nm = 5.0). Five culture suspension samples from each flask were harvested with a disposable syringe and sprayed into a cold methanol solution (−40°C) in order to quench the cellular metabolism. The cell pellets were separated from the extracellular medium by centrifugation at low temperature (−20°C). Additional three samples were harvested and filtered using Millipore membrane (0.45 μm) and the filtrate was stored at −20°C for analysis of extracellular metabolites. The intracellular metabolites were extracted from the cell pellets using a mixture of chloroform, methanol, and buffer at low temperature (−40 to −20°C). The upper polar phase from a three-phase mixture was used for the analysis of intracellular metabolites. Both samples containing intra- and extracellular metabolites were freeze-dried prior to chemical derivatization. Since the intracellular extracts contained large amount of organic solvent, distillated water was added to the samples in order to keep them frozen during the lyophilization process. The dried samples were resuspended in sodium hydroxide solution and derivatized using methylchloroformate (MCF). Fifteen samples of intracellular metabolites and nine of extracellular medium were analyzed for each condition tested. This figure was designed and kindly donated by Joel F. Moxley (Dept. of Chemical Eng./MIT/USA).

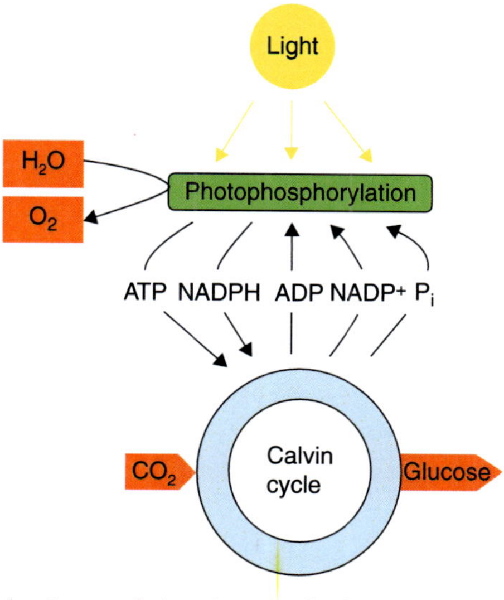

Figure 8.1 Simplified scheme of the photosynthetic process. Light energy is used for photophosphorylation using water, ADP, P_i, and $NADP^+$ producing O_2, ATP, and NADPH. These are further used in the dark reaction (Calvin cycle) for carbon fixation producing glucose.

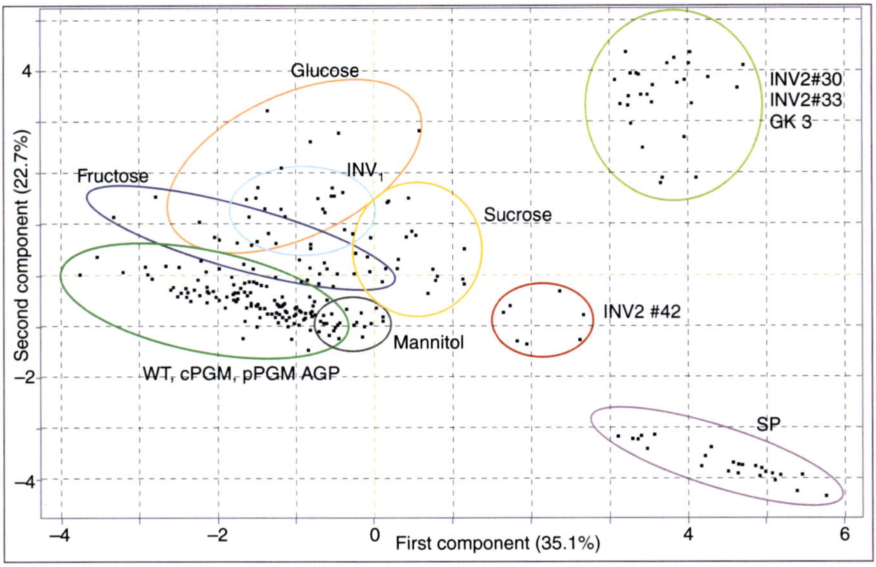

Figure 8.5 Principal component analysis (PCA) of metabolite profiles of environmentally and genetically modified potato tubers (Roessner et al. 2001b). Samples representing wild-type tubers and tubers incubated in buffer alone, plastidial (pPGM) and cytosolic (cPGM) phosphoglucomutase antisense tubers; ADP-glucose pyrophosphorylase (AGP) antisense tubers (dark green circle), mannitol-fed tubers (black circle), fructose-fed tubers (dark blue circle), sucrose-fed tubers (yellow circle), glucose-fed tubers (light red circle), apoplastic invertase (INV1) expressing tubers (light blue circle), cytosolic invertase (INV2) expressing tubers line #30; #33 and cytosolic invertase and glucokinase (GK3) expressing tubers (light green circle), cytosolic invertase (INV2) expressing tubers line #42 (dark red circle), and sucrose phosphorylase (SP) expressing tubers (lilac circle) are marked as described for ease of comparison. PCA Vectors 1 and 2 were chosen for best visualization of differences between experimental treatments and include 57.8% of the information derived from metabolic variances. © American Society of Plant Biologists.

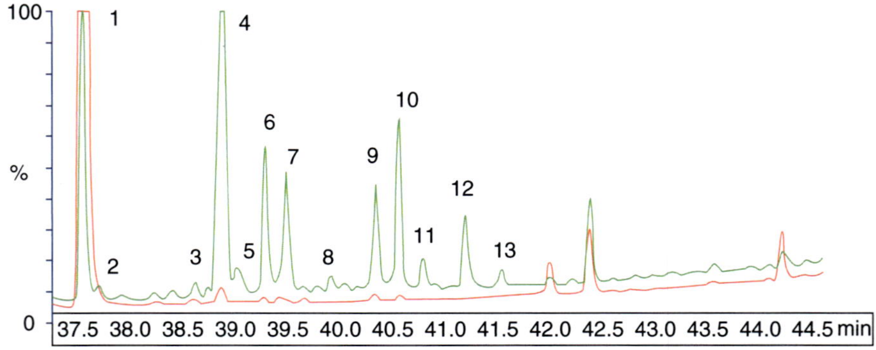

Figure 8.6 Comparison of a specific region of a GC–MS chromatogram of wild-type potato tuber (WT, lower line) compared to tubers expressing a yeast invertase in the cytosol (INV, upper line). 1: sucrose; 3: maltose TMS; 4: maltose MEOX1; 5: trehalose TMS; 6: maltose MEOX2; 7: maltitol TMS; 12: isomaltose MEOX1; 13: isomaltose MEOX2, 2, 8, 9, 10, 11, 14, 15 and 16 are not identified, mass spectra suggest they are sugars or sugar derivatives.

TABLE 10.4. (*Continued*)

Metabolic disorder	Abnormal metabolites	Reference
Glutaric aciduria type I	Glutarc acid; 3-Hydroxyglutaric acid; Glutaconic acid	Matsumoto (1996)
Glutaric aciduria Type II	Glutaric acid; Ethylmalonic acid; Adipic acid; Suberic acid; 2-Hydroxyglutaric acid	Matsumoto (1996)
Glycerol kinase deficiency	Glycerol	Moolenar (2003)
Hawkinsinuria	4-Hydroxycyclohexylacetic acid; Hawkinsin	Moolenar (2003)
Histidinemia	Histidine; *N*-Acetylhistidine	Moolenar (2003)
Homocystinuria	Homocyteine; Methionine; Homocystine	Matsumoto (1996)
Hyperglycemia	Glycine	Matsumoto (1996)
Hyperphenylalaninemia	Phenylalanine	Matsumoto (1996)
Iminoglycinuria	Glycine; Proline; Hydroxyproline	Moolenar (2003)
Isobutyryl-CoA dehydogenase deficience	C4-acylcarnitine	Rinaldo (2004)
Isovaleric academia or Isovleric aciduria	Isovalerylglycine; 3-Hydroxyisovaleric acid	Moolenar (2003)
Isovaleryl-CoA dehydrogenase deficiency	Isovaleric acid; Iso-C5 acylcarnitine	Rinaldo (2004)
Krabbe disease	Galactocerebroside	Rinaldo (2004)
Lactic acidemia	Lactic acid; Alanine	Moolenar (2003)
Lysinuria	Lysine	Matsumoto (1996)
Malonic aciduria	Malonic acid	Moolenar (2003)
Maple syrup urine disease	Leucine; Isoleucine; Valine; 2-Hydroxyisocaproic acid; 2-Hydroxy-3methylvaleric acid; 2-Hydroxyisovaleric acid	Matsumoto (1996)
Medium-chain acyl-CoA dehydrogenase deficiency	Octanoylcarnitine; Hexanoylcarnitine; Decanoylcarnitine; Decenoylcarnitine Hexanoyl-glycine; Suberyl-glycine; Phenylpropionyl-glycine; Cis-4-decenoic acid	Rinaldo (2004)
Methionine adenosyltransferase deficiency	Methionine sulfoxide; Methionine	Moolenar (2003)
Methylmalonic aciduria	Methylmalonic acid; Methylcitric acid	Matsumoto (1996)
Methylmalonic aciduria	Methylmalonic acid; 3-Hydroxypropionic acid	Moolenar (2003)
Mevalonic aciduria	Mevalonic acid; Mevalonolactone	Moolenar (2003)
Molybdenum cofactor deficiency	Xanthine; Hypoxanthine; Uric acid; Sulfite	Moolenar (2003)

(*Continued*)

TABLE 10.4. *(Continued)*

Metabolic disorder	Abnormal metabolites	Reference
Multiple acyl-CoA dehydrogenase deficiency	Cis-4-decenoic acid	Rinaldo (2004)
Multiple carboxylase deficiency	3-Methylcrotonylglycine; Methylcitric acid; 3-Hydroxyisovaleric acid	Matsumoto (1996)
Neuroblastoma	Homovanillic acid; Vanillylmandelic acid	Matsumoto (1996)
Ornithine transcarbamylase deficiency	Orotic acid; Uridine; Uracil	Moolenar (2003)
Oxoprolinuria	5-Oxoproline	Moolenar (2003)
Phenylketonuria	Phenylalanine; Phenyllactic acid; 2-Hydroxyphenylaceitc acid; Phenylpyruvic acid	Matsumoto (1996)
Polyol disease	Arabinitol; Ribotol; Arabinose	Moolenar (2003)
Prolinemia type II	Pyrrole-2-Carboxylglycine; Proline	Moolenar (2003)
Propionic acidemia	Methylcitric acid; Propionylglycine; Tiglyglycine; 3-hydroxy-n-valeric acid; 3-hydroxypropionic acid; 2-Methyl-3-hydroxyvaleric acid	Matsumoto (1996)
Propionic aciduria	Acetona; 3-Hydroxybutyric acid; 3-Hydroxypropionic acid; Acetoacetic acid	Moolenar (2003)
Purine nucleioside phosphorylase deficiency	Inosine; Deoxyinosine; Deoxyguanosine; Guanosine	Moolenar (2003)
Sarcosinemia	Sarcosine	Moolenar (2003)
Short/branched chain acyl-CoA dehydrogenase deficiency	C5-Acylcarnitine; 2-Ethylhydracrylic acid	Rinaldo (2004)
Short-chain acyl-CoA dehydrogenase deficiency	Ethylmalonic acid	Moolenar (2003)
Short-chain acyl-CoA dehydrogenase deficiency	C4-acylcarnitine; Ethylmalonic acid	Rinaldo (2004)
Trimethylaminuria	Trimethylamine *N*-Oxide; Trimethylamine	Moolenar (2003)
Tyrosinemia	Tyrosine; 4-Hydroxyphenyllactic acid; Succinylacetone; 4-Hydroxyphenlpyruvic acid; 4-Hydroxyphenylacetic acid	Matsumoto (1996)
UMP synthase deficiency	Orotic acid; Orotidine; Uracil	Moolenar (2003)
Ureidoproprionase deficiency	3-Ureidopropionic acid; 3-Ureidoisobutyric acid	Moolenar (2003)
Very long-chain acyl-CoA dehydrogenase deficiency	Tetradecenoylcarnitine	Rinaldo (2004)

are clear, it is unlikely that we will see a widespread adoption of metabolomics technology in clinical testing laboratories for quite some time to come.

10.6 FUTURE OUTLOOK

These are exciting times for metabolomics. The field is experiencing a stage of unprecedented growth and excitement. New societies are being established, new journals are appearing on the subject, and major efforts are being made to standardize reporting and data sharing. Likewise, new hardware and new software is being designed, built, and sold by major manufacturers that is specially designed for metabolomics studies. It seems as if new advances are being reported almost every week. However, metabolomics is really at an embryonic stage of development. Indeed, in terms of maturity, it is probably not much different than what genomics was like in the early 1990s. Recall that 15 years ago no living organism had yet been fully sequenced, and the human genome project was only twinkling in a few scientists' eyes. In those early days, we had only wildly incorrect estimates (150,000 vs. 23,000) of the number of genes that might be found in the human genome and a very poor understanding of the complexity of most other genomes. Today the same is true for the human metabolome. We have only best-guess estimates of its size and diversity. Indeed, trying to do human metabolomics today is like trying to do human genetics without the sequence (or even a map) of the human genome! The ironic twist to the situation for metabolomics is the fact that the technology to read metabolite data has effectively jumped far ahead of the knowledge of what those metabolites really are. Therefore the task ahead for metabolomics is quite clear: *we need to complete the human metabolome.* Only by having a list of what constitutes the normal human metabolome can we be in a position to say what is normal and what is abnormal.

Recently, the government of Canada, through a research funding organization called Genome Canada, announced the support for such an undertaking called the Human Metabolome Project (http://www.metabolomics.ca). In these 3 years, $7.5 million project is mandated to identify, quantify, catalog, and store all metabolites that can potentially be found in human tissues and biofluids at concentrations greater than one micromolar. The project is further required to make all these data freely accessible in an electronic format to all researchers through the Human Metabolome Database (www.hmdb.ca). In addition, all compounds synthesized, isolated, or acquired will be made publicly available through the Human Metabolome Library (www.hml.ca). The project itself will employ all the technologies described here, including GC–MS, LC–MS, FT–MS, and NMR and will apply these tools to measure and identify metabolites in urine, blood, CSF, and cell cultures. The project will also depend heavily on using text and data mining tools to track, compile, and consolidate nearly 100 years worth of published metabolite data into a single electronic repository. When the project is completed in early 2008, it is expected that more than 1500 endogenous metabolites and more than 300 exogenous metabolites will be formally identified, and the "normal" concentrations for at least half of these will be known. If and when this goal is reached, then I believe the field of metabolomics will finally have the necessary "legs" to move from a slow walk to a full speed gallop.

REFERENCES

Andreasen N, Blennow K. 2005. CSF biomarkers for mild cognitive impairment and early Alzheimer's disease. Clin Neurol Neurosurg 107:165–173.

Applegarth DA, Toone JR, Lowry RB. 2000. Incidence of inborn errors of metabolism in British Columbia, 1969–1996. Pediatrics 105:E10.

Bamforth FJ, Dorian V, Vallance H, Wishart DS. 1999. Diagnosis of inborn errors of metabolism using 1H NMR spectroscopic analysis of urine. J Inherit Metab Dis 22:297–301.

Bollard ME, Holmes E, Lindon JC, Mitchell SC, Branstetter D, Zhang W, Nicholson JK. 2001. Investigations into biochemical changes due to diurnal variation and estrus cycle in female rats using high-resolution (1)H NMR spectroscopy of urine and pattern recognition. Anal Biochem 295:194–202.

Bollard ME, Stanley EG, Lindon JC, Nicholson JK, Holmes E. 2005. NMR-based metabonomic approaches for evaluating physiological influences on biofluid composition. NMR Biomed 18:143–162.

Braddock DL. 2002. Public financial support for disability at the dawn of the 21st century. Am J Ment Retard 107:478–489.

Brown SC, Kruppa G, Dasseux JL. 2005. Metabolomics applications of FT-ICR mass spectrometry. Mass Spectrom Rev 24:223–231.

Burton BK. 1998. Inborn errors of metabolism in infancy: a guide to diagnosis. Pediatrics 102:E69.

Cohen SM, Ogawa S, Shulman RG. 1979. 13C NMR studies of gluconeogenesis in rat liver cells: utilization of labeled glycerol by cells from euthyroid and hyperthyroid rats. Proc Natl Acad Sci USA 76:1603–1609.

Coley NG. 2004. Medical chemists and the origins of clinical chemistry in Britain (circa 1750–1850). Clin Chem 50:961–972.

Cromwell WC, Otvos JD. 2004. Low-density lipoprotein particle number and risk for cardiovascular disease. Curr Atheroscler Rep 6:381–387.

Daykin CA, Foxall PJ, Connor SC, Lindon JC, Nicholson JK. 2002. The comparison of plasma deproteinization methods for the detection of low-molecular-weight metabolites by (1)H nuclear magnetic resonance spectroscopy. Anal Biochem 304:220–230.

Dickman, SR. 1953. A metabolic cage. Science 1117:284–285.

Dunn WB, Bailey NJ, Johnson HE. 2005. Measuring the metabolome: current analytical technologies. Analyst 130:606–625.

Eckburg PB, Bik EM, Bernstein CN, Purdom E, Dethlefsen L, Sargent M, Gill SR, Nelson KE, Relman DA. 2005. Diversity of the human intestinal microbial flora. Science 308: 1635–1638.

Fandek N, Moreau D, Newell KC, Ofner A. 1995. *Clinical Laboratory Tests: Values and Implications* (2nd edition), Springhouse Press, Springhouse, PA.

Farkas D, Tannenbaum SR. 2005. *In vitro* methods to study chemically-induced hepatotoxicity: a literature review. Curr Drug Metab 6:111–125.

Fauler G, Leis HJ, Huber E, Schellauf C, Kerbl R, Urban C, Gleispach H. 1997. Determination of homovanillic acid and vanillylmandelic acid in neuroblastoma screening by stable isotope dilution GC-MS. J Mass Spectrom 32:507–514.

Forster J, Famili I, Fu P, Palsson BO, Nielsen J. 2003. Genome-scale reconstruction of the Saccharomyces cerevisiae metabolic network. Genome Res 13:244–253.

Freedman DS, Otvos JD, Jeyarajah EJ, Barboriak JJ, Anderson AJ, Walker JA. 1998. Relation of lipoprotein subclasses as measured by proton nuclear magnetic resonance spectroscopy to coronary artery disease. Arterioscler Thromb Vasc Biol 18:1046–1053.

Gates SC, Sweeley CC. 1978. Quantitative metabolic profiling based on gas chromatography. Clin Chem 24:1663–1673.

Guarner F, Malagelada JR. 2003. Gut flora in health and disease. Lancet 361:512–519.

Gupta A, Dwivedi M, Nagana Gowda GA, Ayyagari A, Mahdi AA, Bhandari M, Khetrapal CL. 2005. (1)H NMR spectroscopy in the diagnosis of *Pseudomonas aeruginosa*-induced urinary tract infection. NMR Biomed 18:293–299.

Guttman A, Varoglu M, Khandurina J. 2004. Multidimensional separations in the pharmaceutical arena. Drug Discov Today 9:136–144.

Hall R, Beale M, Fiehn O, Hardy N, Sumner L, Bino R. 2002. Plant metabolomics: the missing link in functional genomics strategies. Plant Cell 14:1437–1440.

Hamamah S, Seguin F, Bujan L, Barthelemy C, Mieusset R, Lansac J. 1998. Quantification by magnetic resonance spectroscopy of metabolites in seminal plasma able to differentiate different forms of azoospermia. Hum Reprod 13:132–135.

Hoffmann G, Aramaki S, Blum-Hoffmann E, Nyhan WL, Sweetmann L. 1989. Quantitative analysis for organic acids in biological samples. Clin Chem 35:587–595.

Hoffmann GF, Surtees RA, Wevers RA. 1998. Cerebrospinal fluid investigations for neurometabolic disorders. Neuropediatrics 29:59–71.

Holmes E, Nicholls AW, Lindon JC, Connor SC, Connelly JC, Haselden JN, Damment, SJ, Spraul M, Neidig P, Nicholson JK. 2000. Chemometric models for toxicity classification based on NMR spectra of biofluids. Chem Res Toxicol 13:471–478.

Idborg-Bjorkman H, Edlund PO, Kvalheim OM, Schuppe-Koistinen I, Jacobsson SP. 2003. Screening of biomarkers in rat urine using LC/electrospray ionization-MS and two-way data analysis. Anal Chem 75:4784–4792.

Jackson M, Mansfield JR, Dolenko B, Somorjai RL, Mantsch HH, Watson PH. 1999. Classification of breast tumors by grade and steroid receptor status using pattern recognition analysis of infrared spectra. Cancer Detect Prev 23:245–253.

Kaiser LG, Schuff N, Cashdollar N, Weiner MW. 2005. Age-related glutamate and glutamine concentration changes in normal human brain: 1H MR spectroscopy study at 4 T. Neurobiol Aging 26:665–672.

Keseler IM, Collado-Vides J, Gama-Castro S, Ingraham J, Paley S, Paulsen IT, Peralta-Gil M, Karp PD. 2005. EcoCyc: a comprehensive database resource for Escherichia coli. Nucleic Acids Res 33(Database issue):D334–337.

Kuhara T. 2005. Gas chromatographic-mass spectrometric urinary metabolome analysis to study mutations of inborn errors of metabolism. Mass Spectrom Rev 24:814–827.

Le Belle JE, Harris NG, Williams SR, Bhakoo KK. 2002. A comparison of cell and tissue extraction techniques using high-resolution 1H-NMR spectroscopy. NMR Biomed 15:37–44.

Leavell MD, Leary JA, Yamasaki R. 2002. Mass spectrometric strategy for the characterization of lipooligosaccharides from *Neisseria gonorrhoeae* 302 using FTICR. J Am Soc Mass Spectrom 13:571–576.

Lenz EM, Bright J, Wilson ID, Hughes A, Morrisson J, Lindberg H, Lockton A. 2004. Metabonomics, dietary influences and cultural differences: a 1H NMR-based study of urine samples obtained from healthy British and Swedish subjects. J Pharm Biomed Anal 36:841–849.

Lutz NW, Maillet S, Nicoli F, Viout P, Cozzone PJ. 1998. Further assignment of resonances in 1H NMR spectra of cerebrospinal fluid (CSF). FEBS Lett 425:345–351.

Matsumoto I, Kuhara T. 1996. A new chemical diagnostic method for inborn errors of metabolism by mass spectrometry–rapid, practical and simultaneous urinary metabolites analysis. Mass Spectrom Rev 15:43–57.

Moolenaar, SH, Engelke UFH, Wevers RA. 2003. Proton nuclear magnetic resonance spectroscopy of body fluids in the field of inborn errors of metabolism. Ann Clin Biochem 40:16–24.

Mountford CE, Somorjai RL, Malycha P, Gluch L, Lean C, Russell P, Barraclough B, Gillett D, Himmelreich U, Dolenko B, Nikulin AE, Smith IC. 2001. Diagnosis and prognosis of breast cancer by magnetic resonance spectroscopy of fine-needle aspirates analysed using a statistical classification strategy. Br J Surg 88:1234–1240.

Mueller P, Schulze A, Schindler I, Ethofer T, Buehrdel P, Ceglarek U. 2003. Validation of an ESI-MS/MS screening method for acylcarnitine profiling in urine specimens of neonates, children, adolescents and adults. Clin Chim Acta 327:47–57.

Nicholson JK, Lindon JC, Holmes E. 1999. 'Metabonomics': understanding the metabolic responses of living systems to pathophysiological stimuli via multivariate statistical analysis of biological NMR spectroscopic data. Xenobiotica 29:1181–1189.

Nicholson JK, Connelly J, Lindon JC, Holmes E. 2002. Metabonomics: a platform for studying drug toxicity and gene function. Nat Rev Drug Discov 1:153–161.

Nicholson JK, Holmes E, Wilson ID. 2005. Gut microorganisms, mammalian metabolism and personalized health care. Nat Rev Microbiol 3:431–438.

Paczkowska A, Toczylowska B, Nyckowski P, Patkowski W, Kanski A, Krawczyk M, Oldakowska-Jedynak U. 2003. High-resolution 1H nuclear magnetic resonance spectroscopy analysis of bile samples obtained from a patient after orthotopic liver transplantation: new perspectives. Transplant Proc 35:2278–2280.

Piotto M, Saudek V, Sklenar V. 1992. Gradient-tailored excitation for single-quantum NMR spectroscopy of aqueous solutions. J Biomol NMR 2:661–665.

Pitt JJ, Egginton M, Kahler SG. 2002. Comprehensive screening of urine samples for inborn errors of metabolism by electrospray tandem mass spectrometry. Clinical Chem 48:1970–1980.

Plumb RS, Granger JH, Stumpf CL, Johnson KA, Smith BW, Gaulitz S, Wilson ID, Castro-Perez J. 2005. A rapid screening approach to metabonomics using UPLC and oa-TOF mass spectrometry: application to age, gender and diurnal variation in normal/Zucker obese rats and black, white and nude mice. Analyst 130:844–849.

Prost E, Sizun P, Piotto M, Nuzillard JM. 2002. A simple scheme for the design of solvent-suppression pulses. J Magn Reson 159:76–81.

Rinaldo P, Tortorelli S, Matern D. 2004. Recent developments and new applications of tandem mass spectrometry in newborn screening. Curr Op Pediatr 16:427–433.

Robosky LC, Wells DF, Egnash LA, Manning ML, Reily MD, Robertson DG. 2005. Metabonomic identification of two distinct phenotypes in Sprague-Dawley (Crl : CD(SD)) rats. Toxicol Sci 87:277–284.

Rosenfeld L. 2002. Clinical chemistry since 1800: growth and development. Clin Chem 48:186–197.

Shockor JP, Unger SE, Wilson ID, Foxall PJD, Nicholson JK, Lindon JC. 1996. Combined HPLC, NMR spectroscopy and ion-trap mass spectrometry with application to the detection and characterization of xenobiotic and endogenous metabolites in human urine. Anal Chem 68:4431–4435.

Silwood CJ, Lynch E, Claxson AW, Grootveld MC. 2002. 1H and (13)C NMR spectroscopic analysis of human saliva. J Dent Res 81:422–427.

Simpson AJ, Brown SA. 2005. Purge NMR: effective and easy solvent suppression. J Magn Reson 175:340–346.

Sklenar V. 1990. Selective excitation techniques for water suppression in one- and two-dimensional NMR spectroscopy. Basic Life Sci 56:63–84.

Smith IC, Baert R. 2003. Medical diagnosis by high resolution NMR of human specimens. IUBMB Life 55:273–277.

Stanley EG, Bailey NJ, Bollard ME, Haselden JN, Waterfield CJ, Holmes E, Nicholson JK. 2005. Sexual dimorphism in urinary metabolite profiles of Han Wistar rats revealed by nuclear-magnetic-resonance-based metabonomics. Anal Biochem 343:195–202.

Suh JW, Lee SH, Chung BC. 1997. GC–MS determination of organic acids with solvent extraction after cation-exchange chromatography. Clin Chem 43:2256–2261.

Sweeley CC, Young ND, Holland JF, Gates SC. 1974. Rapid computerized identification of compounds in complex biological mixtures by gas chromatography-mass spectrometry. J Chromatogr 99:507–517.

Takanashi J, Kurihara A, Tomita M, Kanazawa M, Yamamoto S, Morita F, Ikehira H, Tanada S, Kohno Y. 2002. Distinctly abnormal brain metabolism in late-onset ornithine transcarbamylase deficiency. Neurology 59:210–214.

Takesada H, Ebisawa K, Toyosaki H, Suzuki EI, Kawahara Y, Kojima H, Tanaka T. 2000. A convenient NMR method for in situ observation of aerobically cultured cells. J Biotechnol 84:231–236.

Tanaka, K.; Budd, M. A.; Efron, M. L.; Isselbacher, K. J. 1966. Isovaleric acidemia: a new genetic defect of leucine metabolism. Proc Natl Acad Sci USA 56:236–242.

Tanaka K, Hine DG. 1982. Compilation of gas chromatographic retention indices of metabolically important organic acids and their use in the detection of patients with organic acidurias. J Chromatogr 239:301–322.

Terabe S, Markuszewski MJ, Inoue N, Otsuka K, Nishioka T. 2001. Capillary electrophoretic techniques toward the metabolome analysis. Pure Appl Chem 73:1563–1572.

Tietz NW. 1995. *Clinical Guide to Laboratory Tests*, (3rd edition), WB Saunders Press, Philadelphia, PA.

Trethewey RN. 2004. Metabolite profiling as an aid to metabolic engineering in plants. Curr Op Plant Biol 7:196–201.

van der Graaf M, Janssen SW, van Asten JJ, Hermus AR, Sweep CG, Pikkemaat JA, Martens GJ, Heerschap A. 2004. Metabolic profile of the hippocampus of Zucker Diabetic Fatty rats assessed by in vivo 1H magnetic resonance spectroscopy. NMR Biomed 17:405–410.

Van QN, Chmurny GN, Veenstra TD. 2003. The depletion of protein signals in metabonomics analysis with the WET-CPMG pulse sequence. Biochem Biophys Res Commun 301:952–959.

Verhaeghe BJ, Lefevere MF, De Leenheer AP. 1988. Solid extraction with strong anion exchange column for selective isolation and concentration of urinary organic acids. Clin Chem 34:1077–1083.

Wevers RA, Engelke U, Heerschap A. 1994. High-resolution 1H-NMR spectroscopy of blood plasma for metabolic studies. Clin Chem 40:1245–1250.

Wevers RA, Engelke U, Wendel U, de Jong JG, Gabreels FJ, Heerschap A. 1995. Standardized method for high-resolution 1H-NMR of cerebrospinal fluid. Clin Chem 41:744–751.

Wilson ID, Plumb R, Granger J, Major H, Williams R, Lenz EM. 2005. HPLC–MS-based methods for the study of metabonomics. J Chromatogr B Analyt Technol Biomed Life Sci 817:67–76.

Wishart DS, Querengesser LMM, Lefebvre BA, Epstein NA, Greiner R, Newton JB. 2001. Magnetic resonance diagnostics: a new technology for high-throughput clinical diagnostics. Clin Chemistry 47:1918–1921.

Wishart DS. 2005. Metabolomics: the principles and potential applications to transplantation. Am J Transplant 5:2814–2820.

Wishart DS, Knox C, Guo AC, Shrivastava S, Hassanali M, Stothard P, Chang Z, Woolsey J. 2006. DrugBank: a comprehensive resource for in silico drug discovery and exploration. Nucleic Acids Res 34(Database issue):D668–672.

Zuppi C, Messana I, Forni F, Rossi C, Pennacchietti L, Ferrari F, Giardina B. 1997. 1H NMR spectra of normal urines: reference ranges of the major metabolites. Clin Chim Acta 265:85–97.

INDEX

C

WILEY-INTERSCIENCE SERIES IN MASS SPECTROMETRY

Series Editors

Dominic M. Desiderio
Departments of Neurology and Biochemistry
University of Tennessee Health Science Center

Nico M. M. Nibbering
Vrije Universiteit Amsterdam, The Netherlands

John R. de Laeter • *Applications of Inorganic Mass Spectrometry*
Michael Kinter and Nicholas E. Sherman • *Protein Sequencing and Identification Using Tandem Mass Spectrometry*
Chhabil Dass • *Principles and Practice of Biological Mass Spectrometry*
Mike S. Lee • *LC/MS Applications in Drug Development*
Jerzy Silberring and Rolf Eckman • *Mass Spectrometry and Hyphenated Techniques in Neuropeptide Research*
J. Wayne Rabalais • *Principles and Applications of Ion Scattering Spectrometry: Surface Chemical and Structural Analysis*
Mahmoud Hamdan and Pier Giorgio Righetti • *Proteomics Today: Protein Assessment and Biomarkers Using Mass Spectrometry, 2D Electrophoresis, and Microarray Technology*
Igor A. Kaltashov and Stephen J. Eyles • *Mass Spectrometry in Biophysics: Confirmation and Dynamics of Biomolecules*
Isabella Dalle-Donne, Andrea Scaloni, and D. Allan Butterfield • *Redox Proteomics: From Protein Modifications to Cellular Dysfunction and Diseases*
Silas G. Villas-Boas, Ute Roessner, Michael A. E. Hansen, Jorn Smedsgaard, and Jens Nielson • *Metabolome Analysis: An Introduction*
Mahmoud H. Hamdan • *Cancer Biomarkers: Analytical Techniques for Discovery*
Chabbil Dass • *Fundamentals of Contemporary Mass Spectrometry*